GABA—Biochemistry and CNS Functions

ADVANCES IN EXPERIMENTAL MEDICINE AND BIOLOGY

Recent Volumes in this Series

Volume 117
STEROID HORMONE RECEPTOR SYSTEMS
Edited by Wendell W. Leavitt and James H. Clark

Volume 118
CELL SUBSTRATES: Their Use in the Production of Vaccines and Other Biologicals
Edited by John C. Petricciani, Hope E. Hopps, and Paul J. Chapple

Volume 119
TREATMENT OF EARLY DIABETES
Edited by Rafael A. Camerini-Davalos and Bernard Hanover

Volume 120A
KININS – II: Biochemistry, Pathophysiology, and Clinical Aspects
Edited by Setsuro Fujii, Hiroshi Moriya, and Tomoji Suzuki

Volume 120B
KININS – II: Systemic Proteases and Cellular Function
Edited by Setsuro Fujii, Hiroshi Moriya, and Tomoji Suzuki

Volume 121A
MACROPHAGES AND LYMPHOCYTES: Nature, Functions, and Interaction, Part A
Edited by Mario R. Escobar and Herman Friedman

Volume 121B
MACROPHAGES AND LYMPHOCYTES: Nature, Functions, and Interaction, Part B
Edited by Mario R. Escobar and Herman Friedman

Volume 122A
PURINE METABOLISM IN MAN – III: Clinical and Theoretical Aspects
Edited by Aurelio Rapado, C. H. M. M. de Bruyn, and R. W. E. Watts

Volume 122B
PURINE METABOLISM IN MAN – III: Biochemical, Immunological, and Cancer Research
Edited by Aurelio Rapado, C. H. M. M. de Bruyn, and R. W. E. Watts

Volume 123
GABA – Biochemistry and CNS Functions
Edited by Paul Mandel and Francis V. DeFeudis

Volume 124
THE ENDOCRINE PANCREAS AND JUVENILE DIABETES
Edited by David M. Klachko, Ralph R. Anderson, Thomas W. Burns, and Harold V. Werner

GABA–Biochemistry and CNS Functions

Edited by

Paul Mandel and Francis V. DeFeudis

Université de Strasbourg
Strasbourg, France

Springer Science+Business Media, LLC

Library of Congress Cataloging in Publication Data

International Symposium on GABA – Biochemistry and CNS Functions, Strasbourg, 1978.
GABA – biochemistry and CNS functions.

"Proceedings of the International Symposium on GABA – Biochemistry and CNS Functions, held in Strasbourg, France, July 24–25, 1978."
Includes bibliographies and index.
1. Aminobutyric acid – Congresses. 2. Central nervous system – Congresses.
I. Mandel, Paul, 1908- II. DeFeudis, F. V. III. Title.
QP364.7.I58 1978 596'.01'88 79-22019

ISBN 978-1-4899-5201-1 ISBN 978-1-4899-5199-1 (eBook)
DOI 10.1007/978-1-4899-5199-1

Proceedings of the International Symposium on GABA – Biochemistry and CNS Functions, held in Strasbourg, France, July 24–25, 1978.

ACKNOWLEDGMENTS

We gratefully acknowledge the Centre National de la Recherche Scientifique, 15 Quai Anatole-France, 75700 Paris, for sponsoring this symposium.

The editors are also most grateful to Ms. Suzanne Miller for typing the manuscript for this volume and for her many helpful comments regarding the style of its presentation.

Professor Paul Mandel

Professor Francis V. DeFeudis

PREFACE

In less than 30 years after its discovery in the brain, γ-aminobutyric acid (GABA) has become generally accepted as being one of the major inhibitory neurotransmitters of the vertebrate central nervous system. During the past few years, research on mechanisms associated with central GABA-ergic systems has been conducted at an increasing rate, and therefore it was considered necessary to organize a symposium on this subject.

Recently, three other related meetings have been held: a workshop in California in February 1975, "GABA in Nervous System Function," edited by E. Roberts, T. N. Chase and D. B. Tower, and published in 1976; a NATO Advanced Study Institute in Norway in August 1977, "Amino Acids as Chemical Transmitters," edited by F. Fonnum, and published in 1978; and an Alfred Benzon Symposium in Copenhagen in May 1978, "GABA-Neurotransmitters," edited by P. Krogsgaard-Larsen, J. Scheel-Krüger and H. Kofod, and published in 1979.

The present volume represents the proceedings of an international symposium on "GABA – Biochemistry and CNS Functions," held in the Palais des Congrès, Strasbourg, during July 24-25, 1978. The book is divided into four sections, the articles of which represent reviews and/or new findings on the enzymology, metabolism, uptake, release, receptor-interaction and behavioral implications of central GABA-ergic systems. Biochemical, physiological, pharmacological and morphological approaches are discussed. The use of enzymological, metabolic, and receptor-binding approaches to develop new drugs for modulating central GABA-ergic systems on a biochemical basis is emphasized. Such approaches have already increased our arsenal of anti-epileptic drugs and our understanding of some basic GABA-ergic mechanisms, and should permit further rigorous study of certain neurologic-psychiatric disorders of man (e.g., convulsions, aggressive behavior, Huntington's chorea, Parkinson's disease) which appear to involve altered central GABA-ergic systems.

Taken together, the new information gained from the studies presented in this book provide further evidence for the concept that the regulation of CNS functions occurs by inhibitory-disinhibitory mechanisms, as has been shown for some other biological systems (e.g., transcription of the genetic code, enzyme biosynthesis, enzymatic activities, endocrine function). Thus, it seems important now to develop new drugs that can influence such inhibitory-disinhibitory mechanisms.

In addition to thanking the participants of this meeting, we are grateful to Prof. J. Mark and Dr. M. Maitre for help with its organization and to Dr. Lucienne Ossola, Mme. Henriette Urban, Mlle. Martine Knaeble, Mlle. Brigitte Fennebresque, Mlle. Simone Ott and Mme. Françoise Herth, who also contributed to its success.

P. Mandel and F. V. DeFeudis

CONTENTS

SECTION III

GABA-RECEPTOR

SECTION IV

BEHAVIORAL CORRELATES

SECTION I

ENZYMOLOGY OF THE GABA SYSTEM

GABA-TRANSAMINASE OF MAMMALIAN BRAIN

M. Maitre, L. Ossola and P. Mandel

Centre de Neurochimie, et Institut de Chimie Biologique, Faculté de Médecine, 67085 Strasbourg Cedex, France

INTRODUCTION

GABA-transaminase (4-aminobutyrate: 2-oxoglutarate transaminase; EC 2.6.1.19; GABA-T) is a key enzyme involved in regulating the GABA-ergic system. This is why, during the last 5 years, many studies have been focussed on this enzyme in the mammalian central nervous system (CNS). These studies have been stimulated by the finding that the cerebral GABA level can be modified by controlling GABA-T activity. Moreover, GABA-T is a relatively abundant enzyme in the CNS, and due to its stability it has been isolated from the brains of several animal species (1-5). From a practical point of view, the major points of interest at present concerning this enzyme are the possibilities of inhibiting it specifically, the mechanism of its *in vivo* regulation, and the possible existence of isoenzymes with a well-defined compartmentalization in the brain.

PURIFICATION OF GABA-T FROM MAMMALIAN BRAIN

As GABA-T is a mitochondrial enzyme, certain purification procedures commence with isolation of mitochondria. However, most workers have prepared the enzyme from whole brain homogenates. In our laboratory, GABA-T of rat and human brain has been totally purified. An example of the procedures used to purify about 1 mg

Table I - Purification of Human GABA-T

Stage	Total activity (μ-mole/min)	Total protein (mg)	Specific activity (μ-mole/min/mg)	Purification	Activity yield (%)	Protein yield (%)
Homogenate	65	32.15×20^3	0.0020	1	100	100
Supernatant (20,000 g)	62	10.2×10^3	0.0061	3	96	32
Ammonium sulfate fractionation (45-70%)	30	2.5×10^3	0.012	6	47	7
Sephadex G-200	25	416	0.060	30	39	1.3
DEAE cellulose	14	42	0.34	169	22	0.13
DEAE Sephadex	7.5	22	0.44	212	12	0.068
Hydroxyapatite	5.4	1.1	4.91	2455	8	0.003

Table II - Comparison of Some Data Obtained for Human and Rat Brain GABA-T

Parameter	Rat GABA-T	Human GABA-T
Molecular weight	105,000 ± 5,000	109,000 ± 5,000
Subunits	Dimer	Dimer
Molecular weight of subunits	57,000 ± 5,000	57,000 ± 5,000
Ratio N-terminal/lysine	1/40 - 1/46	1/50
Disulfide bridges	No	No
Optimum pH	8.5 - 8.6	8.5
Thermostability	Rapid denaturation at 55°C	Slow denaturation at 55°C

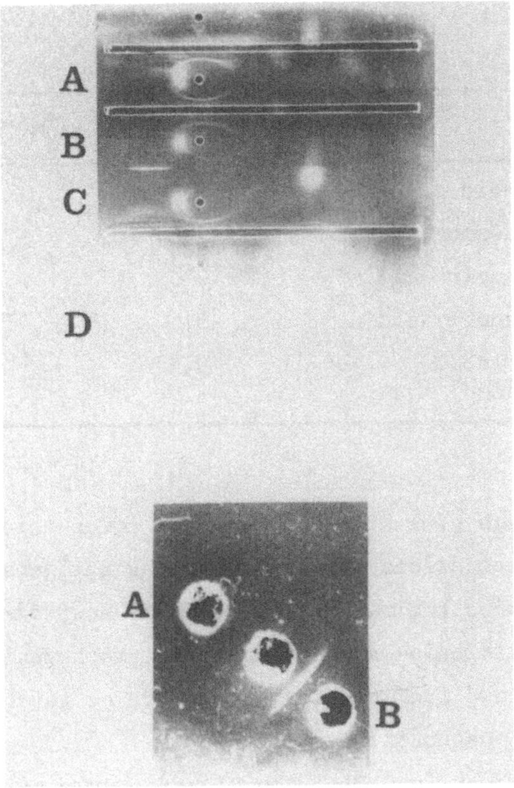

Fig. 1

of human brain GABA-T is shown in Table I. Purification factors were around 2,500-fold.

Chromatography on DEAE-cellulose regularly gave two distinct activity peaks (4), one of which appeared to consist of an aggregate of GABA-T and a contaminating protein. At present, there is no conclusive evidence regarding the existence of isoenzymes (e.g., differences in mapping of tryptic peptides). Most workers have obtained a single protein band by SDS-polyacrylamide gel electrophoresis (2-5). The presence of two different subunits described by Schousboe et al. (1) for the mouse brain enzymes has not been confirmed (L. Ciesielski, unpublished results). In both rat brain and human brain enzymes, dansylation of the pure preparation re-

Table III — K_i Values of the Branched Fatty Acids for Rat and Human GABA-T

Fatty acid	K_i (mM) rat	K_i (mM) human
2-propylpenten-2-oic acid	0.5	4.5
2-methyl-2-ethylhexanoic acid	1.5	6.5
2,2-dimethylpentanoic acid	2	9.5
5,5-dipropylpentanoic acid	5	20
2-propylpentanoic acid (dipropylacetic acid)	9.5	40

vealed the presence of a single N-terminal amino acid (isoleucine) (2-5). These preparations have been used as antigens to prepare GABA-T antibodies in the rabbit. When these antibodies were submitted to immunodiffusion and to immunoelectrophoresis against a supernatant obtained from a water homogenate of rat brain, a single precipitation arc was obtained (Fig. 1).

Rat and human brain GABA-T show considerable similarity in their quaternary structures (Table II). Nevertheless, the K_m (limit) for GABA (i.e., the K_m extrapolated to infinite 2-oxoglutarate concentration) is about 4 mM for rat brain enzyme but only about 0.9 mM for human enzyme. But, for the two enzymes under *in vivo* conditions the 2-oxoglutarate level is always very low, and therefore the affinity of GABA-T for GABA is always maximal, according to the ping-pong Bi Bi mechanism. This means that the apparent K_m for GABA under *in vivo* conditions is much lower than the calculated GABA K_m limit.

Thus it appears that human GABA-T has about four times greater affinity for GABA than the rat enzyme. This difference should be taken into account with regard to the mechanism of regulation of the GABA level in the two species. This result is also significant in relation to the intensity of the competitive inhibition vis à vis GABA, induced by certain branched-chain fatty acids which are

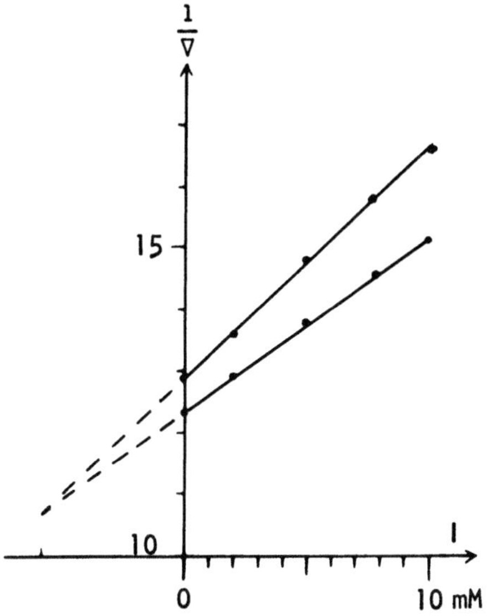

Fig. 2. Graphic determination of the K_i values by Dixon's method. The figure shows an example for 2-propylpenten-2-oic acid.

structural analogues of di-*n*-propylacetate. All these compounds show anticonvulsant properties concerning audiogenic seizure development by mutant mice, in parallel with their effect of increasing cerebral GABA levels in these animals (6-8). The various K_i values of the rat and human enzymes for the different branched-chain fatty acids (BFA) are shown in Table III. As the inhibition is competitive vis à vis GABA (Fig. 2), the product of the limit for GABA and the K_i of each fatty acid is essentially the same for the human and rat enzymes (Table IV). This suggests that the action of the BFA is greater when the enzyme shows less affinity for its substrate, GABA. Thus the actions of these anticonvulsants on GABA-T will be species dependent.

In light of the kinetic differences observed between rat and human brain GABA-T, the possibility of structural differences between the two enzymes was investigated. A comparison of the amino

Table IV — Values of the Product of the K Limit for GABA and the K_i for Each Inhibitor

Inhibitor	Rat		Human	
	K GABA (4 mM)	K GABA x K_i	K GABA (0.9 mM)	K GABA x K_i
2-methyl-2-ethylhexanoic acid	1.5 mM	6	6.5 mM	5.8
2,2-dimethylpentanoic acid	2 mM	8	9.5 mM	8.5
2-propylpenten-2-oic acid	0.5 mM	2	4 mM	3.6
5,5-dipropylpentanoic acid	6 mM	24	20 mM	18
2-propylpentanoic acid (dipropylacetic acid)	9.5 mM	38	40 mM	36

Table V — Comparison of Amino Acid Composition of Rat and Human GABA-T

Amino acid	Rat GABA-T		Human GABA-T	
	Residues/ molecule	Percentage of total	Residues/ molecule	Percentage of total
Glutamic acid	75	9	75	8.1
Aspartic acid	81	9.7	110	11.9
Threonine	30	3.6	31	3.3
Serine	60	7.2	69	7.5
Glycine	84	10	57	6.2
Cysteine	12	1.4	4	0.4
Methionine	8	1	8	0.9
Valine	42	5	46	5
Proline	30	3.6	50	5.4
Arginine	72	8.6	37	4
Histidine	39	4.7	43	4.7
Lysine	90	10.8	89	9.6
Phenylalanine	27	3.2	43	4.6
Tyrosine	12	1.4	30	3.2
Leucine	63	7.5	108	11.7
Isoleucine	30	3.6	47	5
Alanine	78	9.3	75	8.2
Tryptophan	-	-	-	-
Total	833	100	922	100

acid composition of the two enzymes is shown in Table V. There are notable differences in the levels of glycine, cysteine, proline, arginine, phenylalanine, tyrosine and leucine, but the levels of the remaining amino acids are very similar. In particular, the finding that the percentage of methionine residues is the same in both cases is in agreement with the finding that nine peptides are formed from each enzyme after reaction with cyanogen bromide, followed by separa-

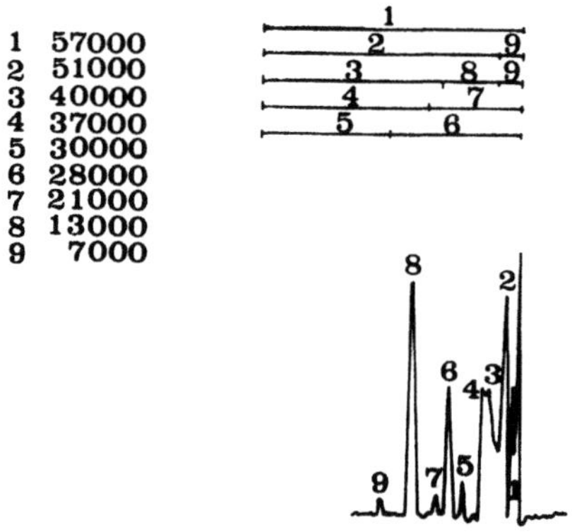

Fig. 3. Molecular weights by polyacrylamide gel electrophoresis of the peptides obtained after cyanogen bromide cleavage of human 4-aminobutyrate transaminase. The scheme proposed to reconstitute the initial monomer is an hypothesis.

tion on polyacrylamide gel electrophoresis using the method of Swank and Munkres (9). However, the molecular weights of the peptides produced is different in each case, which indicates that the methionine residues are distributed differently in each polypeptide chain (Figs. 3 and 4).

Tryptic hydrolysis of the two enzymes, followed by dansylation, and then separation of the peptides by thin layer chromatography (10) gave rise to 19 spots which migrated identically in each case. Moreover, there existed an additional 11 spots which were specific for the human enzyme, and 6 spots specific for the rat enzyme (Fig. 5). These results indicated that there are differences in the primary structure of the two enzymes, as well as differences in their kinetic properties (see above). These structural variations explain the interspecific antigenic differences demonstrated by Saito et al. (11) for GABA-T from various sources. The antigenic sites of

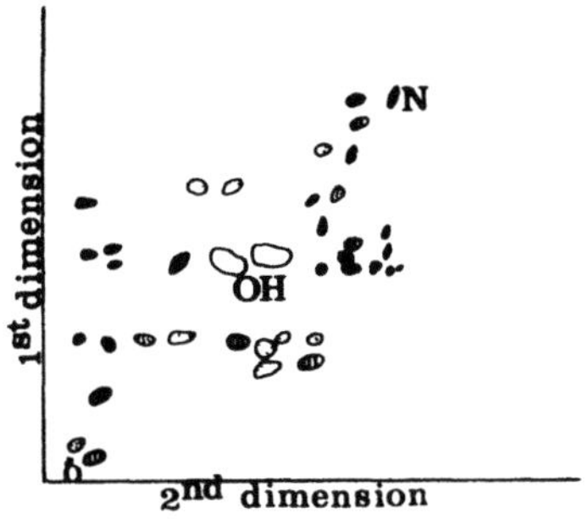

Fig. 4. Molecular weights by polyacrylamide gel electrophoresis of the peptides obtained after cyanogen bromide cleavage of rat 4-aminobutyrate transaminase. The scheme proposed to reconstitute the initial monomer is an hypothesis.

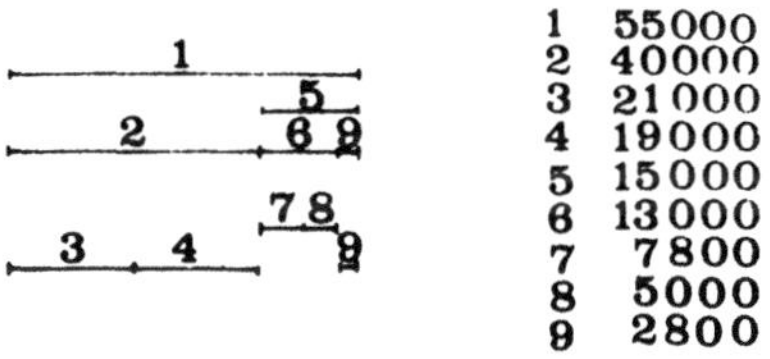

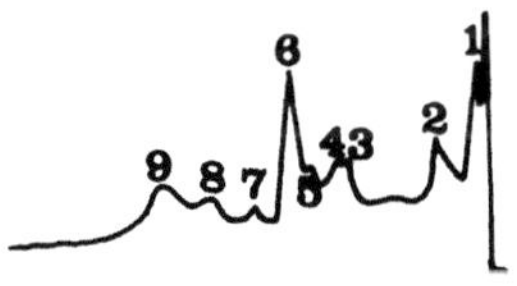

Fig. 5. Superimposition of the two peptide maps in order to identify the common spots. ●, Common spots; ⊙, specific to human 4-aminobutyrate transaminase; ◍, specific to rat 4-aminobutyrate transaminase.

rat and mouse GABA-T are very similar, but larger differences exist among the enzymes isolated from guinea pig, sheep and man (11).

MOLECULAR ACTIVITY OF RAT BRAIN GABA-T

A variation of GABA-T specific activity as a function of its concentration has been reported (12). The same phenomenon was observed with rat brain homogenate (13) using a different method to measure GABA-T activity. This phenomenon was not detected with a much purer preparation of mouse brain GABA-T, but the enzyme concentrations used did not exceed 100 μg/ml (1). To explain the decreased specific activity with high concentrations of homogenates a possible aggregation of the enzyme, giving rise to an inactive form, was suggested. However, measurement of relaxation times (14) showed that pig brain GABA-T behaves as a dimer over a wide concentration range (0.2 – 4 μM). This variation in specific activity in homogenates might involve one or several isoenzymes of GABA-T (15-18). The possible low stability of one such isoenzyme and/or its presence in limited quantity could explain why all purifications of GABA-T yield a single molecular species (2-5). A third hypothesis can be formulated, assuming the existence of one or several effectors of the enzymatic activity. An effector, non-uniformly distributed in the brain, could activate or de-activate the enzyme, perhaps only after homogenization. Molecular activities of GABA-T might also be modulated according to the brain region studied, as a function of variations in GABA-T and effector. This modulation has been verified using a GABA-T radioimmunoassay. Adult male August rats were used to reduce inter-individual variations. After decapitation, the brains were rapidly dissected; various regions were homogenized at 0°C in 6 volumes (w/v) of phosphate buffer (pH 7.4) containing 0.5% Triton X-100 (v/v), 0.5 mM pyridoxal phosphate, 1 mM 2-aminoethylisothiouronium bromide hydrobromide and 0.1 mM EDTA. GABA-T activities were measured in homogenates using a method by which the transamination products of [2,3-^{3}H]GABA are separated

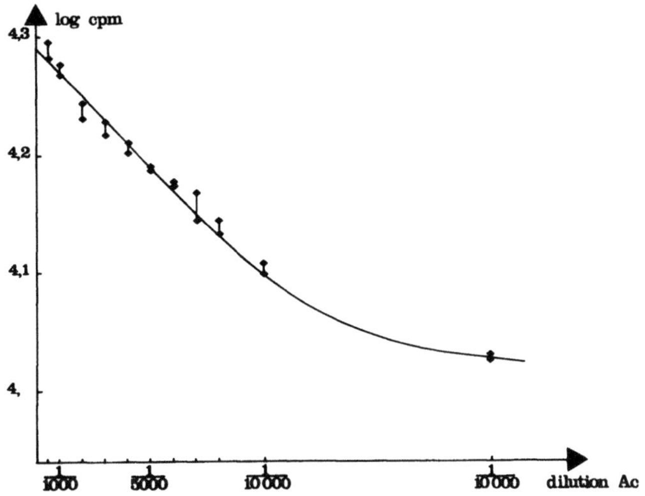

Fig. 6. Precipitation curve of ^{125}I-GABA-T by a series of dilutions of GABA-T antibodies.

on a column of Dowex 50W X8, H^+ form. The concentrations of GABA and 2-oxoglutarate in the incubation medium (Tris-HCl buffer, pH 8.6) were respectively 15 and 7 mM. GABA-T activities were measured with a constant amount of homogenate (0.33 mg protein/ml) in order to obtain valid comparisons.

In parallel, the amount of enzyme was measured in the same homogenate by radioimmunoassay using antibodies to GABA-T that had been prepared from rabbit. These antibodies gave rise to a single precipitation band by immunodiffusion or immunoelectrophoresis against rat brain homogenate. They precipitated all GABA-T activity in this same homogenate. The pure GABA-T was labelled with ^{125}I by the method of Hunter (20). Previous studies on titration of the antiserum showed that the optimum dilution was 1/2000 of the antibodies to be used (Fig. 6). A constant concentration of ^{125}I-GABA-T was incubated with either cold GABA-T (0 — 0.04 μg/400 μl) or with a rat brain homogenate (5 — 20 μg protein) (Fig. 7) in 20 mM phosphate buffer (pH 7.4) containing 1% bovine serum albumin,

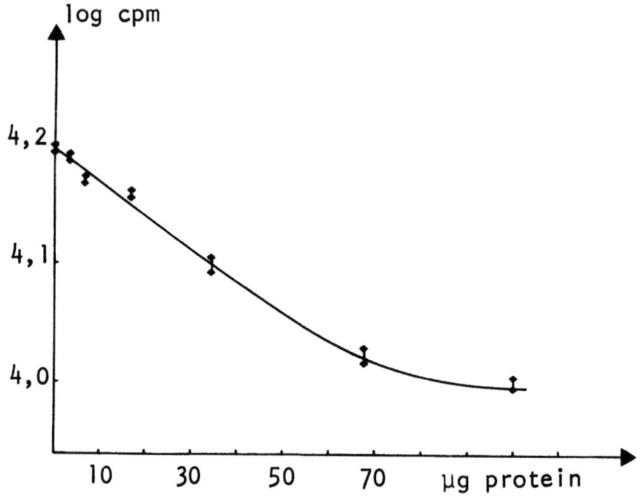

Fig. 7. Inhibition of precipitation of ^{125}I-GABA-T by an increasing amount of protein from a rat brain homogenate.

0.1% sodium azide, 0.5% Triton X-100 and 1/500 rabbit serum (immunized-plus-normal). The incubation was carried out for 6 hours at 37°C. Then, 0.2 ml of 1/300 diluted goat anti-rabbit γ-globulin serum was added to each tube. A second incubation was carried out for 15 hours at 4°C. The precipitate formed was isolated by filtration on a Millipore filter (GS, 0.22 µ). The results are shown in Table VI. Levels of apoenzyme, expressed in n-moles/mg protein, were calculated using a molecular weight of 105,000 for rat brain GABA-T.

Under these conditions, the hypothalamus and substantia nigra showed the highest specific activities. The highest enzyme levels were found in the hypothalamus and substantia nigra and also in the cerebellum. The turnover number of GABA-T, expressed as the number of moles transformed per second per mole of enzyme, differed from one region to another. These were significantly higher in thalamus, geniculate body, and especially in internal capsule and corpus callosum. These latter regions are rich in white matter and oligoden-

Table VI - Activities, Levels of Apoenzyme, and Turnover Numbers for GABA-T in Different Regions of Rat Brain

Region	GABA-T activity (10^{-9} moles/min/mg)	GABA-T apoenzyme (10^{-9} moles/mg)	Turnover number (sec^{-1})
Hypothalamus	13.1 ± 1.1	0.041 ± 0.005	5.3 ± 1.0
Substantia nigra	11.8 ± 0.9	0.033 ± 0.004	5.9 ± 2.0
Septum	11.3 ± 0.7	0.027 ± 0.007	6.8 ± 1.7
Posterior thalamus	11.0 ± 0.6	0.025 ± 0.003	7.3 ± 1.7
Caudate nucleus-plus-putamen	11.0 ± 0.4	0.024 ± 0.003	7.5 ± 1.9
Olfactory bulbs	10.6 ± 0.8	0.031 ± 0.007	5.6 ± 1.4
Geniculate body	10.0 ± 0.7	0.019 ± 0.003	8.6 ± 1.3
Cerebellum	9.8 ± 0.1	0.046 ± 0.008	3.5 ± 0.1
Anterior thalamus	9.7 ± 0.8	0.020 ± 0.002	7.9 ± 1.3
Total brain	9.4 ± 0.5	0.032 ± 0.006	4.8 ± 1.2
Amygdala	8.6 ± 0.4	0.028 ± 0.003	5.1 ± 1.1
Internal capsule	8.1 ± 0.6	0.004 ± 0.001	30.6 ±10.4
Hippocampus	6.4 ± 0.6	0.015 ± 0.005	7.2 ± 1.9
Piriform cortex	5.1 ± 0.5	0.019 ± 0.003	4.4 ± 1.0
Corpus callosum	4.3 ± 0.4	0.005 ± 0.001	12.7 ± 2.7

drocytes. Obviously, we cannot predict whether the effectors are close to the enzyme _in vivo_ or if this occurs during brain homogenization.

TURNOVER NUMBER OF GABA-T DURING CEREBRAL DEVELOPMENT IN THE RAT

The existence of an effector is also suggested when the evolution of GABA-T activity during brain development is considered. Determinations of GABA-T activity and of GABA-T apoenzyme were performed on the male Wistar rat as a function of cerebral development. These assays were carried out on animals ranging from 1 to about 30

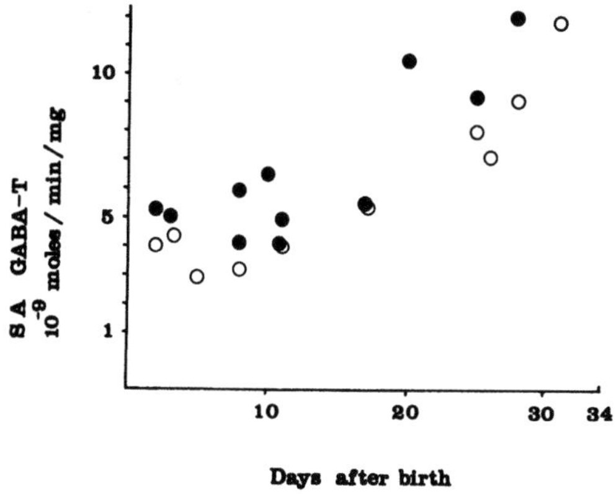

Fig. 8. Specific activity of GABA-T in the developing rat brain. Measurements were made with a constant dilution of protein (1.42 mg/ml) at pH 8.6 and 37°C. ○, Brain hemispheres; ●, cerebellum.

days of age. Litter sizes were always reduced to six individuals.

For animals of each age, the cerebral hemispheres and cerebella were homogenized separately as described above. For GABA assays, the animals were killed in a microwave oven and the brains were homogenized in 10 volumes of 80% ethanol at 0°C. After centrifugation for 30 minutes at 10,000 g and re-extraction of the pellet, the supernatants were pooled and evaporated at low pressure. GABA levels were determined by measuring fluorimetrically the NADH formed in the presence of Gabase (21). Measurements of GABA-T activity and the radioimmunoassay were carried out as described above. The results of these determinations are shown in Figures 8 and 9. Brain GABA concentration remained constant during development of the Wistar rat (Fig. 10); i.e., from birth to maturity. Although there occurred a slight decrease in level of GABA per mg of protein in the cerebellum from the first week after birth to maturity, the level was constant with respect to the wet weight. The specific activities

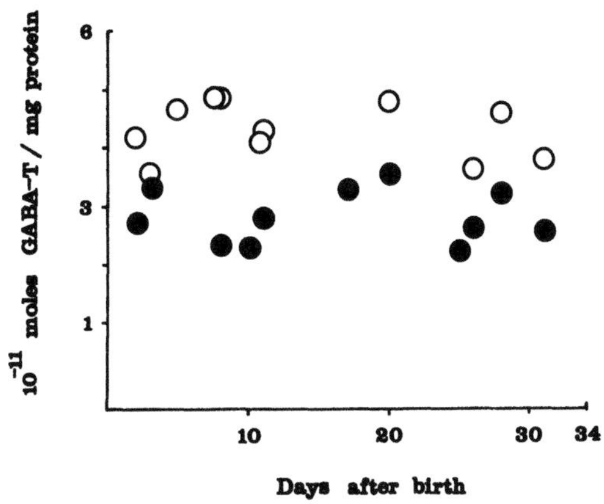

Fig. 9. GABA-T molecules per mg of protein in the developing rat brain. Measurements were made by radio-immuno technique. ●, Brain hemispheres; ○, cerebellum.

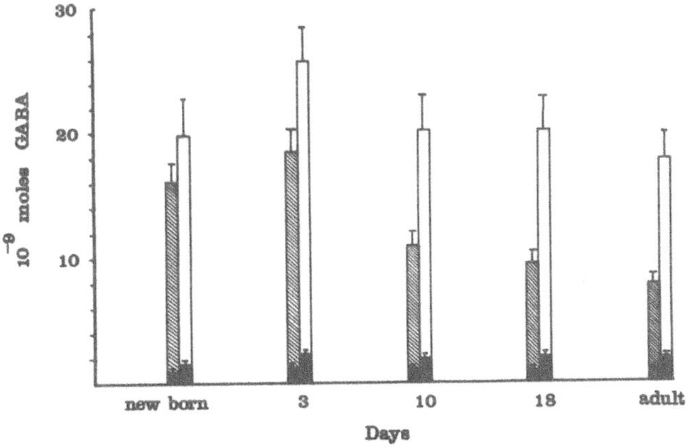

Fig. 10. Brain level of GABA in the developing rat brain. ■, Per mg wet weight; □, brain hemispheres; ▨, cerebellum.

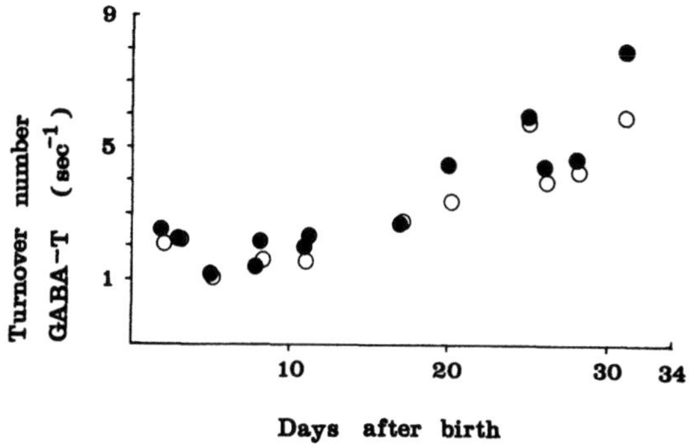

Fig. 11. Molecular activity of GABA-T in the developing rat brain. Results expressed as number of GABA molecules transformes at 37°C and pH 8.6 in one second per molecule of GABA-T. ●, Brain hemispheres; ○, cerebellum.

of GABA-T (Fig. 8), measured at a constant protein concentration (1.42 mg/ml), increased during development in both the cerebellum and cerebral hemispheres. These activities are about three times higher at the age of 30 days than at birth. However, the level of the apoenzyme (per mg of protein) measured in the same brains remained quite constant from birth up to the age of 30 days in both whole brain and cerebellum. This suggests that the molecular activities of GABA-T, expressed as the number of molecules of GABA transformed per second at 37°C and at pH 8.6 per molecule of GABA-T, increased with cerebral development, from 2 to 8 sec^{-1}, and 2 to 6 sec^{-1} for the cerebral hemispheres and the cerebellum, respectively (Fig. 11).

We found that the GABA levels of the cerebral hemispheres and the cerebellum were quite constant from birth up to the age of 30 days, in contrast to a previously published report (22). This discrepancy in the results is probably due to the method used for sacrificing the animals. In our experiments, the animals were killed in a microwave oven, whereas in the previous work (22) the animals were killed by decapitation and the brains were then rapidly frozen. It is possible that the post-mortem increase in GABA level is proportional to the volume of tissue frozen and that this volume increases with the age of the animal. In addition, the specific activities of GAD and GABA-T, which are higher in the adult brain, favorize a post-mortem increase in GABA level. Concerning the turnover numbers of GABA-T, our results show regional heterogeneity in the brain of the adult August rat and an increase during brain development in the Wistar rat. Taking this finding into consideration, one can suggest the presence of a modulation of GABA-T activity which varies depending on the stage of development and on the brain area examined.

Thus, GABA-T activity seems to be determined either by qualitative changes of the apoenzyme or by effectors, namely activators or inhibitors. However, the alternative explanation involving the presence of two or more isoenzymes of GABA-T having distinct immunological specificities and/or kinetic properties, modulated during ontogenesis, and having a well defined compartmentation in the adult brain, cannot be excluded. More information is required to obtain an unambiguous concept of the mechanisms involved in GABA-T modulation.

REFERENCES

(1) Schousboe, A., Wu, J. Y. and Roberts, E. (1973) Biochemistry, 12, 2868-2873.
(2) Maitre, M., Ciesielski, L., Cash, C. and Mandel, P. (1975) Eur. J. Biochem. 52, 157-169.
(3) John, R. A. and Fowler, L. J. (1976) Biochem. J. (Lond.) 155, 645-651.
(4) Bloch Tardy, M., Rolland, B. and Gonnard, P. (1974) Biochimie, 56, 823-832.
(5) Maitre, M., Ciesielski, L., Cash, C. and Mandel, P. (1978) Biochim. Biophys. Acta, 522, 385-399.
(6) Godin, Y., Heiner, L., Mark, J. and Mandel, P. (1969) J. Neurochem. 16, 869-873.
(7) Simler, S., Ciesielski, L., Maitre, M., Randrianarisoa, H. and

Mandel, P. (1973) Biochem. Pharmacol. 22, 1701-1708.
(8) Wood, J. D., Durham, J. S. and Peesker, S. J. (1977) Neurochem. Res. 2, 707-715.
(9) Swank, R. T. and Munkres, K. D. (1971) Anal. Biochem. 39, 462-477.
(10) Zanetta, J. P., Vincendon, G., Mandel, P. and Gombos, G. (1970) J. Chromatogr. 51, 441-458.
(11) Saito, K., Schousboe, A., Wu, J. Y. and Roberts, E. (1974) Brain Res. 65, 287-296.
(12) Waksman, A. and Roberts, E. (1965) Biochemistry, 4, 2132-2139.
(13) De Boer, T. H. and Bruinvels, J. (1977) J. Neurochem. 28, 471-478.
(14) Beeler, T. and Churchich, J. E. (1978) Eur. J. Biochem. 85, 365-371.
(15) Salganicoff, L. and DeRobertis, E. (1965) J. Neurochem. 12, 287-309.
(16) Waksman, A., Rubinstein, M. K., Kuriyama, K. and Roberts, E. (1968) J. Neurochem. 15, 351-357.
(17) Waksman, A. and Bloch, M. (1968) J. Neurochem. 15, 99-105.
(18) Buu, N. T. and Van Gelder, N. M. (1974) Can. J. Physiol. Pharmacol. 52, 674-680.
(19) Dye, D. J. and Taberner, P. V. (1975) J. Neurochem. 24, 997-1001.
(20) Hunter, W. M. (1974) Brit. Med. Bull. 30, 18-23.
(21) Graham, L. T. and Aprison, M. H. (1966) Anal. Biochem. 15, 487-497.
(22) Coyle, J. T. and Enna, S. J. (1976) Brain Res. 111, 114-133.

GABA TRANSAMINASE INHIBITORS

L. Ciesielski[a], S. Simler[a], C. Gensburger[a], P. Mandel[a], G. Taillandier[b], J. L. Benoit-Guyod[b], A. Boucherle[b], C. Cohen-Addad[c] and J. Lajzerowicz[c]

[a]Centre de Neurochimie du CNRS, et Institut de Chimie Biologique, Faculté de Médecine, 67085 Strasbourg Cedex, France

[b]UER des Sciences Pharmaceutiques et Biologiques, Avenue de Verdun, 38240 Meylan, France

[c]Laboratoire de Spectrometrie Physique Associé au CNRS, Université Scientifique et Médicale de Grenoble, B.P. 53, Centre de tri, 38041 Grenoble, France

INTRODUCTION

Owing to the role of γ-aminobutyric acid (GABA) as an inhibitory neurotransmitter (for reviews, see refs. 1-3), the production of an increase in its level in the brain has been the goal of many research teams for nearly twenty years. Many methods have been tried, such as: (a) modification of the blood-brain barrier by physical or chemical means to allow entry of exogenous GABA (4); (b) administration of lipophilic GABA derivatives such as 2-pyrrolidone (5), piracetam (2-pyrrolidone acetamide; ref. 6), β-(p-chlorophenyl)GABA (7) or GABA-cetyl ester (8); see review in ref. 9; (c) blocking the re-uptake of GABA by agents such as nipecotic acid, L-2,4-diaminobutyric acid and 1,3-diaminocyclohexane carboxylic acid which do not cross the blood-brain barrier (10); (d) reinforcement of precursor or cofactor availability. Glutamate, the direct precursor of GABA, does not cross the blood-brain barrier, and no mechanisms exist for producing an increase in glutamate by transamination of α-ketoglutarate or by deamination of glutamine. Administration of pyridoxine can block seizures induced by dietary deficiency or by familial pyridoxine deficiency (11,12) but cannot increase the brain GABA level; (e) interaction with GABA-receptors. Muscimol, a naturally-occurring toxin, is a potent GABA-agonist and has some anticonvulsant as well as convulsant properties (10,13); (f) production of an intraneuronal storage by inhibition of GABA catabolism. On the basis that an inhibition of GABA catabolism can produce, through the increase of intraneuronal GABA, an increase of GABA released into the synaptic cleft, the effect of GABA-transaminase (GABA-T) inhibitors, have been tested. GABA is catabolized successively by two enzymes, GABA-transaminase (E.C. 2.6.1.19) and succinic semialdehyde dehydrogenase (E.C. 1.2.1.24, SSADH), located in mitochondria. Experiments by Baxter and Roberts (14) have shown that inhibition of SSADH does not increase the brain GABA level.

The present paper is focussed on some aspects of GABA-T inhibitors which, by producing increases in the brain GABA level, are potential anticonvulsants.

CLASSIFICATION OF GABA-TRANSAMINASE INHIBITORS

The inhibitors of GABA-T can be classified in several groups (Table I).

Carbonyl Trapping Agents

Carbonyl trapping agents which link to the cofactor pyridoxal phosphate were the first used, but they lack specificity since they act on both the anabolizing enzyme and the catabolizing enzyme.

GABA-Analogues

This group is composed of amino acids which compete with GABA for the active site of the enzyme.

Table I - Inhibitors of GABA-Transaminase

Compound	Reference
Carbonyl Trapping Reagents	
Hydroxylamine	15
Aminooxyacetic acid	16
Hydrazines	17
Carbazides	18
Hydrazinopropionic acid	19
GABA-Analogues	
β-Chloro-GABA	20
β-Phenyl-GABA	20
4-Aminotetrolic acid	21
<u>Trans</u>-4-aminocrotonic acid	22
α-Amino-β-chloropropionic acid	23
Substrate-Coenzyme Analogues	
N (Pyridoxyl-5'-phosphate)GABA	24
N (Pyridoxyl-5'-phosphate)glutamate	24
Irreversible (K Cat) Inhibitors	
Ethanolamine-O-sulphate	26
Acetylenic-GABA	27
Vinyl-GABA	28
Gabaculine	30
Isogabaculine	31
Threo-α-cycloglutamate	32
Cycloserine	33
3-Isoxazolidone	34

Substrate-Coenzyme Analogues

Substrate-coenzyme analogues which may be irreversible inhibitors (24) have been studied only *in vitro*.

"K Cat" Inhibitors, or Suicide Compounds

"K cat" inhibitors (25) are metabolized by GABA-T to form an acylating molecule which acts at the active site. Theoretically, these should be highly specific, but experimentally they are not so specific (e.g., γ-acetylenic-GABA is also an inhibitor of glutamate decarboxylase (GAD), though γ-vinyl-GABA is more specific (29).

Branched-Chain Fatty Acids

This group is composed of acids without amino groups such as di-*n*-propylacetate (DPA), 2-2,dimethylvalerate (DMV), 2-methyl-2-ethylcaproate (MEC), tripropylacetate (TPA) and 2-propyl-2-pentenoate (PPΔ). DPA's anticonvulsant properties were discovered serendipitously in 1963 by Meunier et al. (35). Inhibition of GABA-T by DPA was described in 1969 (36) and analyzed in 1973 (37). DPA is now used extensively in the clinical treatment of epilepsy (for review see refs. 38,39).

Anticonvulsant properties of DMV and MEC were described by Lespagnol et al. (40) and their inhibitory effects on GABA-T were analyzed by Maitre et al. (41). The crystal structures of DPA and tripropylacetate (TPA) were shown to have some analogy with the non-planar folded form of GABA (42). A large number of branched-chain fatty acids have been synthesized in Professor Boucherle's laboratory (43) and have been tested for their inhibition of GABA and for their antiepileptic properties in our laboratory.

BRANCHED-CHAIN FATTY ACIDS

Inhibition of GABA-Transaminase

All branched-chain fatty acids tested were inhibitors of GABA-T prepared from rat or mouse brain. Table II shows the inhibition of purified rat brain GABA-T in a standard assay for the most representative branched-chain fatty acids, in correlation with their structure. An important point is the increase in the inhibition of GABA-T

Table II - Percentage of Inhibition by Branched-Chain Fatty Acids of Purified Rat Brain GABA-T

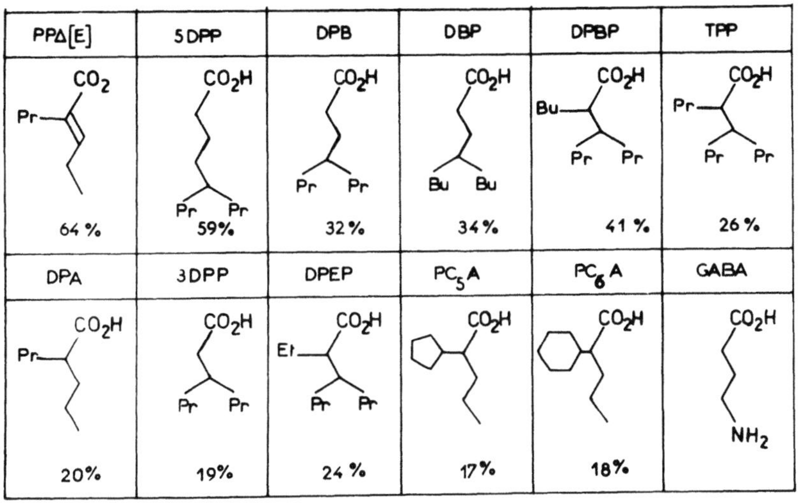

PPΔ[E]	5DPP	DPB	DBP	DPBP	TPP
64 %	59%	32 %	34 %	41 %	26 %
DPA	**3DPP**	**DPEP**	**PC_5A**	**PC_6A**	**GABA**
20%	19%	24 %	17%	18%	

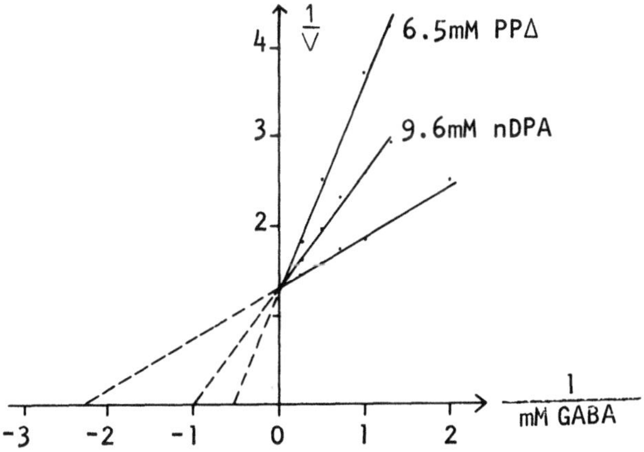

Fig. 1. Mouse GABA-T; DPA-GABA and PPΔ-GABA interactions. Lineweaver-Burk plot; 1/V = f(1/GABA).

Table III - Comparison of K_i Values for Branched-Chain Fatty Acids on Purified Brain GABA-T from Different Species

Fatty acid	Rat K_i (mM)	Human K_i (mM)	Mouse K_i (mM)
2-Propyl-2-pentenoic	0.5	4.5	1.75
2-Methyl-2-ethylcaproic	1.5	6.5	2.9
2,2-Dimethylvaleric	2	9.5	3.7
5,5-Dipropylpentanoic	5	20	5.7
2,3,3-Tripropylpropionic	7	-	6.6
Di-n-propylacetic	7.5	40	7.2
K GABA (mM)	4	0.9	2.3

activity brought about by desaturation of DPA to 2-propyl-2-pentenoate (PPΔ).

All of the branched-chain fatty acids studied are competitive inhibitors of GABA-T with respect to GABA (e.g., see Fig. 1). K_i values have been determined for these compounds on purified GABA-T from three species (44). The results shown in Table III reveal that the classification according to the K_i is the same for the three species and that PPΔ is the most potent inhibitor of the group.

Inhibition of Succinic Semialdehyde Dehydrogenase

Tests on brain homogenates have suggested that DPA is a competitive inhibitor of SSADH (45,46). Our experiments on purified SSADH showed that the inhibition of human, mouse and rat enzymes is non-competitive with any of the substrates (44) (Figs. 2 and 3).

An hypothesis for the increase in GABA level through dehydrogenase inhibition has been suggested. However, several sets of investigations are against this hypothesis: (a) The rate of synthesis of GABA from [^{3}H]succinic semialdehyde is not altered significantly by

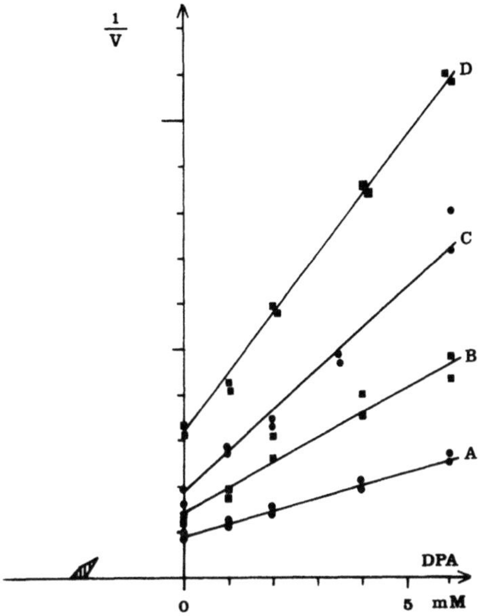

Fig. 2. Inhibition of mouse SSADH by DPA. Dixon plot, 1/V = f(DPA) for several SSA concentrations: A, 10^{-4} M; B, 0.4×10^{-4} M; C, 0.2×10^{-4} M; D, 10^{-5} M. Hatched area: confidence limit for intersection point.

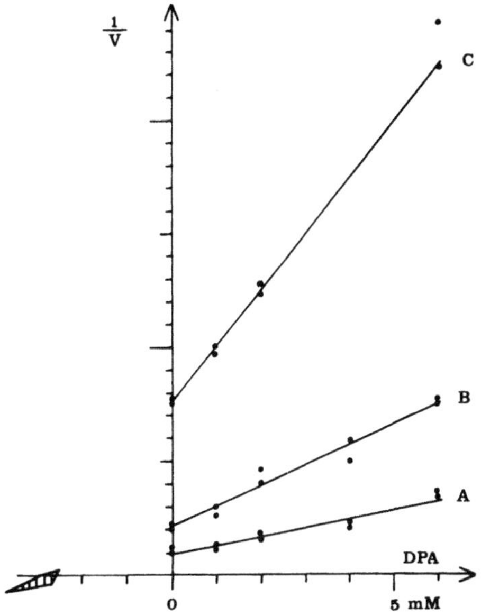

Fig. 3. Inhibition of mouse SSADH by DPA. Dixon plot, 1/V = f(DPA) for several NAD concentrations: A, 3×10^{-4} M; B, 3.75×10^{-5} M; C, 1.87×10^{-5} M. Hatched area: confidence limit for intersection point.

branched-chain fatty acids or by 4-hydroxybenzaldehyde, a potent SSADH inhibitor (47); (b) No correlation exists between the K_is for SSADH and the protecting dose against audiogenic seizures (44); (c) Computation of the ratio of SSADH activity to GABA-T activity as a function of the inhibitor concentration shows that the dehydrogenase is not the limiting enzyme in the catabolism of GABA in the presence of PPA, DMV, MEC or DPA (Fig. 4), the ratio of SSADH activity to GABA-T activity being always greater than unity. Experimental data on homogenates by Sawaya et al. (48) have indicated that the ratio

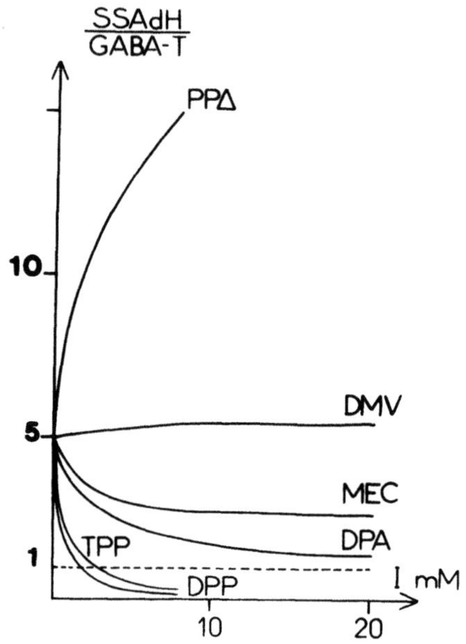

Fig. 4. Ratio of SSADH activity to GABA-T activity as a function of branched-chain fatty acid concentration; theoretical computation (the higher the ratio, the higher the specificity for GABA-T).

of SSADH activity to GABA-T activity is 1.9 for 1 mM DPA and 1.3 for 10 mM DPA, confirming the theoretical calculation; (d) Elevation of the brain GABA level by inhibitors of SSADH has been studied without success by Baxter and Roberts (14); the inhibitors used were azaserine, oxamate, barbiturate, arsenite and malonate.

Structural Studies

Crystal Structure of DPA. The crystal structure of DPA has already been published (42,49,50) and has been compared with that of GABA. GABA crystallizes in two forms: a folded gauche-trans conformation

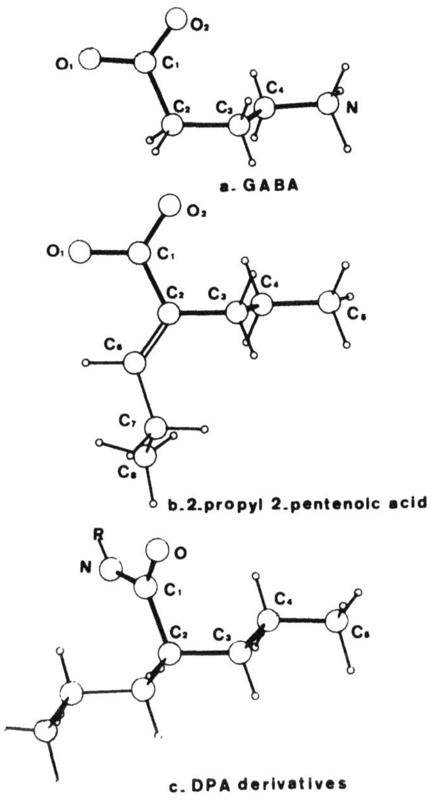

Fig. 5. Crystal structures of GABA, 2-propyl-2-pentenoic acid and DPA-derivatives.

for the (C_1-C_2-C_3-C_4-N) chain is observed in GABA (and in a GABA complex (51-53; see Fig. 5) and another one, an <u>all-trans</u> conformation, is observed for GABA-chlorohydrate (54). The crystal structures of DPA derivatives have some analogy with the non-planar folded form of GABA: the dipropyl part is a planar, fully extended chain, symmetrically oriented with respect to the central amide or acid group (Fig. 5).

<u>Determination of the Crystal Structure of PPΔ</u>. Parallelepipedic, colorless single crystals of suitable quality for diffraction

Table IV - Atomic Coordinates in Å of 2-Propyl-2-Pentenoic Acid; Comparison with GABA

PPΔ	x	y	z	GABA (from ref. 51)	x	y	z
O_1	-1.942	1.371	-0.029	O_1	-1.864	1.460	-0.241
O_2	0.044	2.367	0.04	O_2	0.070	2.365	0.224
C_1	-0.638	1.332	0.0	C_1	-0.630	1.392	0.0
C_2	0.0	0.0	0.0	C_2	0.0	0.0	0.0
C_3	1.511	0.0	0.0	C_3	1.519	0.0	0.0
C_4	2.077	0.169	-1.406	C_4	2.107	0.394	-1.326
C_5	3.602	0.192	-1.414	N	3.572	0.495	-1.266
C_6	-0.768	-1.081	-0.026				
C_7	-0.320	-2.511	-0.056				
C_8	-0.567	-3.122	-1.419				

Coordinate system: origin on C_2, C_2C_3 = X axis, $(C_2-C_3-C_1)$ = XY-plane (see Fig. 5).

experiments (size 0.2 — 0.4 mm) are obtained by slow evaporation from a solution in ethanol. The crystals are triclinic, space group $P\bar{1}$ with:

$a = 8.023\ (5)$ $b = 11.991\ (6)$ $c = 4.933\ (3)$ Å

$\alpha = 99.70\ (5)$ $\beta = 104.40\ (5)$ $\gamma = 103.70\ (5)$ d°

Z = 2 molecules per unit cell which correspond to a center of symmetry.

Three-dimensional X-ray data were collected on a four-circle Siemens diffractometer using Mo-K_α radiation with a five-point method (55) and resulted in 949 significant diffracted intensities. Measurements were made at Laüe-Langevin Institut (Grenoble). The coordinates of the atoms in a Cartesian coordinate system are listed in Table IV. Torsion angles are compared to those of GABA and DPA derivatives in Table V.

Table V - Torsion Angles in 2-Propyl-2-Pentenoic Acid; Comparison with GABA and DPA Derivatives

Atoms	Angles (d°) in PPΔ	Angles in GABA	Angles in DPA derivatives
O_1-C_1-C_2-C_3	178.6	167.6 (180)	118.6 (N)
O_2-C_1-C_2-C_3	-2.2	-12.2 (0)	-61.5 (0)
C_1-C_2-C_3-C_4	-83.2	-73.4 (180)	-60.9
C_2-C_3-C_4-N	179.1	175.2 (180)	181.2
O_1-C_1-C_2-C_6	-0.1		
O_2-C_1-C_2-C_6	179.1		
C_1-C_2-C_6-C_7	178.8		
C_2-C_6-C_7-C_8	-107.7		
C_3-C_2-C_6-C_7	0.2		

The ideal torsion angles for the extended form of GABA are given in parentheses.

There exists a good analogy between the folded form of GABA and the (O_1-O_2-C_1-C_2-C_3-C_4-C_5) chain of PPΔ. Torsion angles are nearly the same in the two chains, even for the carboxylic groups. Like the NH_3^+ group of GABA, the methyl group around C_5 is found in a staggered position (see Fig. 5). The chain lengths can be compared and the following values are observed:

4.418 and 5.841 Å for O_2-C_5 and O_1-C_5 in PPΔ

4.328 and 5.612 Å for O_2-N and O_1-N in GABA.

On the other hand, the O_1-O_2-C_1-C_2-C_6-C_7-C_8 chain of PPΔ cannot be compared to the folded or extended form of GABA, as shown from torsion angles. .In this case, the C_2-C_6 double bond involves a planar group, and the symmetry observed in the case of the dipropyl chain of DPA cannot occur; it is this structural constraint which leads to a conformation which is, for a part of the molecule, very similar to the folded form of GABA. The structural analogy is better than with other branched-chain fatty acids, and this result is in agree-

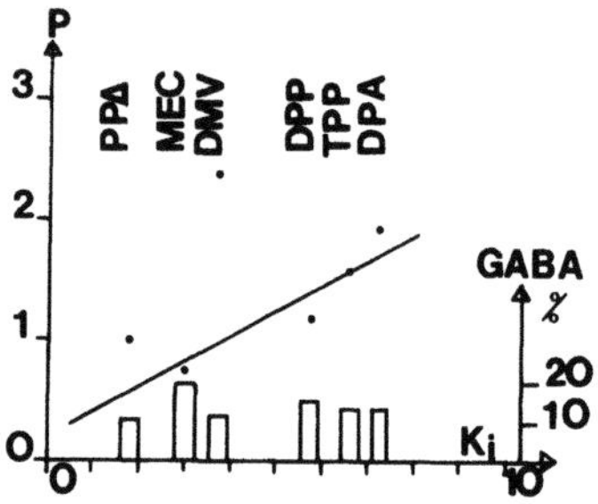

Fig. 6. P, Protecting dose (m-mole/kg) as a function of K_i for mouse GABA-T (line). Corresponding GABA increases are indicated by vertical bars.

ment with biochemical studies which showed that PPΔ was, among the compounds tested, the most specific for GABA-T. It should be noted that trans-aminocrotonic acid which has a structure similar to that of the extended form of GABA is a potent inhibitor of GABA-T, whereas cis-aminocrotonic acid is not (22).

Anticonvulsant Properties of Branched-Chain Fatty Acids

All of the branched-chain fatty acids tested have anticonvulsant properties which have been correlated with their abilities to produce increases in brain GABA content (37,56,57). More convincing evidence for the role of GABA is shown in Figure 6 which indicates the existence of a correlation between the K_i for GABA-T and the threshold dosage which protects mice against audiogenic seizures.

Regional Increases in Brain GABA Level

The increases in brain GABA level produced by treatment with branched-chain fatty acids were determined on whole brain at the threshold dosage. The increase in GABA level was the same (about

15%) for all the compounds tested. The time course of the GABA increase in different mouse brain regions has been investigated following administration of DPA and PPΔ. Figure 7 shows that after DPA treatment (56) the increase in brain GABA is not homogeneous and is not a function of the initial GABA level; the increase is maximal in the olfactory bulbs and in the hypothalamus. A similar experiment with PPΔ (57) has shown that the increase is maximal in the substantia nigra and in the frontal cortex. The more selective increase of GABA in the substantia nigra might be of therapeutic use in the therapy of Parkinson's disease. The 15% increase in brain GABA level following a treatment with branched-chain fatty acids should be compared to the 300% increase produced by "k cat" inhibitors at the ED_{50} dose; however, protection afforded by "k cat" inhibitors lasts longer (29).

Effect of DPA on Cells in Culture

The effect of DPA treatment on the GABA level of primary cultures of astroblasts and neuronal (enriched) cells has been investigated. An increase in GABA level has been observed only in astroblasts prepared either from rat enbryo or newborn rat (Table VI). No increases in GABA level have been found in C_6 astrocytoma cells or in C-1300 neuroblastoma cells (58) after DPA treatment.

Effects of Branched-Chain Fatty Acids on the High-Affinity Uptake of GABA

The actions of several branched-chain fatty acids were tested on the high-affinity uptake of GABA by brain slices and synaptosomes (59). They were found to have very weak or negligible inhibitory actions on the uptake system.

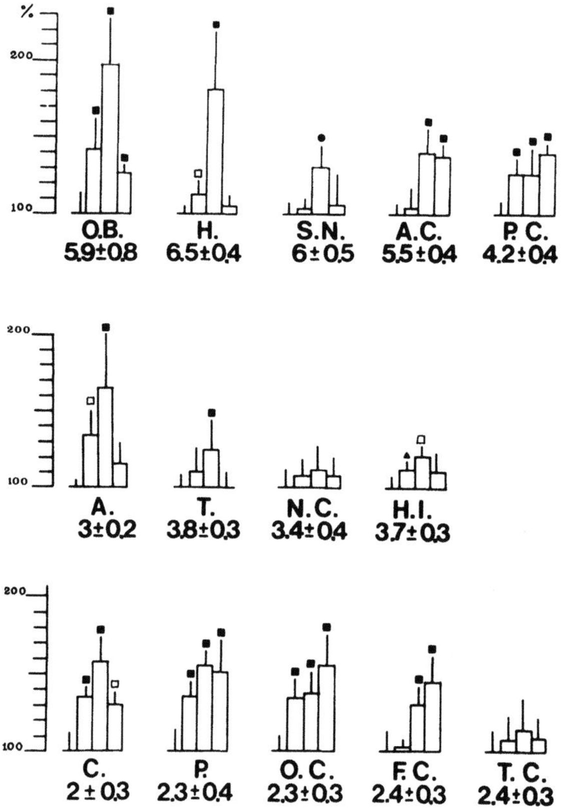

Fig. 7. GABA level in different brain areas (μ-mole/g ± S.D.; $6 < n < 10$) and its evolution (percentage of control value ± S.D.) after an i.p. (400 mg/kg) DPA treatment. Abbreviations: O.B., olfactory bulbs; H, hypothalamus; S.N., substantia nigra; A.C., anterior colliculus; P.C., posterior colliculus; A, amygdala; T, thalamus; N.C., nucleus caudatus; H.I., hippocampus; C, cerebellum; P, pons-medulla; O.C., occipital cortex; F.C., frontal cortex; T.C., temporal cortex. Time-course: 0, 15, 45, 75 min. Significance: ■ $p < 0.001$; □ $p < 0.005$; ● $p < 0.01$.

Table VI - Effects of DPA on the GABA Levels of Primary Cultures of Astroblasts and Neuronal Cells

Primary Culture \ DPA	Control	20 μM	1 mM	5 mM	10 mM
Neurones (enriched); 7-day culture of 10-13-day rat embryo	10.5 ± 0.6 (6)	10.3 ± 1.8 (10)	10.3 ± 0.7 (4)	10.4 ± 0.8 (4)	10.3 ± 0.5 (4)
Astroblasts; 14-day culture of 10-13-day rat embryo	3.0 ± 0.2 (4)	-	2.9 ± 0.2 (4)	3.0 ± 0.3 (4)	4.7 ± 0.4**(6)
Astroblasts; 2-day culture of newborn rat	3.3 ± 0.4 (10)	3.6 ± 0.4*(10)	-	-	-

GABA is expressed in n-mole/mg protein ± standard deviation; numbers of determinations in parentheses. Cells were incubated for one day with DPA in the culture medium. * and ** indicate $p < 0.05$ and $p < 0.001$ between these values and their respective controls.

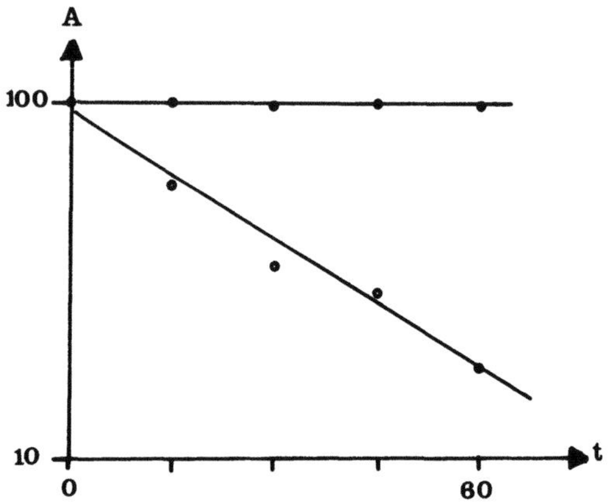

Fig. 8. Kinetics of inactivation of purified human brain GABA-T by acetylenic-GABA. Residual activity is plotted on a logarithmic scale as a function of time. Open circles, acetylenic-GABA (0.5 mM); closed circles, GABA (5 mM) added. Inactivation: 24 ± 1 min (r = 0.99).

A New "K Cat" GABA-Transaminase Inhibitor

A common feature of all known "k cat" inhibitors of GABA-T is that the enzyme is protected from irreversible inactivation by GABA (Fig. 8), although if α-ketoglutarate is added inactivation resumes. Owing to the high ratio in the brain of GABA to α-ketoglutarate, it would be of interest to have a "k cat" inhibitor whose action is not affected by GABA. One molecule has this property: 1-propyl-3-5-dicyano 4-phenyl-6-hydroxy 2-pyridone (Py4; Fig. 9). This compound has anticonvulsant properties (60). Inactivation of GABA-T by this compound and protection by α-ketoglutarate are shown in Figures 10 and 11.

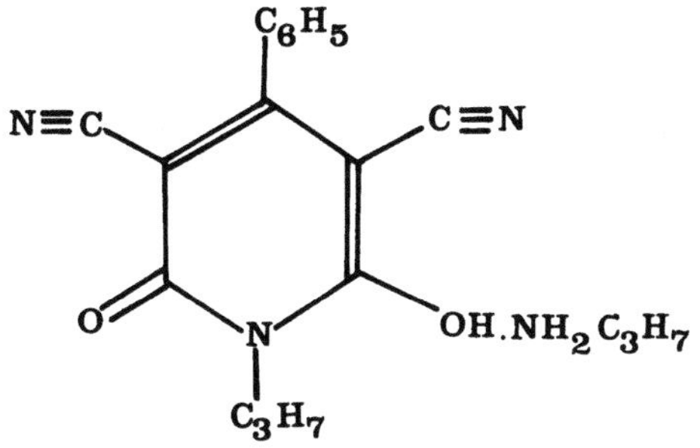

Fig. 9. Structure of an irreversible inhibitor of GABA-T with anticonvulsant properties (Py4).

Table VII - Omega-Amino Acid Transaminase Activity in the Purification of Mouse Brain GABA-T

Amino acid	Homogenate	Ammonium sulfate	G-200	DEAE cellulose	Hydroxy-apatite
β-Alanine	177%	196%	167%	163%	126%
GABA	100%	100%	100%	100%	100%
δ-Aminovaleric	114%	153%	150%	133%	114%
ε-Aminocaproic	28%	40%	33%	20%	21%

Table VIII - Inhibition of Mouse Brain GABA-T

Amino acid	DPA (20 mM) inhibition	AOAA (60 mM) inhibition	Acetylenic-GABA (0.5 mM) inactivation period (min)	
			without AA	5 mM AA
β-Alanine	15%	15%	3	6
GABA	14%	13%	3	9
δ-Aminovaleric	10%	10%		
ε-Aminocaproic	38%	48%		

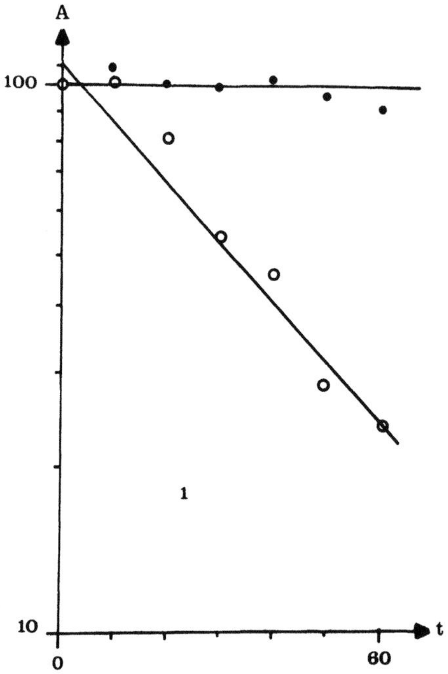

Fig. 10. Kinetic inactivation of purified rat brain GABA-T by Py4. Residual activity is plotted on a logarithmic scale as a function of time. Closed circles, control; open circles, 1.3 mM Py4. Inactivation period: 26 ± 4 min (r = 0.98).

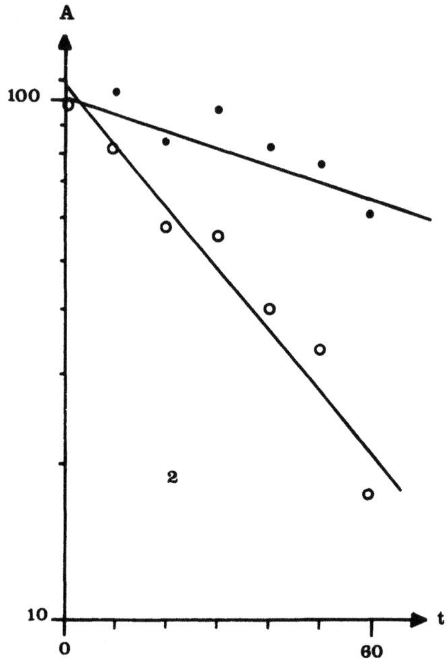

Fig. 11. Kinetics of inactivation of purified rat brain GABA-T by Py4; effects of substrates. Residual activity is plotted on a logarithmic scale as a function of time. Closed circles, 1.3 mM Py4 + 5 mM α-ketoglutarate, period = 26 ± 4 min; open circles, 1.3 mM Py4 + 25 mM GABA, period = 92 ± 30 min.

β-Alanine Aminotransferase

During the purification of GABA-T there is always co-purification of ω-amino acid aminotransferases. The ratio of the activity of any ω-amino acid aminotransferase to GABA-T activity is constant (see Table VII). GABA-T inhibitors, DPA, aminooxyacetic acid, and γ-acetylenic GABA, have the same inhibitory properties on the aminotransferase activities, suggesting that there is a single protein with ω-amino acid aminotransferase activity (Table VIII).

CONCLUSION

In summary, several branched-chain fatty acids appeared to be competitive inhibitors of GABA-T and non-competitive inhibitors of SSADH. These compounds produce an increase in brain GABA level, and for two of these it was shown that the increase differs among various brain areas. An increase of GABA cannot be obtained by inhibition of SSADH. The increase in brain GABA seems to correlate with the anticonvulsant activity of branched-chain fatty acids.

ACKNOWLEDGEMENTS

This work was supported by a grant from the CNRS; ATP Pharmacodynamie et Chimiotherapie (no. 1-655-1805).

REFERENCES

(1) Curtis, D. R. and Johnston, G. A. R. (1974) Ergebn. Physiol. 69, 97-188.
(2) Krnjević, K. (1974) Physiol. Rev. 54, 518-540.
(3) Roberts, E., Chase, T. N. and Tower, D. B. (eds.) (1976) GABA in Nervous System Function. Raven Press, New York.
(4) Strasberg, P., Krnjević, K., Schwartz, S. and Elliott, K. A. C. (1967) J. Neurochem. 14, 755-760.
(5) Hawkins, J. E., Jr. and Sarett, L. H. (1957) Clin. Chim. Acta, 2, 481-484.
(6) Giurgea, C. E., Moyersoons, F. E. and Evraerd, A. C. (1967) Arch. Int. Pharmacodyn. Therap. 166, 238-251.
(7) Khaunina, R. (1968) Farmak. Toks. 31, 202-204.
(8) Ostrovskaya, R. U. and Parin, V. V. (1972) Byulleten Eksperimental'noi Biologii, 75, 47-49.
(9) Meldrum, B. S. (1975) Int. Rev. Neurobiol. 17, 1-36.
(10) Meldrum, B. S. (1979) In: GABA-Neurotransmitters (P. Krogsgaard-

Larsen, J. Scheel-Krüger and H. Kofod, eds.) pp. 390-405. Munksgaard, Copenhagen.
(11) Coursin, D. B. (1955) Amer. J. Dis. Child. 90, 344-348.
(12) Hunt, A. D., Stokes, J., McCrory, W. W. and Stroud, H. H. (1954) Pediatrics, 13, 140-145.
(13) Anlezark, G., Collins, J. and Meldrum, B. S. (1978) Neurosci. Lett. 7, 337-340.
(14) Baxter, C. F. and Roberts, E. (1961) J. Biol. Chem. 236, 3287-3294.
(15) Baxter, C. F. and Roberts, E. (1959) Proc. Soc. Exp. Biol. Med. 101, 811-815.
(16) Wallach, D. P. (1961) Biochem. Pharmacol. 15, 533-539.
(17) Medina, M. A. (1961) J. Pharmacol. Exp. Therapeut. 140, 133-137.
(18) Killam, K. F. and Bain, J. A. (1957) J. Pharmacol. Exp. Therapeut. 119, 255-262.
(19) van Gelder, N. M. (1967) J. Neurochem. 14, 747-757.
(20) Buu, N. T. and van Gelder, N. M. (1974) Brit. J. Pharmacol. 52, 401-406.
(21) Beart, P. M., Uhr, M. L. and Johnston, G. A. R. (1972) J. Neurochem. 19, 1849-1854.
(22) Johnston, G. A. R., Curtis, D. R., Beart, P. M., Game, C. J. A., McCulloch, R. M. and Twitchin, B. (1974) J. Neurochem. 24, 157-160.
(23) Tsui, D., Gorecki, D. K., Charington, C. B. and Wood, J. D. (1976) Eur. J. Pharmacol. 38, 189-192.
(24) Severin, E. S., Gulyaev, N. N., Khurs, E. N. and Khomutov, R. M. (1969) Biochem. Biophys. Res. Commun. 35, 318-323.
(25) Rando, R. R. (1974) Nature (Lond.) 250, 586-587.
(26) Fowler, L. J. and John, R. A. (1972) Biochem. J. (Lond.) 130, 569-573.
(27) Jung, M. J. and Metcalf, B. W. (1975) Biochem. Biophys. Res. Commun. 67, 301-306.
(28) Lippert, B., Metcalf, B. W., Jung, M. J. and Casara, P. (1977) Eur. J. Biochem. 74, 441-445.
(29) Schechter, P. J., Tranier, Y., Jung, M. J. and Böhlen, P. (1977) Eur. J. Pharmacol. 45, 319-328.
(30) Kobayashi, K., Miyazawa, S. and Endo, A. (1977) FEBS Lett. 76, 207-210.
(31) Schechter, P. J., Tranier, Y. and Grove, J. (1979) this volume.
(32) Severin, E. S., Sashchenko, L. P., Kovaleva, G. K. and Khomutov, R. M. (1968) Biokhimiya, 33, 1210-1213.
(33) Dann, O. T. and Carter, C. E. (1964) Biochem. Pharmacol. 13, 677-684.
(34) Gorecki, D. K. J., Peesker, S. J. and Wood, J. D. (1976) Biochem. Pharmacol. 25, 1653-1657.
(35) Meunier, H., Carraz, G. and Eyhar, M. (1963) Thérapie, 18, 435-438.
(36) Godin, Y., Heiner, L., Mark, J. and Mandel, P. (1969) J. Neurochem. 16, 869-873.
(37) Simler, S., Ciesielski, L., Maitre, M., Randrianarisoa, H. and

Mandel, P. (1973) Biochem. Pharmacol. 22, 1701-1708.
(38) Simon, D. and Penry, J. K. (1975) Epilepsia, 16, 549-573.
(39) Pinder, R. M., Brogden, R. N., Speight, T. M. and Avery, G. S. (1977) Drugs, 13, 81-123.
(40) Lespagnol, A., Mercier, J., Erb-Debruyne, F. and Dessaigne, S. (1972) Annales Pharm. Franc. 30, 193-198.
(41) Maitre, M., Ciesielski, L. and Mandel, P. (1974) Biochem. Pharmacol. 23, 2363-2368.
(42) Ferrandes, B., Cohen-Addad, C., Benoit-Guyod, J. L. and Eymard, P. (1974) Biochem. Pharmacol. 23, 3363-3376.
(43) Taillandier, G. (1974) Thesis, Université de Grenoble, 155 pp.
(44) Mandel, P., Ciesielski, L. and Maitre, M. (1977) J. Medicinal Chem. 5, 155-171.
(45) Harvey, P. K. P., Bradford, H. F. and Davison, A. N. (1975) FEBS Lett. 52, 251-254.
(46) Anlezark, G., Horton, R. W., Meldrum, B. S. and Sawaya, M. C. (1976) Biochem. Pharmacol. 25, 413-417.
(47) Maitre, M., Ossola, L. and Mandel, P. (1976) FEBS Lett. 72, 53-57.
(48) Sawaya, M. C. B., Horton, R. W. and Meldrum, B. S. (1975) Epilepsia, 16, 649-655.
(49) Grand, A. and Cohen-Addad, C. (1973) Acta Cryst. B29 , 1149-1151.
(50) Cohen-Addad, C. and Grand, A. (1974) Acta Cryst. B30, 1342-1346.
(51) Tomita, K., Higashi, H. and Fujiwara, J. (1973) Bull. Chem. Soc. Japan, 7, 2199-2204.
(52) Steward, E. G., Player, R. B. and Warner, D. (1973) Acta Cryst. B29, 2038-2040.
(53) Takenaka, A., Oshima, E., Yamada, S. and Watanabe, T. (1973) Acta Cryst. B29, 503-513.
(54) Steward, E. G., Player, R. B. and Warner, D. (1973) Acta Cryst. B29, 2825-2826.
(55) Troughton, P. G. H. (1969) Ph.D. Thesis, Imperial College, London.
(56) Simler, S., Gensburger, C., Ciesielski, L. and Mandel, P. (1978) Commun. Psychopharmacol. 2, 123-130.
(57) Weissman, D., Simler, S., Ciesielski, L. and Mandel, P. (1978) C. R. Soc. Biol. Paris, 172 (4), 707-712.
(58) Gensburger, C., Simler, S., Ciesielski, L. and Mandel, P. (1978) C. R. Soc. Biol. Paris, 172 (4), 701-706.
(59) Balcar, V. J. and Mandel, P. (1976) Experientia, 32, 904-905.
(60) Pera, M. H. (1974) Thesis, Université de Grenoble, 150 pp.

ATTEMPTS TO CORRELATE ALTERATIONS IN BRAIN GABA METABOLISM BY GABA-T INHIBITORS WITH THEIR ANTICONVULSANT EFFECTS

Paul J. Schechter, Yvette Tranier and Jeffrey Grove

Centre de Recherche Merrell International, 16 rue d'Ankara, 67084 Strasbourg Cedex, France

INTRODUCTION

It has been known for the last 20 years that decreases in brain GABA concentrations are associated with convulsions in animals (1). Furthermore, many agents which increase brain GABA concentrations have anti-convulsant activity (2,3). On the other hand, it is clear that the relationship between brain GABA and convulsions is not a simple one. Convulsions can occur even in the presence of elevated brain GABA concentrations (4-6) and the time course of seizure incidence does not necessarily correlate with the time course of changes in brain GABA (7).

Enzyme-activated irreversible inhibitors of GABA-transaminase (4-aminobutyrate: 2-oxoglutarate aminotransferase; EC 2.6.1.19; GABA-T), recently developed in our laboratories and in Japan, represent the most specific method to date for increasing brain GABA concentrations. We therefore attempted to correlate the increases in brain GABA caused by these inhibitors with protection against a variety of seizures in mice. In addition, a similar correlation was attempted with di-n-propylacetate, an effective clinical anti-epileptic agent (8) reported to act by increasing brain GABA concentrations (9).

MATERIALS AND METHODS

Structural formulas of the four enzyme-activated GABA-T inhibitors are given below (Fig. 1).

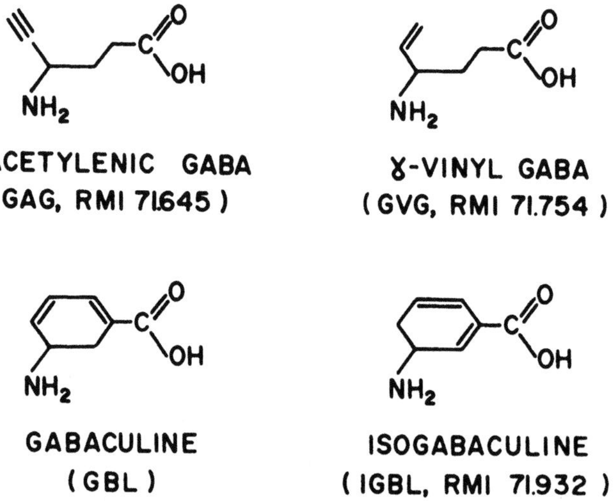

Fig. 1

γ-Acetylenic GABA (GAB, RMI 71.645), γ-vinyl GABA (GVG, RMI 71.754), and isogabaculine trifluoromethane-sulfonate (IGBL, RMI 71.932) were synthesized by previously-described methods (10-12). Gabaculine HCl (GBL) was synthesized by a modification of the published procedure (13). All doses of GBL and IGBL were calculated as the base. Inhibitors were dissolved in water, the pH adjusted when necessary, and injected intraperitoneally into mice in a volume of 10 ml/kg body weight. Control mice received saline, or in the case of IGBL, an equivalent dose of trifluoromethane-sulfonic acid, pH-adjusted in saline.

Di-*n*-propylacetate (DPA) was purchased as Depakine solution (Berthier-Derol, Paris) and diluted in deionized water to yield a concentration of 40 mg/ml. Mice were injected with a volume of 10 ml/kg (i.p.). Control mice received this same volume of saline.

Male CD_1 albino mice (Charles River, France) were used throughout these studies, except for audiogenic seizures and for biochemical effects of DPA where DBA/2 mice (breeding stock from CNRS, Orléans, France) were used.

RESULTS AND DISCUSSION

Enzyme-Activated GABA-T Inhibitors

γ-Acetylenic GABA, γ-vinyl GABA, gabaculine and isogabaculine are all irreversible inhibitors of mammalian brain GABA-T (11,12, 14,15). With GAG and GVG, peak brain GABA concentrations in mice are reached by 4 hours after intraperitoneal administration; these concentrations are sustained for approximately 4 hours with GAG (16) and for 16 - 24 hours with GVG (17). In the cases of GBL and IGBL, peak GABA levels in mouse brain are reached by 16–24 hours following intraperitoneal administration and remain at this maximum for at least 24 hours (Schechter *et al*., unpublished results, 1978). GVG, GBL and IGBL do not influence the activity of glutamate-decarboxylase (L-glutamate-1-carboxylase, EC 4.1.1.15; GAD) *in vitro*, whereas GAG irreversibly inhibits GAD although to a lesser extent than GABA-T. *In vivo*, however, GVG, GBL and IGBL produce slight decreases in GAD activity, probably via a feed-back type of inhibition.

Pharmacological effects of the four enzyme-activated GABA-T inhibitors on several laboratory seizure models were compared at the time of maximum effects on brain GABA concentrations; i.e., 4 hours after administration for GAG and GVG and 24 hours for GBL and IGBL.

Audiogenic Seizures. Audiogenic seizures in genetically-susceptible mice are very reproducible, and protection against such seizures has been associated with elevations in GABA content of the central nervous system (CNS) (18). GAG, GVG, GBL and IGBL all produce time- and dose-related protection against audiogenic seizures. The duration of anti-seizure activity parallels the duration of elevations in whole brain GABA concentrations. ED_{50} estimates for

Table I - ED_{50} (CL_{95}) Estimates[1] for Audiogenic Seizure Protection

	Complete protection[2]	Attenuation of seizure intensity[2]
γ-Acetylenic GABA	41 (33-51)	25 (18-33)
γ-Vinyl GABA	990 (750-1307)	540 (367-794)
Gabaculine	17 (16-18)	15 (13-17)
Isogabaculine	16 (14-19)	14 (12-16)

Values represent mg/kg i.p., 4 hrs post-injection for GAG and GVG and 24 hrs post-injection for GBL and IGBL.

[1] Determined by method of Litchfield and Wilcoxon (19).

[2] Methods as described by Schechter *et al.* (17).

the four agents for seizure protection (Table I) follow the same order as their relative potency as GABA-T inhibitors.

Linear regression analyses relating brain GABA increase and degree of seizure protection (Table II) indicated significant correlations between brain GABA levels and both audiogenic seizure incidence and attenuation of seizure intensity for all four GABA-T inhibitors. In addition, the brain GABA concentrations associated with 50% protection and 50% seizure attenuation were remarkably similar for the four inhibitors. The close association between brain GABA concentrations and blockade of seizures induced by sound stimuli in susceptible mice strongly suggests that in this seizure model these inhibitors act as anti-convulsants via a GABA-ergic mechanism.

Chemical Convulsants. GAG has been shown to have varying anti-seizure activity when tested against a variety of chemically-induced convulsions in mice (20). It protects mice against seizures and mortality induced by thiosemicarbazide, isoniazid and strychnine. It does not protect against the incidence of seizures and mortality induced by pentylenetetrazol. GVG has a spectrum of anti-seizure activity essentially identical to that of GAG, albeit at higher

Table II - Correlations Between Brain GABA Increase in Mice and Audiogenic Seizure Protection

GABA - T INHIBITOR	COMPLETE PROTECTION			ATTENUATION OF SEIZURE INTENSITY		
	CORRELATION COEFFICIENT	P	% GABA INCREASE CORRESPONDING TO 50 % PROTECTION	CORRELATION COEFFICIENT	P	% GABA INCREASE CORRESPONDING TO 50 % ATTENUATION
γ-ACETYLENIC GABA	0.974	<0.005	322	0.991	<0.001	265
γ-VINYL GABA	0.834	<0.005	324	0.962	<0.001	264
GABACULINE	0.901	<0.005	311	0.851	<0.01	284
ISOGABACULINE	0.884	<0.005	317	0.890	<0.01	201

doses, consistent with its lower biochemical potency (17). Hence, 1500 to 2000 mg/kg of GVG (Table III) significantly protect mice against seizures and mortality induced by isoniazid and strychnine but have no influence on the effect of pentylenetetrazol.

GAG and GVG do not influence the frequency of seizures and death after picrotoxin or bicuculline treatment, but in the case of picrotoxin produce a dose-dependent prolongation of the time to onset of seizures and death (21).

The pharmacological anti-seizure profiles of GBL and IGBL are very similar to each other but differ somewhat from those of GAG and GVG. Hence, whereas GAG and GVG provide almost complete protection against seizure and mortality frequencies induced by isoniazid, thiosemicarbazide and strychnine, GBL and IGBL provide a lesser degree of protection or no protection (Table IV). On the other hand, GBL and IGBL significantly attenuate pentylenetetrazol seizures (Table IV) in contrast to GAG and GVG which are without significant effect on these seizures.

GBL and IGBL also do not affect the frequency of seizures or mortality after picrotoxin or bicuculline (Table V). However, the times required for tonic convulsions (CT_{50}) or respiratory arrest

Table III - Effect of γ-Vinyl GABA (GVG) Pretreatment on Chemically-Induced Seizures and Mortality in Mice[1]

TREATMENT	DOSE (MG/KG)	TOTAL	TONIC SEIZURES NO. CONVULSING	NO. NOT CONVULSING	P*	MORTALITY NO. DEAD	NO. ALIVE	P*
ISONIAZID	250	10	10	0	< 0.005	9	1	< 0.005
+ GVG	1500	10	0	10		0	10	
STRYCHNINE	0.5	8	7	1	< 0.005	4	4	< 0.025
+ GVG	2000	10	1	9		0	10	
PENTYLENETETRAZOL	60	27	23	4	N.S.	18	9	N.S.
+ GVG	1500	18	12	6		11	7	

[1] GVG administered intraperitoneally 3 hrs (for isoniazid) or 4 hrs prior to intravenous administration of convulsant.

* Calculated using Chi^2 analysis (uncorrected) of 2 x 2 contingency table.

Table IV - Effect of Gabaculine and Isogabaculine Pretreatments[1] on Chemically-Induced Seizures and Mortality in Mice

CONVULSANT - MG/KG	GABACULINE (40 MG/KG) NO.	% CONVULSING	% DEAD	ISOGABACULINE (40 MG/KG) NO.	% CONVULSING	% DEAD
ISONIAZID - 250	10	100	100	10	100	100
+ GABA-T INHIBITOR	10	60*	50*	10	40*	40*
THIOSEMICARBAZIDE - 25	20	100	100	20	100	100
+ GABA-T INHIBITOR	20	65*	100	20	25*	45*
STRYCHNINE - 0.6	8	100	100	10	100	90
+ GABA-T INHIBITOR	8	100	75	10	100	70
PENTYLENETETRAZOL - 70	20	100	95	10	100	100
+ GABA-T INHIBITOR	20	70*	60*	10	40*	40*

[1] GABA-T inhibitors were administered intraperitoneally 24 hrs prior to intravenous challenge with convulsants.

* $p < 0.05$ compared to convulsant alone; Chi^2 analysis (uncorrected) of 2 x 2 contingency table.

Table V - Effects of Gabaculine and Isogabaculine Pretreatments on Picrotoxin- and Bicuculline-Induced Tonic Seizures and Mortality

TREATMENT	GABACULINE (40 mg/kg)							ISOGABACULINE (40 mg/kg)						
	NO. TESTED	% CONVULSING	p*	CT_{50}** (sec)	% DEAD	p*	MT_{50}** (sec)	NO. TESTED	% CONVULSING	p*	CT_{50}** (sec)	% DEAD	p*	MT_{50}** (sec)
PICROTOXIN - 6 mg/kg	10	90	N.S.	305	100	N.S.	355	9	100	N.S.	295	100	N.S.	336
+ GABA-T INHIBITOR	8	100		850	100		880	9	100		770	88.9		990
BICUCULLINE - 3 mg/kg	9	100	N.S.	318	100	N.S.	402	11	100	N.S.	294	100	N.S.	372
+ GABA-T INHIBITOR	9	100		288	100		372	9	100		485	100		564

Mice were pre-treated 24 hrs prior to challenge with picrotoxin (i.v.) or bicuculline (s.c.)

* Determined using Chi^2 test (uncorrected); ** estimated using method of Paton (22).

(MT_{50}) to occur in 50% of the animals were significantly increased by GBL and IGBL (using a doubling of the value as indicating significance; ref. 22) after picrotoxin but not after bicuculline. This is similar to our findings with GAG and GVG (21).

Di-n-propylacetate

DPA is an effective anti-convulsant when tested against a range of experimentally-induced seizures in laboratory animals (23). The mechanism for its anti-convulsant effect in laboratory animals, as well as in man, is thought to be inhibition of GABA-T resulting in an increase in brain GABA concentrations (9,24). This conclusion is based primarily on a close time-related similarity between brain GABA concentrations in mice and protection against audiogenically-induced seizures following DPA administration (9). Of the principal amino acids, only GABA was found to be altered by DPA treatment (24). The increase in brain GABA associated with complete protection

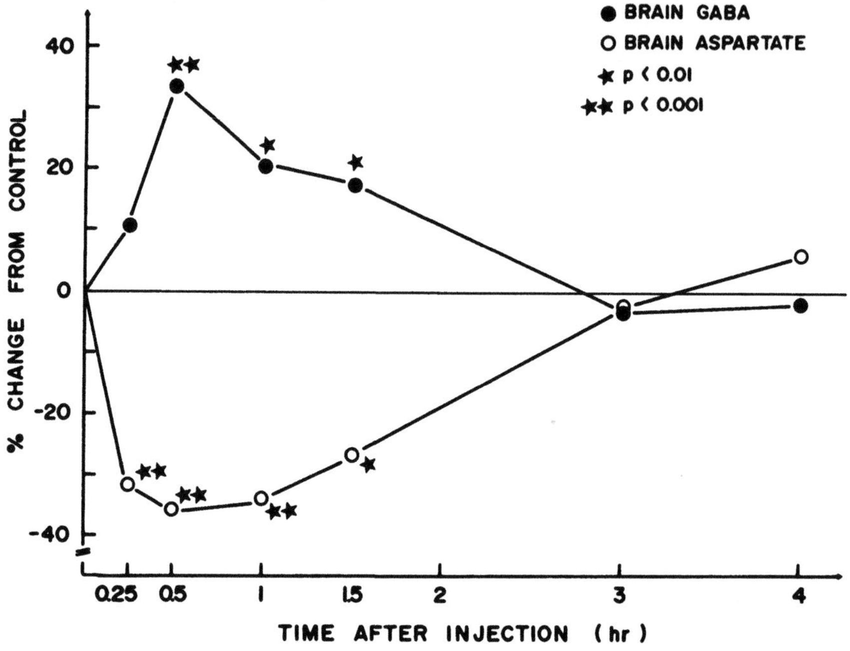

Fig. 2. Effect of DPA (400 mg/kg, i.p.) on brain GABA and aspartate concentrations of DBA/2 mice as a function of time after injection. Each value represents the percent change (n = 6 mice/point) from control values. The S.E.M. of each value never exceeded 8% of the mean. Control values: GABA concentrations, 2.11 ± 0.03 μ-mole/g (mean ± S.E.M.; n = 42); aspartate concentrations, 3.82 ± 0.04 μ-mole/g (n = 42). Significant differences from control values are indicated.

against audiogenic seizures (approximately 6%; ref. 9) was, however, considerably less than that found in the present study with the four irreversible GABA-T inhibitors. In an attempt to clarify these findings, DBA/2 mice were injected with DPA (400 mg/kg, i.p.) and sacrificed at various times after treatment, and different amino acid concentrations in whole brain were estimated using high performance liquid chromatography (25). In a parallel experiment, susceptibility of DBA/2 mice to audiogenic seizures was determined as a function of time after treatment with DPA (400 mg/kg, i.p.).

GABA concentrations of mouse brain (Fig. 2), as previously shown (9), increased within 15 minutes of DPA administration, reached a maximum by 30 minutes and returned to control values by 3 hours. Conversely, brain aspartate concentrations decreased significantly within 15 minutes to approximately 68% of control values, remained depressed for up to 1½ hours after treatment and by 3 hours returned to a level not different from control mice. The symmetry and duration of the changes in GABA and aspartate concentrations were remarkably similar. Cerebral concentrations of threonine, serine, glutamate, glutamine, glycine, β-alanine and β-isobutyrate following DPA (400 mg/kg i.p.) did not differ from control values at any point.

DPA, at a dose of 400 mg/kg (i.p.), protected sensitive DBA/2 mice against the occurrence of audiogenically-induced seizures within 30 minutes after injection (Fig. 3). Significant protection lasted up to 2 hours, but by 3 hours following a single treatment the incidence of seizures was not different from that of control mice ($p > 0.05$, Chi^2 analysis (uncorrected) of 2 x 2 contingency tables). At no time were mice completely protected against seizures. Intensity of seizure response followed a similar pattern, but even 4 hours after injection the intensity of seizure response in treated animals was less severe than in saline-injected controls ($p < 0.05$, Mann-Whitney U test using values from individual mice).

These results indicate that the parallelism between anti-convulsant activity of DPA and decreases in brain aspartate concentrations as a function of time after treatment is as good as that between anti-seizure effects and elevated brain GABA concentration. Furthermore, the magnitude of the changes in aspartate are as great or greater than the GABA alterations. Hence, a decrease in aspartate, a putative excitatory neurotransmitter in the CNS (26), is as plausible an explanation for the anti-convulsive effects of DPA in experimental seizures as is an increase in brain GABA.

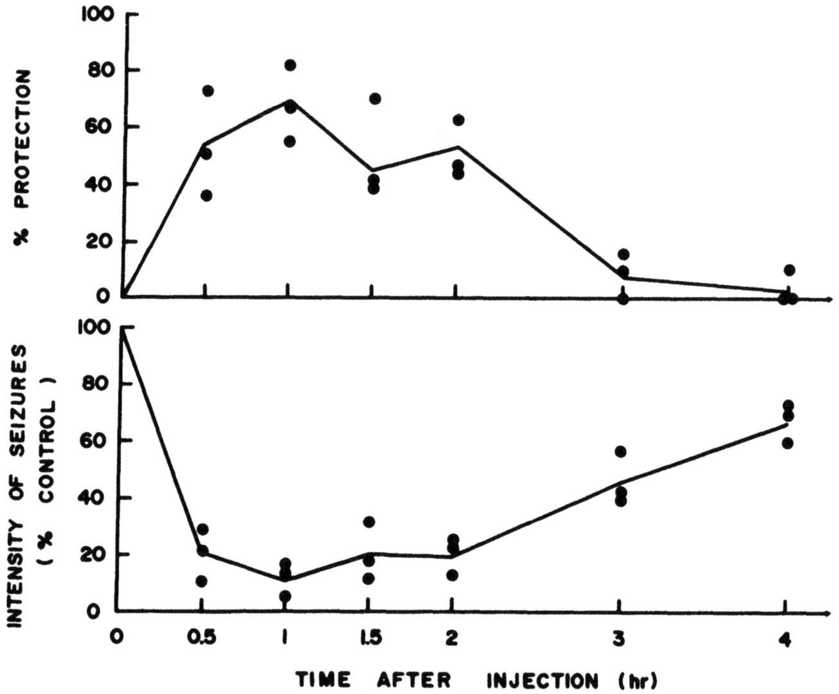

Fig. 3. Effect of DPA (400 mg/kg, i.p.) on audiogenic seizures in DBA/2 mice as a function of time after injection (methods as described by Schechter _et al._; ref. 17). The upper panel illustrates % absolute protection relative to controls at various times after injection (control frequency of seizures = 99.1 ± 0.6%, mean ± S.E.M.; for 18 groups of 8 - 14 mice each (average = 11 mice/group)). The lower panel illustrates the intensity of seizures expressed as % control values (intensity of seizures in control mice = 62.9 ± 2.4 for 18 groups where a maximum possible score is 80 or all animals dead). Each point represents a different experiment using 8 — 14 mice/group with lines drawn through the average of the 3 values at each time.

Table VI - Summary of Anti-Seizure Activity of Four Enzyme-Activated GABA-T Inhibitors

Test Protection against:	γ-Acetylenic GABA	γ-Vinyl GABA	Gabaculine	Iso-gabaculine
Audiogenic seizures	Active	Active	Active	Active
Picrotoxin seizures	Inactive	Inactive	Inactive	Inactive
Bicuculline seizures	Inactive	Inactive	Inactive	Inactive
Isoniazid seizures	Active	Active	Slightly active	Slightly active
Thiosemicarbazide seizures	Active	—	Slightly active	Slightly active
Strychnine seizures	Active	Active	Inactive	Inactive
Pentylenetetrazol seizures	Inactive	Inactive	Slightly active	Slightly active

CONCLUSION

GAG, GVG, GBL and IGBL share the ability to inhibit brain GABA-T activity irreversibly and increase whole brain GABA concentrations to approximately the same degree. Furthermore, effects on GAD activity *in vivo* are similar for GVG, GBL and IGBL. Hence, if anticonvulsant effects were related to alterations in GABA metabolism, as previously hypothesized (27-29), one might expect similar pharmacological activities among the inhibitors. This was the case only in certain types of seizures (Table VI). When tested against audiogenic seizures, all four inhibitors protected susceptible mice with protection being related to the absolute increase in brain GABA concentrations. Against picrotoxin and bicuculline, none of the inhibitors influenced the incidence of tonic convulsions or mortality; however, all prolonged latency after picrotoxin but not after bicuculline.

When tested against other laboratory seizure models, dissimilarities among the four inhibitors were apparent, with GAG and GVG comprising one group and GBL and IGBL another. Hence, GAG and GVG are active against isoniazid-, thiosemicarbazide- and strychnine-induced seizures, whereas GBL and IGBL are less active or inactive. The converse holds for seizures induced by pentylenetetrazol.

It is, therefore, obvious that changes in GABA metabolism in whole brain cannot explain all of the pharmacological effects observed with these GABA-T inhibitors. It is possible that regional biochemical differences within the brain or differential effects on other enzyme or neurotransmitter systems of the various inhibitors may provide clearer explanations for the differences in their actions.

In the case of DPA, there is considerable controversy as to how this branched-chain fatty acid alters brain GABA metabolism (30-32). Irrespective of the means of influencing brain GABA, our results indicate that aspartate is as likely a candidate as GABA to explain the anti-seizure effects of DPA in animals.

On the basis of pharmacokinetic data, however, it would appear that neither a GABA-ergic nor an aspartate mechanism is responsible for the clinical anti-convulsant properties of DPA. Plasma concentrations of DPA associated with clinical anti-epileptic effects are 0.5 - 1.0 mM (33-36). The concentrations of DPA in the CNS at these plasma concentrations are likely to be considerably lower since DPA has been shown to have a low volume of distribution in man (34,37, 38), probably due to a high degree of binding to human serum albumin (37); i.e., 95% bound compared to 12% in the mouse (39). Further, the concentration of DPA in human cerebrospinal fluid is only 10% of corresponding serum levels (40,41). Even in mice, maximum brain concentrations of DPA represent only 1/4 to 1/5 of serum levels (42). Hence, it is likely that maximum brain concentrations of DPA in humans associated with therapeutic plasma concentrations are considerably less than 0.2 mM. Since the K_I of DPA for inhibition of human brain GABA-T is 40 mM or about four times greater than that of rat brain enzyme (43), it is doubtful that brain GABA is affected in man.

If the clinical anti-epileptic activity of DPA is not due to alterations in GABA metabolism, it remains uncertain whether an elevation in brain GABA might have any eventual therapeutic usefulness. On theoretical and neurochemical grounds, it has been suggested that Huntington's disease (44-46), tardive dyskinesias (47,48), Parkinson's disease (44), schizophrenia and other psychoses (49) and epilepsy may be among the clinical syndromes amenable to treatment with effective GABA-T inhibitors. Due to their inherent potency and specificity, the enzyme-activated inhibitors of GABA-T are the most promising candidates available to validate some of these suggestions in clinical trials and may represent a new approach to the treatment of these and other neuropsychiatric disorders.

REFERENCES

(1) Killam, K. F. and Bain, J. A. (1957) J. Pharmacol. Exp. Therapeut. 119, 255-262.
(2) DaVanzo, J. P., Greig, M. E. and Cronin, M. A. (1961) Amer. J. Physiol. 201, 833-837.
(3) Baxter, M. G., Fowler, L. J., Miller, A. A. and Walker, J. M. G. (1973) Brit. J. Pharmacol. 47, 681P.
(4) Baxter, C. F. and Roberts, E. (1960) Proc. Soc. Exp. Biol. Med. 104, 426-427.
(5) Maynert, E. W. and Kaji, H. K. (1962) J. Pharmacol. Exp. Therapeut. 137, 114-121.
(6) Tapia, R., De la Mora, M. P. and Massieu, H. (1969) Ann. N.Y. Acad. Sci. 166, 257-266.
(7) Kuriyama, K., Roberts, E. and Rubinstein, M. K. (1966) Biochem. Pharmacol. 15, 221-236.
(8) Pinder, R. M., Brogden, R. N., Speight, T. M. and Avery, G. S. (1977) Drugs, 13, 81-123.
(9) Simler, S., Ciesielski, L., Maitre, M., Randrianarisoa, H. and Mandel, P. (1973) Biochem. Pharmacol. 22, 1701-1708.
(10) Metcalf, B. W. and Casara, P. (1975) Tetrahedron Lett. 3337-3340.
(11) Lippert, B., Metcalf, B. W., Jung, M. J. and Casara, P. (1977) Eur. J. Biochem. 74, 441-445.
(12) Metcalf, B. W., Lippert, B. and Casara, P. (1978) In: Enzyme-Activated Irreversible Inhibitors (N. Seiler, M. J. Jung and J. Koch-Weser, eds.) Elsevier/North-Holland Biomedical Press, Amsterdam, pp. 123-133.

(13) Kobayashi, K., Mijazawa, S., Terahara, A., Mishima, H. and Kurihara, H. (1976) Tetrahedron Lett. 537-540.
(14) Jung, M. J. and Metcalf, B. W. (1975) Biochem. Biophys. Res. Commun. 67, 301-306.
(15) Rando, R. R. and Bangerter, F. W. (1977) Biochem. Biophys. Res. Commun. 76, 1276-1281.
(16) Jung, M. J., Lippert, B., Metcalf, B. W., Schechter, P. J., Böhlen, P. and Sjoerdsma, A. (1977) J. Neurochem. 28, 717-723.
(17) Schechter, P. J., Tranier, Y., Jung, M. J. and Böhlen, P. (1977) Eur. J. Pharmacol. 45, 319-328.
(18) Schlesinger, K., Stavnes, K. L. and Boggan, W. O. (1969) Psychopharmacologia, 15, 226-231.
(19) Litchfield, J. T., Jr. and Wilcoxon, F. (1949) J. Pharmacol. Exp. Therapeut. 96, 99-112.
(20) Schechter, P. J., Tranier, Y., Jung, M. J. and Sjoerdsma, A. (1977) J. Pharmacol. Exp. Therapeut. 201, 606-612.
(21) Schechter, P. J. and Tranier, Y. (1977) Psychopharmacology, 54, 145-148.
(22) Paton, W. D. M. (1967) Brit. J. Pharmacol. 29, 350-366.
(23) Frey, H.-H. and Löscher, W. (1976) Arzneim.-Forsch. 26, 299-301.
(24) Godin, Y., Heiner, L., Mark, J. and Mandel, P. (1969) J. Neurochem. 16, 869-873.
(25) Schechter, P. J., Tranier, Y. and Grove, J. (1978) J. Neurochem. 31, 1325-1327.
(26) Krnjević, K. (1970) Nature (Lond.) 228, 119-124.
(27) Wood, J. D. and Peesker, S. J. (1974) J. Neurochem. 23, 703-712.
(28) Wood, J. D. and Peesker, S. J. (1975) J. Neurochem. 25, 277-282.
(29) Abe, M. and Matsuda, M. (1976) J. Biochem. 80, 1165-1171.
(30) Harvey, P. K. P., Bradford, H. F. and Davison, A. N. (1975) FEBS Lett. 52, 251-254.
(31) Sawaya, M. C. B., Horton, R. W. and Meldrum, B. S. (1975) Epilepsia, 16, 649-655.
(32) Anlezark, G., Horton, R. W., Meldrum, B. S. and Sawaya, M. C. B. (1976) Biochem. Pharmacol. 25, 413-417.
(33) Meinardi, H., Hanke, N. F. J. and Van Beveren, J. (1974) Pharm. Weekbl. 109, 45-47.
(34) Schobben, F., van der Kleijn, E. and Gabreëls, F. J. M. (1975) Eur. J. Clin. Pharmacol. 8, 97-105.
(35) Loiseau, P., Brachet, A. and Henry, P. (1975) Epilepsia, 16, 609-615.
(36) Baruzzi, A., Bordo, B., Bossi, L., Castelli, D., Gerna, M., Tognoni, G. and Zagnoni, P. (1977) Int. J. Clin. Pharmacol. 15, 403-408.
(37) Klotz, U. and Antonin, K. H. (1977) Clin. Pharmacol. Therapeut. 21, 736-743.
(38) Perucca, E., Gatti, G., Frigo, G. M., Crema, A., Calzetti, S. and Visintini, D. (1978) Brit. J. Clin. Pharmacol. 5, 495-499.

(39) Löscher, W. (1978) J. Pharmacol. Exp. Therapeut. 204, 255-261.
(40) Meijer, J. W. A. and Hessing-Brand, L. (1973) Clin. Chim. Acta, 43, 215-222.
(41) Johannessen, S. I. (1977) Arzneim.-Forsch. 27, 1083-1085.
(42) Löscher, W. and Esenwein, H. (1978) Arzneim.-Forsch. 28, 782-787.
(43) Maitre, M., Ciesielski, L., Cash, C. and Mandel, P. (1978) Biochim. Biophys. Acta, 522, 385-399.
(44) Barbeau, A. (1973) Lancet, II, 1499-1500.
(45) Shoulson, I., Kartzinel, R. and Chase, T. N. (1976) Neurology, 26, 61-63.
(46) Perry, T. L., MacLeod, P. M. and Hansen, S. (1977) New Engl. J. Med. 297, 840.
(47) Linnoila, M., Viukari, M. and Hietala, O. (1976) Brit. J. Psychiat. 129, 114-119.
(48) Palfreyman, M. G., Huot, S., Lippert, B. and Schechter, P. J. (1978) Eur. J. Pharmacol., 50, 325-336.
(49) Roberts, E. (1972) Neurosci. Res. Program Bull. 10, 468-482.

L-GLUTAMATE DECARBOXYLASE

Paul Y. Sze

Department of Biobehavioral Sciences, The University of Connecticut, Storrs, Connecticut 06268, U.S.A.

INTRODUCTION

Phylogenetically, enzymatic decarboxylation of L-glutamate to form γ-aminobutyrate (GABA) is found throughout living organisms, from bacteria and fungi to animals. In vertebrate and invertebrate species, L-glutamate decarboxylase (GAD) is confined mainly, though not exclusively, to the nervous system (1,2). Using anti-mouse GAD immunoglobulin, cross-reactivity tests by immunodiffusion and microcomplement fixation methods have indicated that brain GAD is very similar among various mammalian species, whereas the enzymes from frog, bird and fish differ greatly from the mouse enzyme (3). Although immunochemical data on invertebrate GAD are not available, biochemical data have revealed that GAD from crustacean inhibitory neurones differs in some fundamental ways from the mouse brain enzyme; e.g., its behavior toward enzyme inhibitors (4). Bacterial GAD, purified from Escherichia coli (5), is a distinctly different protein from the mammalian brain enzyme. Moreover, a mammalian nonneural GAD, purified from bovine heart, has been shown to be different from both the bacterial and the brain enzymes (2). Even within the same species, e.g., between several inbred strains of mice, microcomplement fixation tests have revealed small immunochemical differences in brain GAD (6), indicating the possible existence of enzyme variants according to genotypes. Thus, there seems to be a high degree of polymorphism among the enzyme proteins that catalyze the α-decarboxylation of glutamate, and GAD appears to be evolutionarily labile. In this article, some of the current issues concerning this enzyme will be examined; emphasis will be placed on the mammalian central nervous system (CNS).

LOCALIZATION OF GAD IN THE MAMMALIAN CNS

In higher nervous systems where heterogeneous types of neurones are interconnected, GAD is a phenotypic protein of the types of neurones that synthesize and release GABA as their neurotransmitter ("GABA-ergic" neurones). In the mammalian CNS, GAD activity has been found in virtually all brain areas examined, as well as in the spinal cord (7,8,9,10), indicating that GAD-containing neurones and their processes are scattered throughout the CNS. The highest specific enzyme activities are localized in substantia nigra, nucleus accumbens, globus pallidus and dentate nucleus. In a recent study (11), GAD activity was measured in 70 discrete brain nuclei diligently dissected from the rat by the "punching" technique. Again, the enzyme activity was found in virtually all of the nuclei examined, the highest activity being present in the reticular part of the substantia nigra. Biochemical measurement of neuronal enzymes such as GAD does not distinguish between neuronal cell bodies or their processes. However, with the availability of specific antibodies to GAD, recent immunocytochemical localization of the enzyme at the electron microscopic level has provided typological information on the organization of GABA-ergic neurones and their targets, particularly in the cerebellum (12,13) and in the substantia nigra (14).

When brain homogenates are subjected to subcellular fractionation (15,16,17,18,19), about 40% of the total GAD in the whole tissue is associated with synaptosomes. A comparable amount is found in the cell cytosol, and a substantial amount can also be found in the mitochondrial fraction. Little or no GAD is associated with microsomes or highly purified nuclei. It is generally believed that in synaptic terminals, GAD exists in soluble and particulate-bound forms. This contention is supported by the biochemical demonstration that GAD can associate with synaptic vesicles *in vitro* through a Ca^{2+}-dependent process (18,20,21) and by immunocytochemical localization of the enzyme on synaptic vesicle membranes (22). The possible alteration of its kinetic properties which could occur as a result of this binding represents a potentially important regulatory process, but it has not yet been explored. To date, knowledge is still lacking concerning the dynamic aspects of GAD translocation; i.e., its rate of axoplasmic flow in maintaining steady-state levels of the enzyme in synaptic terminals; its turnover rate in either nerve terminals or neuronal cell bodies.

ENZYMOLOGICAL PROPERTIES OF GAD

Although mammalian GAD was first described in mouse brain in 1950 (23), total purification of the enzyme was not achieved until two decades later (24). Subsequently, GAD has also been purified from human (25,26), rat (see Blindermann *et al.*, this volume) and bovine (27) brains. In all of these studies, the purification involved conventional procedures such as fractional precipitation by ammonium-sulfate and successive chromatography on Sephadex G-200, calcium phosphate gel and DEAE-Sephadex. The homogeneity of the purified preparations was assessed by polyacrylamide gel electrophoresis and/or sedimentation equilibrium (24,25,27), and in the case of mouse enzyme, also by immunodiffusion and immunoelectrophoresis (3, 6,28). The molecular weight (M.W.) of the purified mouse enzyme was estimated to be about 85,000 daltons by both sedimentation equilibrium and Sephadex gel filtration (24). When such a native enzyme was treated with guanidine-hydrochloride and mercaptoethanol, a homogeneous species with a M.W. of about 44,000 was obtained, indicating two ultracentrifugally-indistinguishable subunits. In SDS-urea gels, electrophoresis revealed multiple bands, with M.W. ranging from 16,000 to 129,000; these were interpreted as dissociation and aggregation (28,29). Since the M.W. of almost all the bands are close to integral multiples of a M.W. of approximately 15,000, this M.W. may represent the smallest dissociable subunit of GAD. Thus, mouse brain GAD might consist of six minimal subunits of 15,000, organized into two larger subunits (29). However, from studies on the human (25) and rat enzyme (see Blindermann _et al._, this volume), the M.W. of purified GAD was estimated to be about 140,000 by polyacrylamide gradient gel electrophoresis. Moreover, in SDS-urea gels, only a single protein band was obtained, corresponding to a M.W. of 67,000. The presence of a single N-terminal amino acid (L-alanine), in the human enzyme (25), further indicated the presence of only two similar, if not identical, subunits. Unless mouse GAD is an entirely different protein from the rat and human enzyme, which appears unlikely from immunochemical data (3), the discrepancies in M.W. and in the electrophoretic patterns between laboratories must be attributed to methodological sources.

The GAD with a M.W. of 85,000 reported for mouse brain was prepared from a synaptic terminal fraction (24). When whole tissue homogenate of mouse brain in hypotonic buffer was used as the starting material, two GAD peaks from gel filtration were found (30). The one with lower M.W. eluted at a volume that would be expected for the GAD of M.W. 85,000. The other eluted at the exclusion volume on Sephadex G-200, indicating a M.W. greater than 200,000. When a cell cytosol fraction prepared from mouse brain was used, only the peak with high M.W. could be obtained from gel filtration (30). From these data, it appears that two forms of GAD may exist in brain: one with a M.W. greater than 200,000 localized in neuronal cell bodies; the other with a M.W. of about 85,000 localized in synaptic terminals. That two major peaks of brain GAD might exist has also been reported for rat brain by other workers (31). In several other laboratories (25,26,27), however, only a single form of GAD was obtained with Sephadex gel filtration using whole tissue of human, bovine or rat brain. The reason for these different results is not clear. Possibly, one form of GAD is labile, or readily convertible into the other, depending on the conditions of tissue preparation. Alternatively, only a single form might exist in situ, but a tight association with another macromolecular substance during tissue preparation could produce an artifactual second form. Whatever the source of the discrepancy, it it noteworthy that even in the case where two forms were observed with gel filtration, the two forms were found to be kinetically and immunologically identical (30). Future work will be required to clarify this issue on molecular forms, as well as the uncertainties concerning molecular weight and subunit structure.

The chemical properties of the active site on mammalian brain GAD are little known. The binding of glutamate to the enzyme for catalytic action appears to be very specific. Other than L-glutamate, only L-aspartate can serve as a substrate to a small extent (24). Judging from the effects of pH on the catalytic activity, it has been postulated (24) that a histidyl imidazole group participates in a proton transfer reaction, as in the case of other decarboxylases. There seems to be at least one sensitive sulfhydryl group on or near the active site that is required for catalytic activity (24, 28,32). Also, there exists an absolute requirement for pyridoxal 5'-phosphate (PLP). In bacterial GAD from Escherichia coli, which is a hexamer with a M.W. of about 300,000 (5), each subunit has a binding site for PLP (33). Whether this also holds for brain GAD is not yet known. Although the brain enzyme has a high content of acidic amino acids and a relatively low pH_i (25), the active site itself may be positively charged, as suspected from the competitive inhibition of the enzyme by a variety of dicarboxylic acids (29,32).

The catalytic mechanism of brain GAD can only be extrapolated from those of other decarboxylases. In several amino acid decarboxylases studied, including bacterial GAD (33,34), PLP is linked to the ε-amino group of a specific lysyl residue by a labile covalent bond. A general mechanism of PLP-dependent decarboxylation is be-

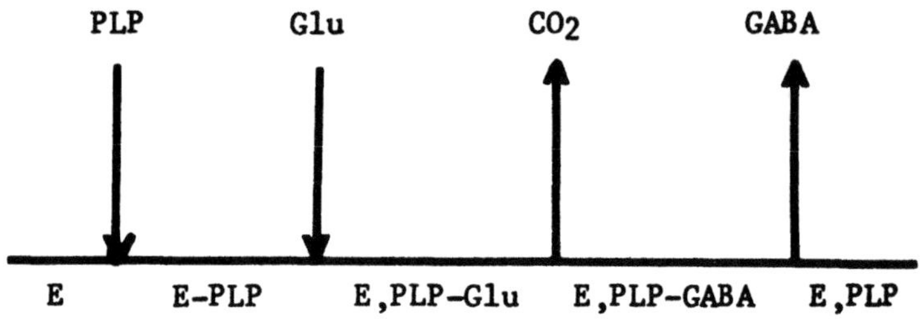

Fig. 1. A hypothetical kinetic scheme of GAD catalysis proposing an "Order Bi Bi" mechanism.

lieved to involve an electrophilic attack from the α-amino nitrogen of the substrate amino acid, displacing the enzyme-PLP bond with the formation of a new Schiff base between PLP and the substrate amino acid (35). After decarboxylation and subsequent release of the amine product, the enzyme-PLP covalent bond is re-established. If brain GAD is not an exception to this general mechanism of PLP-dependent amino acid decarboxylation, its kinetics would be expected to follow an "Order Bi Bi" mechanism, as shown in Figure 1. Purposefully speculative, this scheme is intended to illustrate the need for kinetic information. For example, there is the argument that binding of glutamate or ATP may promote dissociation of PLP from the holoenzyme, thereby modulating its activity (31,36). If the kinetics of enzyme catalysis are known at the basic level, issues of this kind could be readily subjected to an analysis.

Some normal metabolites are known to modify the catalytic activity of GAD. A few of these are shown in Table I. Several dicarboxylic acids derived from carbohydrate metabolism can inhibit GAD activity at millimolar concentrations. Among the neurotransmitter substances tested, norepinephrine exhibits the most potent inhibitory effect (32). In contrast to the crustacean enzyme (4), mammalian GAD is not inhibited by GABA, even at a concentration of 10 mM. Although it has been suggested that GAD activity may be modulated by Cl^- during membrane conduction (37), Cl^- is not likely to be a significant inhibitor of the enzyme. A number of physiological and

Table I - Inhibition of GAD[a]

Inhibitor	K_i (M)
α-Ketoglutarate	1.2×10^{-3}, competitive
Fumarate	1.8×10^{-3}, competitive
Glutarate	3.5×10^{-3}, competitive
Oxaloacetate	6.2×10^{-3}, competitive
GABA	(10% inhibition at 10^{-2} M)
Norepinephrine	1.3×10^{-3}, non-competitive
Zn^{2+}	1.0×10^{-5} ?
Cl^-	1.7×10^{-2}, competitive
ATP[b]	50% inhibition on the order of 10^{-3} M in the presence of low PLP concentrations

[a] From Wu (29,32).
[b] From Seligmann *et al.* (36).

pharmacological substances appear to have inhibitory effects on the catalytic activity of brain GAD *in vitro*, at realistic or unrealistic concentrations. Lists of these substances have been provided in recent reviews (29,38).

CELLULAR REGULATION OF GAD

Enzymes are regulated not only by binding with effectors that alter catalytic activity or by cofactor availability, but more importantly also by a variety of cellular mechanisms that involve macromolecular synthesis or degradation and post-translational modification of protein molecules through covalent bond formation or hydrolysis. These latter issues form the core of current regulatory cell biology. The homeostasis of a neurone, including its levels of enzymes, as in the case of other cell types, is determined by metabolic events within the cell, which are further influenced by extracellular signals such as ions, hormones and neurotransmitters. Such a concept of enzyme regulation in neurones can be best illustrated by tyrosine hydroxylase (TH), perhaps the most intensively studied neuronal enzyme in terms of regulatory phenomena. For example, in the superior cervical ganglion, the level of TH is induced by enhanced cholinergic input from central innervation ("trans-synaptic" regulation) (39,40,41,42), influenced by the adrenal hormones, epinephrine (43,44) and glucocorticoids (45,46,47), and diminished by transection of target organs, such as the salivary gland ("retrograde" regulation; see ref. 48). The cholinergic receptor-mediated induction of TH, at least in adrenal chromaffin cells, may involve cyclic nucleotides in a sequence of intracellular events including de novo RNA synthesis (49,50,51,52). Moreover, in central noradrenergic neurones, neural stimulation may lead to a post-translational modification of the enzyme which involves phosphorylation of the enzyme molecule by a cyclic-AMP-dependent protein kinase (53,54,55). Within this context of enzyme regulation, as illustrated by TH, what is presently known about GAD? Unfortunately, no element of GAD regulation has been systematically studied. Nevertheless, to pave the way for future work, a bold attempt will be made here to assemble some available data on GAD, preliminary or indirect as they may be.

In general, biochemical development provides a dynamic system whereby the underlying regulatory events may be probed. In the case of GAD, developmental changes in enzyme activity have been described in the CNS of several species, including regional examinations in brain (56,57,58,59,60,61). The increase of brain GAD activity which occurs during development, at least in the mouse, is due to an increase in the enzyme protein, as determined recently by immunological quantitation (27). In the optic lobe of the chick embryo, the developing levels of GAD activity could be readily depressed by increasing tissue GABA levels (62,63), as achieved either directly by injecting GABA into the yolk sac or indirectly by administration of aminooxyacetic acid (AOAA), a potent inhibitor of GABA-transaminase

(GABA-T). Similar findings on such an effect of GABA on GAD activity *in vivo* have been reported for developing mouse brain (64,65) and recently in chick embryo retina (66). Cultured astrocytes obtained from dissociated mouse hemispheres have been used to provide a more direct demonstration; GAD levels were similarly depressed in the presence of GABA added to the culture medium (Tardy *et al.*, this volume). Since GABA is not an inhibitor of vertebrate brain GAD, its effect on the development of the enzyme *in vivo* or in culture may be attributable to a feedback repression of the enzyme through cellular mechanisms regulating its synthesis or degradation. A definitive experiment using antibody titration to determine enzyme amounts, however, has not yet been performed. During early development of the nervous system, before significant appearance of GAD, GABA may be formed mainly from putrescine (66) via an alternate pathway involving ornithine decarboxylase in its initial step (67, 68,69). It has been speculated (66) that if GABA does indeed repress GAD synthesis, gene expression for GAD in GABA-ergic neurones during early development may be mediated by regulatory events related to ornithine decarboxylase, an enzyme known to respond to a variety of extracellular signals, including hormones and neurotransmitters (70, 71,72). This developmental-genetic model, although attractive, has not yet been tested. For phylogenetic comparison, it may be noted that in *E. coli* GABA does not affect GAD synthesis in either the constitutive strain H or the inducible strain H/Gl_1, a mutant derived from the wild type H (73,74). L-Glutamate, L-aspartate and α-keto-glutarate can induce GAD synthesis by strain H/Gl_1, whereas α-methyl-DL-glutamate represses enzyme formation by H/Gl_1 but not by strain H. The differences in GAD levels among various strains of *E. coli* K12 have been attributed to different alleles of a regulatory gene, gadR, which may vary in their response to regulatory molecules (75).

Brain GAD may also be inducible. An early report claimed an increase of brain GAD activity following administration of glutamate (76), which was not replicated in our laboratory. However, an on-

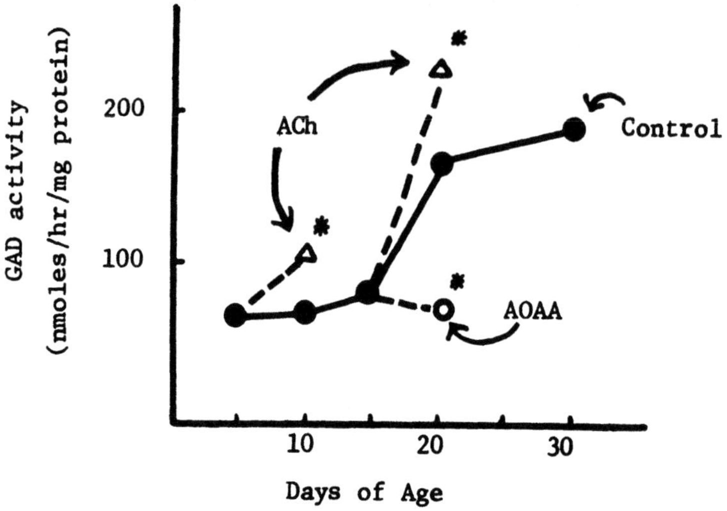

Fig. 2. Effect of ACh treatment on GAD activity in cerebellar cortex. Mice were given daily injections of ACh (1 mg/kg, s.c.) preceded by paraoxon (1 mg/kg) 2 hrs earlier, beginning at 5 or 15 days of age as indicated. For comparison, one 15-day-old group was given AOAA as previously described (64,65). The cerebellar cortex was dissected after 5 days of the treatment, and GAD was measured according to the procedure previously described (24) with some modifications (65). * $p < 0.01$. (P. Y. Sze, unpublished results).

going study has indicated that acetylcholine (ACh) can affect <u>in vivo</u> GAD activity in the cerebellar cortex of developing mouse (Fig. 2). After daily injections of ACh (1.0 mg/kg) together with paraoxon (1.0 mg/kg) to neonatal mice for 5 days, beginning at 5 days of age, GAD activity in cerebellar cortex was found to be significantly elevated. Although antibody titration has not been performed, GAD from treated and control tissues exhibited identical kinetic characteristics (K_m for glutamate and for PLP) and thermostability (at 50°C). These data are in accord with the suggestion that the increased enzyme activity may be due to induction. In parallel experiments for comparison, the

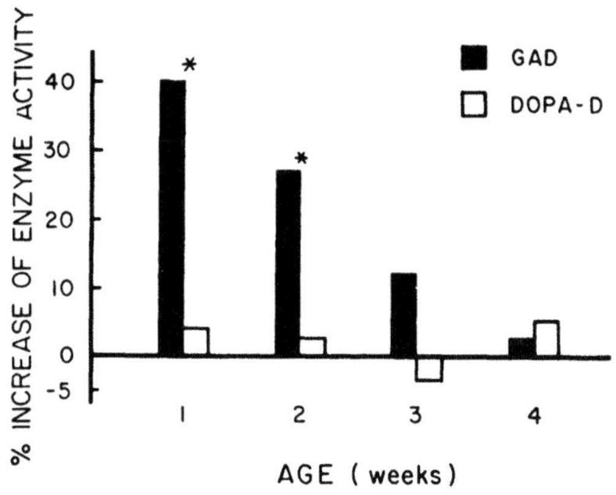

Fig. 3. Responsivity of cerebellar cortical GAD to ACh treatment in mice of various ages. Mice were treated as described in the legend to Figure 2, beginning at the indicated ages. L-Aromatic amino acid decarboxylase (DOPA-D) activity was measured as a reference in the same tissue preparations using [1-^{14}C]dihydroxyphenylalanine as the substrate. * $p < 0.01$ (P. Y. Sze, unpublished results).

effect of AOAA at depressing GAD levels (described in the previous paragraph and in ref. 65) was reproduced in the cerebellar cortex of similar neonatal mice. Further, the responsivity of GAD to chronic ACh treatment was examined at various stages of development (Fig. 3). The response was apparent only during the first two weeks of postnatal life. The lack of response after two weeks could have been due to the full maturation of the blood-brain barrier. The mechanism of this increase in GAD activity, presumably enzyme induction initiated by pharmacologically-enhanced cholinergic activity, is presently not known. It is noteworthy that while the AOAA effect was universal in all brain regions, the increase in GAD activity following cholinergic treatment was found only in the cerebellar cortex. GABA-ergic neurones in the cerebellar cortex are known to receive excitatory inputs extrinsically from the climbing and mossy fibers and intrinsically from the parallel fibers. So far, the evidence suggests that only a minor portion of the mossy fibers may be cholinergic. Thus, although one is tempted to speculate here about a trans-synaptic cholinergic influence on GAD, current understanding is more in favor of other possibilities such as an indirect or metabolic effect of ACh.

However, a trans-synaptic mechanism regulating GAD in the CNS has been implicated in other studies. One study has shown a significant decrease of GAD activity in the cervical spinal cord of rats following section of the dorsal roots one month earlier (77). Supported by autoradiographic and other data, the authors concluded that the reduction of enzyme activity following deafferentation of the spinal cord was restricted to GABA-ergic interneurones located within the superficial layers of the anterior horn. This would be analogous to the reduction of TH activity in the sympathetic ganglion that followed transection of pre-ganglionic fibers, a case of trans-synaptic regulation. In the striatum, where GABA-ergic neurones receive dopaminergic innervation from the nigrostriatal pathway, GAD activity was found to be markedly reduced in Parkinson's disease which involves a degeneration of the dopaminergic pathway (78). In a parallel study on rats, striatal GAD was significantly increased following long-term L-DOPA administration (79). These responses of striatal GAD may also be interpreted as a case of "trophic" effect from presynaptic dopaminergic activity. However, a clear study of trans-synaptic regulation of GAD, in these or other neuronal systems, still remains to be accomplished. In the mammalian CNS, where GABA-ergic neurones are diffusely scattered, mostly as interneurones with multiple connections, a definitive demonstration of trans-synaptic regulation will not be an easy task. A simpler neuronal system, perhaps a peripheral one in a lower species, should be studied.

Not many physiological states or pharmacological treatments are known to modify the steady-state levels of brain GAD activity. This hinders the study of cellular regulation of the enzyme. Obviously, without an experimental perturbation of the steady-state, it would be difficult to identify the regulatory processes. Although an early study indicated that adrenalectomy produced an increase in GAD activity of rat brain (80), various attempts in our laboratory have failed to demonstrate any effect of adrenal-pituitary function. Other early reports indicated that brain GAD activity was increased in animals during hibernation (81,82). More recently, an increase of GAD activity has been localized in the thalamus following an acute dose of morphine (83) and also in the whole brain following an acute dose of phenobarbital (84). The responses of GAD to these sedative drugs are particularly interesting, since in both cases the increase of enzyme activity occurred within an hour after drug administration. The rapidity of these responses suggests that the observed increase of activity may result from an activation of the enzyme.

Can GAD indeed be activated as a short-term response to altered neuronal activity, such as that occurring during sedation, or under any condition? A recent study has provided some in vitro evidence that supports such a possibility (85). In that study, a significant increase of glutamate decarboxylation by striatal slices was found after pre-incubation of the slices in the presence of either veratridine or high K^+ concentration. Since the uptake of glutamate by the brain slices was not affected by the pre-incubation, the authors suggested that depolarization by veratridine or high K^+ might lead to an activation of GAD. Obviously, many explanations are possible for the increased rate of glutamate decarboxylation observed in slice preparations; that enzyme activation is involved must await at least kinetic characterization in soluble preparations. If, however, an activation does indeed occur here, one may recall a similar activation of TH in the hippocampus following electrical stimulation of noradrenergic neurones in the locus coeruleus (86). The molecular mechanism of TH activation in this case involved phosphorylation of the enzyme, perhaps by a cyclic-AMP-dependent protein kinase (53,54, 55). Another neuronal enzyme of the mammalian CNS, tryptophan hydroxylase (TPH), appears also to be activated through a phosphoryla-

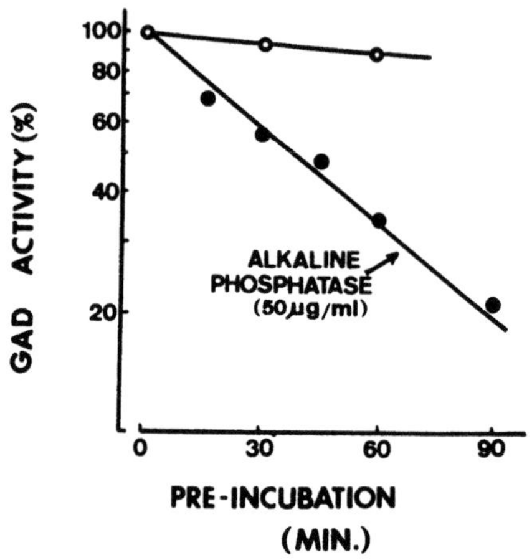

Fig. 4. Effect of alkaline phosphatase on GAD activity. Synaptosomal pellets prepared from mouse whole brains were rehomogenized in hypotonic buffer (50 mM Tris-acetate, pH 7.4) and centrifuged at 100,000 g for 1 hr. The supernatant was pre-incubated at 37°C in the presence of alkaline phosphatase (bovine intestinal source; Sigma Chemical Co.). The pre-incubated preparation was chromatographed on a DEAE-cellulose column. Alkaline phosphatase was removed by elution with 0.4 M NaCl-50 mM Tris-acetate (pH 6.5), and GAD was then eluted with 1 M NaCl-50 mM Tris-acetate. After dialysis, GAD activity was measured by the procedure previously described (24,36), except that 0.2 M phosphate (pH 7.0) was used to inhibit any residual phosphatase activity and the PLP present was increased to 1 mM (from Sze et al., ref. 90).

tion process, perhaps involving a cyclic nucleotide-independent protein kinase (87,88,89).

Regardless of whether or not an activation of GAD can occur in the CNS as a function of neuronal activity, a body of data is emerging in our laboratory which suggests that, as in the case of TH or TPH, brain GAD may be regulated by a protein phosphorylation process. Two lines of evidence are currently accumulating. First, as shown in Figure 4, GAD in a supernatant preparation from mouse brain synaptosomes was found to lose its activity rapidly in the presence of purified alkaline phosphatase (AP), an enzyme known to dephosphorylate a variety of phosphate esters as well as phosphoproteins. Following pre-incubation with AP, GAD activity was measured after the phosphatase had been removed by DEAE-cellulose chromatography using step-wise salt gradients. This inactivation of GAD by AP could be totally prevented by addition of inorganic phosphate (0.1 M), EDTA (5 mM), or arsenate (10 mM), all potent inhibitors of the phosphatase, or by the removal of Zn^{2+} from the phosphatase with o-phenanthroline. Therefore, the inactivation was not due to any possible contaminating proteolytic activity in the phosphatase preparation. From Lineweaver-Burk analysis, the inactivation was characterized by a reduction of V_{max} with no change in K_m for PLP. This irreversibility of the inactivation by high concentrations of PLP in the assay indicated that hydrolysis of the cofactor PLP (which is a substrate for the phosphatase) could not account for the loss of GAD activity. Thus, it is reasonable to suspect that AP may act on protein-bound phosphate, present in GAD itself or in a protein that modulates GAD. A second line of evidence is the demonstration that GAD can be activated by a system that favors phosphorylation. Table II briefly summarizes the key data from an ongoing study. Preincubation of a supernatant preparation from mouse cerebellum in the presence of ATP and Mg^{2+} nearly doubled GAD activity. Once activated, the increased enzyme activity was fully retained after dialysis which removed the added ATP and Mg^{2+}. The non-dialyzability indicated that the increase in enzyme activity was not the result of a direct enzyme-ligand interaction, such as the formation of GAD-Mg^{2+}-ATP complex, but rather, that it may be due to covalent bond formation. Reinforced by the effect of phosphatase, this activation by Mg^{2+}-ATP further strengthens the possibility that brain GAD may be regulated by a phosphorylation process. A detailed characterization of the Mg^{2+}-ATP activating system for GAD is currently underway in our laboratory.

Table II - Activation of GAD by Mg^{2+}-ATP

Addition	% Control
None (control)	100
ATP	97
Mg^{2+}	100
ATP + Mg^{2+}	190

Mouse cerebellum homogenate in 20 mM Tris-acetate (pH 7.5) was centrifuged at 100,000 g for 1 hr. The supernatant was treated with 5 mM Ca^{2+} plus 5 mM mercaptoethanol at 30°C for 10 min. The supernatant was then incubated at 37°C for 15 min with the addition of ATP (1 mM) and/or Mg^{2+} (1 mM) and in the presence of 5 mM mercaptoethanol. GAD activity was determined in aliquots of the incubated supernatant by the method previously described (24,36), except that the PLP present was increased to 1 mM. (R. Alderson and P. Y. Sze, unpublished results).

It may be mentioned, as a historical note, that protein phosphorylation as a regulatory mechanism was first demonstrated in glycogen phosphorylase, which is also a PLP-containing enzyme (91).

SUMMARY

Much progress has been made in recent years regarding enzymological aspects of mammalian brain GAD, such as its purification and characterization, but some uncertainty still remains concerning its molecular weight and forms, and its subunit structure. The availability of antibodies to this enzyme has allowed immunocytochemical studies which have provided important information on the intrinsic organization of GABA-ergic neurones in the CNS, particularly in the cerebellum and nigrostriatal pathway. With the increased understanding of the enzymology of GAD and the distribution of central

GABA-ergic neurones, it is becoming feasible to study the regulatory biochemistry of GAD in terms of control and adaptive mechanisms at the cellular level. In our own laboratory, as well as in others, initial approaches have already begun. Obviously, cellular regulation of this phenotypic enzyme is an important issue for the understanding of GABA-ergic neurones and their functions.

The preparation of this manuscript was supported in part by U.S.P.H.S. grant MH29237-02.

REFERENCES

(1) Wu, J. Y., Chude, O., Wein, J., Roberts, E., Saito, K. and Wong, E. (1978) J. Neurochem. 30, 849-857.
(2) Wu, J. Y. (1977) J. Neurochem. 28, 1359-1367.
(3) Saito, K., Wu, J. Y., Matsuda, T. and Roberts, E. (1974) Brain Res. 65, 277-285.
(4) Molinoff, P. B. and Kravitz, E. A. (1968) J. Neurochem. 15, 391-409.
(5) Strausbauch, P. H. and Fischer, E. H. (1970) Biochemistry, 9, 226-233.
(6) Wong, E., Schousboe, A., Saito, K., Wu, J. Y. and Roberts, E. (1974) Brain Res. 68, 133-142.
(7) Fahn, S. (1976) In: GABA in Nervous System Function (E. Roberts, T. N. Chase and D. B. Tower, eds.) pp. 169-186, Raven Press, New York.
(8) Graham, L. T. and Aprison, M. H. (1969) J. Neurochem. 16, 559-566.
(9) Müller, P. B. and Langemann, H. (1962) J. Neurochem. 9, 399-401.
(10) Albers, R. W. and Brady, R. O. (1959) J. Biol. Chem. 234, 926-928.
(11) Tappaz, M. L., Brownstein, M. J. and Palkovits, M. (1976) Brain Res. 108, 371-379.
(12) McLaughlin, B. J., Wood, J. G., Saito, K., Barber, R., Vaughn, J. E., Roberts, E. and Wu, J. Y. (1974) Brain Res. 76, 377-391.
(13) McLaughlin, B. J., Wood, J. G., Saito, K., Roberts, E. and Wu, J. Y. (1975) Brain Res. 85, 355-371.
(14) Ribak, C. E., Vaughn, J. E., Saito, K., Barber, R. and Roberts, E. (1976) Brain Res. 116, 287-298.
(15) Fonnum, F. (1968) Biochem. J. (Lond.) 106, 401-412.
(16) Balázs, R., Dahl, D. and Harwood, J. R. (1966) J. Neurochem. 13, 897-905.

(17) Whittaker, V. P. (1965) Progr. Biophys. Biophys. Chem. 15, 39-96.
(18) Salganicoff, L. and DeRobertis, E. (1965) J. Neurochem. 12, 287-309.
(19) Van Kempen, G. M. J., Van den Berg, C. J., Van der Helm, H. J. and Veldstra, H. (1965) J. Neurochem. 12, 581-588.
(20) Kuriyama, K., Roberts, E. and Kakefuda, T. (1968) Brain Res. 8, 132-152.
(21) Haber, B. and Kuriyama, K. (1969) Brain Res. 14, 767-771.
(22) Wood, J. G., McLaughlin, B. J. and Vaughn, J. E. (1976) In: GABA in Nervous System Function (E. Roberts, T. N. Chase and D. B. Tower, eds.) pp. 133-148, Raven Press, New York.
(23) Roberts, E. and Frankel, S. (1950) J. Biol. Chem. 187, 55-63.
(24) Wu, J. Y., Matsuda, T. and Roberts, E. (1973) J. Biol. Chem. 248, 3029-3034.
(25) Blindermann, J.-M., Maitre, M., Ossola, L. and Mandel, P. (1978) Eur. J. Biochem. 86, 143-152.
(26) Drummond, G. S. and Goldstein, M. (1978) Trans. Amer. Soc. Neurochem. 9, 183.
(27) Hadjian, R. A. and Stewart, J. A. (1977) J. Neurochem. 28, 1249-1257.
(28) Matsuda, T., Wu, J. Y. and Roberts, E. (1973) J. Neurochem. 21, 159-166.
(29) Wu, Y. Y. (1976) In: GABA in Nervous System Function (E. Roberts, T. N. Chase and D. B. Tower, eds.) pp. 7-55, Raven Press, New York.
(30) Wu, J. Y., Wong, E., Saito, K., Roberts, E. and Schousboe, A. (1976) J. Neurochem. 27, 653-659.
(31) Tursky, T. and Lassanova, M. (1978) J. Neurochem. 30, 903-905.
(32) Wu, J. Y. and Roberts, E. (1974) J. Neurochem. 23, 759-767.
(33) Strausbauch, P. H. and Fischer, E. H. (1970) Biochemistry, 9, 233-238.
(34) Fasella, P. (1967) Ann. Rev. Biochem. 36, 185-210.
(35) Snell, E. E. and DiMari, S. J. (1970) In: The Enzymes (P. D. Boyer, ed.) Vol. II, pp. 335-368, Academic Press, New York.
(36) Seligmann, B., Miller, L. P., Brockman, D. E. and Martin, D. L. (1978) J. Neurochem. 30, 371-376.
(37) Susz, J. P., Haber, B. and Roberts, E. (1966) Biochemistry, 5, 2870-2876.
(38) Tapia, R. (1975) In: Handbook of Psychopharmacology (L. L. Iversen, S. D. Iversen and S. H. Snyder, eds.) Vol. 4, pp. 1-58, Plenum Press, New York.
(39) Thoenen, H. (1970) Nature (Lond.) 228, 861-862.
(40) Thoenen, H., Kettler, R., Burkard, W. and Sauer, A. (1971) Naunyn-Schmiedeberg's Arch. Exp. Path. Pharmak. 270, 146-160.
(41) Black, I. B. and Reis, D. J. (1975) Brain Res. 84, 269-278.
(42) Zigmond, R. E. and Chalagonitis, A. (1977) Trans. Amer. Soc. Neurochem. 8, 146.
(43) Hanbauer, I., Kopin, I. J., Guidotti, A. and Costa, E. (1975) J. Pharmacol. Exp. Therapeut. 193, 95-105.

(44) Markey, K. and Sze, P. Y. (1977) Trans. Amer. Soc. Neurochem. 8, 145.
(45) Hanbauer, I., Guidotti, A. and Costa, E. (1975) Brain Res. 85, 527-531.
(46) Otten, U. and Thoenen, H. (1975) Proc. Nat. Acad. Sci., U.S.A. 72, 1415-1419.
(47) Otten, U. and Thoenen, H. (1976) Mol. Pharmacol. 12, 353-361.
(48) Dibner, M. and Black, I. B. (1976) Brain Res. 103, 93-102.
(49) Guidotti, A. and Costa, E. (1973) Science, 179, 902-904.
(50) Guidotti, A., Hanbauer, I. and Costa, E. (1975) Adv. Cyclic Nucleotide Res. 5, 619-639.
(51) Chuang, D. M. and Costa, E. (1976) Mol. Pharmacol. 12, 514-518.
(52) Costa, E. and Guidotti, A. (1978) In: Psychopharmacology: A Generation of Progress (M. A. Lipton, A. DiMascio and K. F. Killam, eds.) pp. 235-246, Raven Press, New York.
(53) Harris, J. E., Baldessarini, R. J., Morgenroth, V. H., III and Roth, R. H. (1975) Proc. Nat. Acad. Sci., U.S.A. 72, 789-793.
(54) Morgenroth, V. H., III, Hegstrand, L. R., Roth, R. H. and Greengard, P. (1975) J. Biol. Chem. 250, 1946-1948.
(55) Lovenberg, W., Bruckwick, E. A. and Hanbauer, I. (1975) Proc. Nat. Acad. Sci., U.S.A. 72, 2955-2958.
(56) Sisken, B., Sano, K. and Roberts, E. (1961) J. Biol. Chem. 236, 503-507.
(57) Himwich, W. A. (1962) Int. Rev. Neurobiol. 4, 117-158.
(58) Van den Berg, C. J., Van Kempen, G. M. J., Schade, J. P. and Veldstra, H. (1965) J. Neurochem. 12, 863-869.
(59) Bayer, S. M. and McMurray, W. C. (1967) J. Neurochem. 14, 695-706.
(60) Roberts, E. and Kuriyama, K. (1968) Brain Res. 8, 1-35.
(61) Sze, P. Y. (1977) Brain Res. 122, 59-69.
(62) Haber, B., Sze, P. Y., Kuriyama, K. and Roberts, E. (1970) Brain Res. 18, 545-547.
(63) Sze, P. Y., Kuriyama, K., Haber, B. and Roberts, E. (1971) Brain Res. 26, 121-130.
(64) Sze, P. Y. (1970) Brain Res. 19, 322-325.
(65) Sze, P. Y. and Lovell, R. A. (1970) J. Neurochem. 17, 1657-1664.
(66) DeMello, F. G., Bachrach, U. and Nirenberg, M. (1976) J. Neurochem. 27, 847-851.
(67) Seiler, N., Al-Therib, M. J. and Kataoka, K. (1973) J. Neurochem. 20, 699-708.
(68) Seiler, N. and Lamberty, U. (1973) J. Neurochem. 20, 709-717.
(69) Seiler, N. and Al-Therib, M. J. (1974) Biochem. J. (Lond.) 144, 29-35.
(70) Byus, C. V. and Russell, D. H. (1975) Science, 187, 650-652.
(71) Matsuzawa, H. and Nirenberg, M. (1975) Proc. Nat. Acad. Sci., U.S.A. 72, 3472-3476.
(72) Bachrach, U. (1975) Proc. Nat. Acad. Sci., U.S.A. 72, 3087-3091.
(73) Halpern, Y. S. (1962) Biochim. Biophys. Acta, 61, 953-962.
(74) Halpern, Y. S., Even-Shoshan, A. and Cohen, S. (1965) Biochim. Biophys. Acta, 95, 321-328.

(75) Lupo, M. and Halpern, Y. S. (1970) J. Bacteriol. 103, 382-386.
(76) Kraus, P. (1968) Hoppe Seylers Z. Physiol. Chem. 349, 1425-1427.
(77) Kelly, J. S., Gottesfeld, Z. and Schon, F. (1973) Brain Res. 62, 581-586.
(78) Lloyd, K. G. and Hornykiewicz, O. (1973) Nature (Lond.) 243, 521-523.
(79) Hornykiewicz, O., Lloyd, K. G. and Davidson, L. (1976) In: GABA in Nervous System Function (E. Roberts, T. N. Chase and D. B. Tower, eds.) pp. 479-485, Raven Press, New York.
(80) Pandolfo, L. and Macaione, S. (1964) Giorn. Biochim. (Italy) 13, 247-252.
(81) Robinson, J. D. and Bradley, R. M. (1963) Nature (Lond.) 197, 389-390.
(82) Mihailovic, Lj. T., Krzalic, Lj. and Cupic, D. (1965) Experientia, 21, 709-710.
(83) Kuriyama, K. and Yoneda, Y. (1978) Brain Res. 148, 163-179.
(84) Tzeng, S. and Ho, I. K. (1977) Biochem. Pharmacol. 26, 699-704.
(85) Gold, B. I., Simon, J. R. and Roth, R. H. (1978) Life Sci. 22, 187-194.
(86) Roth, R. H., Morgenroth, V. H. and Salzman, P. M. (1975) Arch. Pharmacol. 289, 327-343.
(87) Lysz, T. W. and Sze, P. Y. (1978) J. Neurosci. Res. 3, 411-418.
(88) Lysz, T. W. and Sze, P. Y. (1978) Trans. Amer. Soc. Neurochem. 9, 128.
(89) Kuhn, D. M., Vogel, R. L. and Lovenberg, W. (1978) Trans. Amer. Soc. Neurochem. 9, 184.
(90) Sze, P. Y., Hedrick, B. J. and Alderson, R. (1978) Trans. Soc. Neurosci. 8, 324.
(91) Krebs, E. G. and Fischer, E. H. (1956) Biochim. Biophys. Acta, 20, 150-157.

STUDIES ON GLUTAMATE DECARBOXYLASE OF THE MAMMALIAN BRAIN

J. M. Blindermann, M. Maitre and P. Mandel

Centre de Neurochimie et Institut de Chimie Biologique, Faculté de Médecine, 67085 Strasbourg Cedex, France

INTRODUCTION

Extensive studies have suggested that L-glutamate decarboxylase (GAD; EC 4.1.1.15) may be used as a specific marker-enzyme for GABA-ergic neurones. It has been reported that the activity of GAD correlates well with the distribution of GABA (1-6) and that the enzyme is found mainly in synaptosomes (7,8), suggesting that the formation of the neurotransmitter portion of GABA may occur in nerve endings *in vivo*.

Some neurological diseases, such as Huntington's Chorea, Parkinson's Disease (9) and certain types of convulsions (10-12), for which alterations of GABA-ergic neurotransmission have been postulated, have been correlated with decreases in GAD activity. Decreases in enzyme activity by cellular metabolites have also been described in terms of inhibition of the catalytic process (13-16). One has to distinguish between this direct control of the catalytic reaction and the mechanisms which may influence the synthesis or degradation of the enzyme. For instance, it has been postulated that GABA itself, which is a weak inhibitor of vertebrate GAD activity *in vitro* (13,14) and which has a significant depressant action on GAD activity *in vivo* (17-19), might produce a feedback regulation of GAD synthesis.

Regarding activation of GAD, it has been shown that chronic administration of acetylcholine to neonatal mice produced an increase in GAD activity in the cerebellar cortex (Sze, this volume). Similarly, enhanced GAD activity was found in the thalamus after injection of an acute dose of morphine (20) and in the whole brain after phenobarbital (21).

Although the mechanisms underlying these regulatory processes are still unknown, such findings support the idea that GAD might be inducible. However, studies involving the determination of both enzyme activity and apoenzyme level would offer more convincing evidence for enzyme regulation than studies based only upon enzyme activity. Using this rationale, some aspects of possible regulatory effectors have been examined.

METHODS

Assay for GAD Activity

GAD activities were determined by the radiometric decarboxylation of [U-^{14}C]glutamate, as previously described (22): [U-^{14}C]GABA formation was determined according to Molinoff and Kravitz (24) and $^{14}CO_2$ evolution was measured by a modification of the method of Albers and Brady (23). In a typical experiment, aliquots (20 - 50 µl) of homogenates (representing about 0.1 - 0.5 mg protein) were incubated together with 100 µl of K^+-phosphate buffer (0.1 M; pH 7.0) which contained (as final concentrations) 0.5 mM pyridoxal phosphate, 1.0 mM 2-aminoethylisothiouronium bromide hydrobromide (AET), 0.1 mM EDTA and 15 mM L-glutamate (10.4 mCi/m-mole).

Assay for GAD Protein

GAD apoenzyme levels were determined by a specific radioimmunoassay (25) using specific antibodies. Antiserum to rat brain GAD was obtained in an adult rabbit 60 days after it received four subcutaneous injections of purified enzyme (each of 0.2 mg protein).

This antiserum gave a single precipitation arc by immunoelectrophoresis against pure GAD and against a crude enzyme preparation. By immunoprecipitation, the antibodies inhibited the pure enzyme by about 70%. Radioimmunoassay was performed with a dilution of antiserum (1/1000) that gave 70% precipitation of the antigen. The method is based on competition for the precipitation of a fixed amount of [^{125}I]-labelled enzyme (about 10 ng protein), either by increased levels (0 - 10 ng) of unlabelled purified enzyme (standard curve) or by increased amounts of homogenate (representing 10 - 50 μg protein). Under these conditions the method allowed measurement of GAD levels as low as 1 ng (about 1.5 x 10^{-15} mole).

PURIFICATION OF RAT BRAIN GAD

Although some attempts have been made to purify mammalian GAD by affinity chromatography (26,27) or by preparative electrophoresis (28), the method used here, which consists of conventional techniques in addition to an hydrophobic chromatography step, seems to provide the most reliable results (13,29,30). Different experiments were performed in order to obtain a high extraction of the enzyme with substantial activity stabilization. Thus, Triton X-100 homogenization and $(NH_4)_2SO_4$ fractionation were eliminated from the protocol of purification because these substances produced large losses of enzyme activity. That the partial binding of brain GAD was dependent on Ca^{2+} (6,7,31) was further substantiated by showing that homogenization in the presence of a high concentration (5 mM) of EDTA, a Ca^{2+}-complexing agent, increased the yield of the GAD extracted. Stabilization of the enzyme was obtained by using a combination of 2-aminoethylisothiouronium bromide hydrobromide (1.0 mM) and pyridoxal phosphate (0.5 mM). A step that involved heating of the homogenate was introduced in the sequence of purification. This step, which was performed under acidic conditions (50°C for 5 min; pH 5.4) provided optimal thermal stability of the enzyme as well as precipitating a large amount of contaminant proteins.

Ion exchange chromatography (DEAE-cellulose and QAE-Sephadex A-50), adsorption on calcium phosphate (hydroxyapatite) and gel filtration (Ultrogel ACA 34) were used in addition to an hydrophobic chromatography procedure on phenyl sepharose. These steps provided a 3000-fold purification of the rat brain enzyme (Table I) whereas an 8000-fold purified enzyme was obtained from the human brain (13, 29).

The enzymatic preparation appeared to be homogeneous by the following criteria: (1) By polyacrylamide gel electrophoresis, under denaturing conditions (sodium dodecylsulphate and urea), the purified enzyme migrated as a single molecular species. (2) A similar result was obtained with non-denaturing conditions (on polyacrylamide gradient gel slab electrophoresis; PAA 4/30), the single band corresponding to a peak which contained all the enzyme activity. (3) Isoelectric focussing gave a single peak of activity at pH 5.2. The protein contained in this peak, when applied to sodium dodecylsulphate/polyacrylamide gel (containing 6 M urea), migrated as a single band. (4) N-terminal analysis indicated the presence of a single amino acid in the N-terminal position (α-alanine). (5) The purified rat enzyme and a crude rat brain preparation gave only one precipitation arc by immunoelectrophoresis using specific antibodies against the above-described purified GAD.

The main properties of the enzyme are summarized in Table II. Since the molecular weight of the native GAD (140,000 ± 10,000) was approximately twice the value obtained under denaturing conditions (67,000 ± 5,000) we conclude that the enzyme from rat and human brain is most probably a dimer which consists of two similar, if not identical, subunits. Further support for this hypothesis was provided by the findings that the enzyme contained a single N-terminal amino acid (α-alanine) and lacked disulfide bridges. For mouse brain GAD, Wu (14) estimated a molecular weight of about 90,000 for the native enzyme by gel filtration. By equilibrium sedimentation under denaturing conditions (in the presence of guanidine hydrochloride and mercaptoethanol) two indistinguishable species

Table I - Purification of Rat Brain GAD

Fraction	Total protein (mg)	Specific activity (units/mg prot)	Purification (x)	Total activity (units)	Activity yield (%)
Homogenate	19,850	2.30		45,600	100
Supernatant	2,405	4.93	2.14	11,860	26
DEAE-cellulose	737	14.7	6.40	10,816	23.7
Hydroxyapatite	73	88.4	38.5	6,425	14
Phenyl-sepharose	5.7	713.8	310	4,067	9
QAE-Sephadex	2.1	1,128.6	491	2,370	5.2
Ultrogel	0.200	6,370	2,770	1,274	2.8

1 unit = 10^{-9} moles/min

Table II - Comparison of Some Properties of Human and Rat GAD

Property	Human enzyme	Rat enzyme
Molecular weight:		
native	140,000 ± 15,000	140,000 ± 15,000
subunits	67,000 ± 5,000	67,000 ± 5,000
Disulphide bridged	None	None
N-Terminal amino acid	α-Alanine	α-Alanine
pH_i	5.0 ± 0.2	5.4 ± 0.2
pH optimum	6.8 ± 0.15	7.0 ± 0.15
K_m, glutamate	1.28 ± 0.07 mM	1.10 ± 0.07 mM
K_m, pyridoxal phosphate	0.13 ± 0.09 μM	0.50 ± 0.09 μM

having a molecular weight of about 45,000 were obtained, whereas by polyacrylamide gel electrophoresis four major protein bands, ranging from 60,000 to 120,000 were detected. Since these bands can be taken to represent multiples of 15,000, he postulated that the mouse brain enzyme might be a hexamer.

In an attempt to overcome these discrepancies, the writers will re-interpret the results of the two sets of experiments. First of all, the two equal subunits obtained by equilibrium sedimentation for the mouse enzyme are consistent with our results. A single protein band obtained on SDS/polyacrylamide gel both for the human and rat enzymes has also been reported by other authors (26,32), whereas the major band for the mouse enzyme had a molecular weight of 60,000, suggesting that the other bands represent contaminant proteins. Secondly, the apparent molecular weight of the native enzyme determined by gel filtration for rat and human whole brain (90,000 and 100,000, respectively) and bovine cerebellum (85,000) (33), is also consistent with the result obtained for the synaptosomal mouse brain enzyme (85,000). However, it seems likely that such estimates are not valid since it is well known that sieve chromatography underestimates the molecular weights of some proteins (34); e.g., purified cysteine sulfinate decarboxylase (EC 4.1.1.29) of rat liver (35). Furthermore, no significant differences among mouse, rat and human enzymes could be found by comparing immunochemical data (30,36).

DETERMINATION OF GAD TURNOVER NUMBER IN SOME REGIONS OF RAT BRAIN

The regional variation of GAD activity in rat brain (Table III) is consistent with previous data (4), the highest values being found for the substantia nigra and hypothalamus and low values being found for cerebellum and pons-medulla. Levels of GAD protein followed the variation in its specific activity, as revealed by the relatively constant turnover numbers obtained for the different brain regions. Moreover, the molecular activity of the purified rat brain enzyme, calculated from its specific activity (6370 n-moles of GABA formed/min/mg protein), its protein concentration and its molecular weight (140,000), was about the same as that of the GAD present in whole brain homogenate determined by radioimmunoassay (16.4 and 15 sec^{-1},

Table III - Specific Activities, Levels of Apoenzyme and Turnover Numbers of GAD for Some Regions of Rat Brain, for Whole Rat Brain, and for Purified Rat GAD

	Specific activity (10^{-9} mole min^{-1} mg^{-1})	GAD apoenzyme (10^{-12} mole mg^{-1})	Turnover number (sec^{-1})
Total brain	2.73 ± 0.03 (12)	2.79 ± 0.19 (12)	16.4 ± 0.98 (12)
Cerebellum	2.59 ± 0.11 (6)	2.42 ± 0.21 (6)	17.0 ± 1.67 (6)
Olfactory bulb	3.25 ± 0.59 (6)	2.92 ± 0.32 (6)	18.6 ± 2.55 (6)
Pons-medulla	2.52 ± 0.61 (6)	2.38 ± 0.27 (6)	16.5 ± 1.78 (6)
Substantia nigra	6.06 ± 0.55 (6)	6.32 ± 0.89 (6)	17.2 ± 2.15 (6)
Hypothalamus	6.20 ± 0.30 (6)	7.28 ± 1.17 (6)	14.5 ± 1.46 (6)
Thalamus	4.16 ± 0.34 (6)	4.04 ± 0.55 (6)	17.2 ± 1.79 (6)
Hippocampus	3.06 ± 0.17 (6)	3.25 ± 0.42 (6)	15.7 ± 0.45 (6)
Caudate nucleus and putamen	4.25 ± 0.15 (6)	4.59 ± 0.31 (6)	15.4 ± 0.69 (6)
Amygdala	3.24 ± 0.22 (6)	3.25 ± 0.36 (6)	16.6 ± 0.79 (6)
Frontal cortex	3.52 ± 0.13 (6)	3.22 ± 0.89 (6)	18.2 ± 2.67 (6)
Purified enzyme	6370 ± 230	7100 ± 71	15 ± 0.7

Experiments were performed on Wistar rats. The whole brain (including the brain stem and cerebellum) was homogenized in 4 volumes (w/v) of water containing 0.5 mM pyridoxal phosphate, 1.0 mM AET, 5 mM EDTA and 0.5% Triton X-100 (v/v). Results are expressed as means ± S.D.; numbers of determinations in parentheses.

respectively), indicating that our preparation was quite homogeneous. These results indicated also that regulatory effectors of GAD activity were not detectable in brain homogenate. However, processes involving such effectors could have been destroyed by the conditions of tissue dilution and homogenization employed. In this regard, recent experiments have revealed that the degree of saturation of GAD by its cofactor, pyridoxal phosphate, in vivo, might be determined by a balance between the rate of dissociation of the cofactor from the enzyme and the rate of association of the cofactor with the apoenzyme, the latter of which is inhibited strongly by ATP and ADP (15). It has been further postulated that these adenine nucleotides might be involved in the mechanism controlling the rate of GABA synthesis. Thus, the high concentration of exogenous pyridoxal phosphate (0.5 mM) employed in the above-mentioned experiments might alter this steady state and destroy the possible in vivo regulation of the enzyme activity.

GAD IN DEVELOPING RAT BRAIN

The developmental changes in cerebral GAD activity of various species has been fully described (28,37-39). For mouse brain GAD, a semi-quantitation of enzyme protein has been proposed (28). In this study, residual GAD activities were determined after immunoprecipitation, using antibodies against GAD from bovine cerebellum. From birth to day 28 the protein level increased by a factor of 12, in parallel with the enzymatic activity. The amount of GAD enzyme protein seemed to decline after 28 days and to remain at a constant level after 40 days for at least 50 days longer. In order to evaluate quantitatively GAD protein versus its activity during ontogenesis, we have performed parallel determinations of GAD apoenzyme by radioimmunoassay and GAD activity on rat brains over a period of 60 days. The data of Figure 1 show that the increase in GAD activity (6-fold, during the period studied) was maximal during the second and the third postnatal weeks (from 10 to 26 days), reflecting

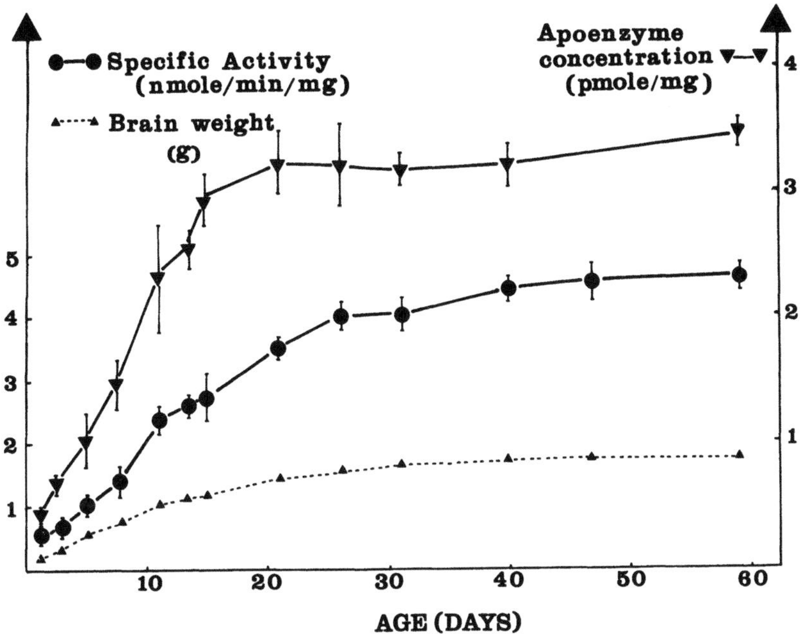

Fig. 1. Specific activity and protein concentration of rat brain GAD during ontogenesis. Results are given as means ± S.D. (n = 12).

whole brain protein synthesis (40). The adult value for GAD activity (about 4 n-moles of GABA formed/min/mg protein) seemed to be reached after 30 days of age.

The developmental profile of GAD apoenzyme (Fig. 1) was similar to that of the enzyme activity. From birth to day 30 apoenzyme increased from 0.70 to 3.82 p-moles/mg protein and then remained constant for at least 30 days longer. Determination of turnover numbers, expressed in p-moles of GABA formed/sec/molecule of enzyme (Fig. 2), suggested relatively constant values, confirming the finding that the increase in GAD activity is due to a parallel increase of the specific protein, at least for the period studied (60 days). There is a notable disparity with GABA-T, the GABA degradative enzyme, for which a constant protein concentration was deter-

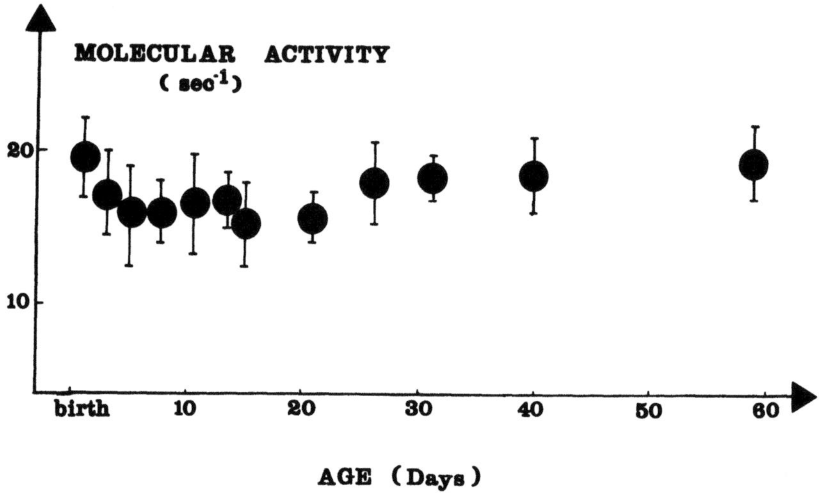

Fig. 2. Turnover numbers of GAD during ontogenesis. Results are given as means ± S.D. (n = 12).

mined while the enzyme activity increased (Maitre et al., this volume). Endogenous GABA levels remained quite constant during the postnatal maturation of rat brain, and values for GABA uptake were close to those of adult brain at this early postnatal period (37). On the contrary, GABA-receptor sites, as determined by binding experiments, increased during the first 30 postnatal days (37). Thus, among the different parameters of the GABA-ergic system during ontogeny the closest relationship seems to exist between the presumed postsynaptic GABA-receptors and the GAD protein and activity.

CEREBRAL GAD ACTIVITIES IN DIFFERENTIALLY-HOUSED MICE

Recent results have suggested that GAD is inducible (for review see Sze, this volume) and that the enzyme activity may be regulated by genetically-mediated mechanisms (41,42). Modifications in the GABA-ergic system have also been produced by environmental changes. For example, GABA binding to synaptosome-enriched fractions occurred

Table IV - GAD Activities in Homogenates of Brain Regions of Differentially-Housed Mice

Region	GAD activity: n-moles $^{14}CO_2$ formed/min/mg protein	GAD activity: n-moles ^{14}C-GABA formed/min/mg protein
	Isolated mice	
Olfactory bulb	4.96 ± 0.15 (12)[a]	4.29 ± 0.16 (12)[b]
Cerebral cortex	3.47 ± 0.13 (6)	2.62 ± 0.10 (6)
Hippocampus	2.89 ± 0.10 (6)	2.07 ± 0.06 (6)
Striatum	4.81 ± 0.09 (6)	3.48 ± 0.27 (6)
Hypothalamus	9.50 ± 0.34 (6)	8.21 ± 0.32 (6)
Cerebellum	3.43 ± 0.08 (12)[c]	2.68 ± 0.09 (12)
Pons-medulla	3.42 ± 0.13 (6)	2.43 ± 0.07 (6)
Midbrain	5.67 ± 0.16 (6)	4.15 ± 0.14 (6)
	Grouped mice	
Olfactory bulb	5.53 ± 0.10 (12)	4.82 ± 0.18 (12)
Cerebral cortex	3.68 ± 0.18 (6)	2.59 ± 0.24 (6)
Hippocampus	2.91 ± 0.19 (6)	2.10 ± 0.11 (6)
Striatum	4.71 ± 0.12 (6)	3.80 ± 0.30 (6)
Hypothalamus	9.40 ± 0.12 (6)	8.82 ± 0.14 (6)
Cerebellum	3.97 ± 0.08 (12)	2.57 ± 0.10 (12)
Pons-medulla	3.35 ± 0.12 (6)	2.39 ± 0.11 (6)
Midbrain	5.89 ± 0.18 (6)	4.21 ± 0.20 (6)

Means ± S.E.M.; numbers in parentheses indicate numbers of determinations (12 determinations on 6 homogenates or 6 determinations on 3 homogenates, each homogenate representing regions from two brains. [a], [b], and [c] indicate, respectively, $p < 0.05$, $p < 0.025$, and $p < 0.001$ when these values were compared with corresponding values for grouped mice (Student's *t*-test; two-tailed.) Unpublished data of J. M. Blindermann, M. Maitre, F. V. DeFeudis, R. Misslin and P. Mandel (1978); see also ref. 22.

to a lesser extent in "isolated" (aggressive) mice and cerebral GABA content was lower in "isolated" than in "grouped" mice (43,44). Furthermore, GABA levels in the olfactory bulbs is lower in spontaneous "killer" rats than in non-killers (45). Muricidal behavior could be inhibited by bilateral injections of GABA into the olfactory bulbs (46). These results have led us to compare GAD activities in the whole brains (22) and in several brain regions of "isolated" and "grouped" mice.

Post-weanling male mice of the C3H/He/orl strain were used. "Isolated" mice were kept singly in small cages and "grouped" mice were kept in groups of 10 - 12 in large cages. After 4 - 5 weeks of differential housing, the mice were decapitated and their brains were rapidly removed, dissected, and homogenized in ten volumes (w/v) of water containing 0.5 mM pyridoxal phosphate, 1 mM AET and 5 mM EDTA. GAD activities were determined as described in "Materials and Methods" (see above). Among the different regions studied (Table IV), GAD activity was significantly lower in the olfactory bulbs of "isolated" mice than in "grouped" mice. The difference found for the whole forebrain (3.87 ± 0.09 (12) n-moles [^{14}C]GABA formed/min/mg protein for the "isolated" mice versus 4.18 ± 0.09 (12) n-moles [^{14}C]GABA formed/min/mg protein for the "grouped" mice) indicated that some regions not examined in the present study (e.g., thalamus) could also be involved in this environmentally-induced behavioral change. These results supported the notion that lower GABA levels might be responsible for aggressive behavior and that the cerebral GABA-ergic system is adaptable and plastic.

REFERENCES

(1) Fahn, S. (1976) In: GABA in Nervous System Function (E. Roberts, T. N. Chase and D. B. Tower, eds.), pp. 169-186, Raven Press, New York.
(2) Ben Ari, Y., Kanazawa, I. and Zigmond, R. E. (1976) J. Neurochem. 26, 1279-1283.
(3) Yoneda, Y. and Kuriyama, K. (1978) J. Neurochem. 30, 821-825.
(4) Tappaz, M. L., Browstein, M. J. and Palkovits, M. (1976) Brain

Res. 108, 371-379.
(5) Balcom, G. J., Lenox, R. H. and Meyerhoff, J. L. (1975) J. Neurochem. 24, 609-615.
(6) Fonnum, F. and Walberg, F. (1973) Brain Res. 54, 115-127.
(7) Whittaker, V. P., Michaelson, I. A. and Kirkland, R. J. A. (1964) Biochem. J. (Lond.) 90, 293-303.
(8) Fonnum, F. (1968) Biochem. J. (Lond.) 106, 401-412.
(9) McGeer, P. L. and McGeer, E. G. (1976) J. Neurochem. 26, 65-76.
(10) Horton, R. W. (1978) Biochem. Pharmacol. 27, 1471-1477.
(11) Sawaya, C., Horton, R. and Meldrum, B. (1978) Biochem. Pharmacol. 27, 475-481.
(12) Adcock, T. and Taberner, P. V. (1978) Biochem Pharmacol. 27, 246-248.
(13) Blindermann, J. M., Maitre, M., Ossola, L. and Mandel, P. (1978) Eur. J. Biochem. 86, 143-152.
(14) Wu, J. Y. (1976) In: GABA in Nervous System Function (E. Roberts, T. N. Chase and D. B. Tower, eds.) pp. 7-55, Raven Press, New York.
(15) Seligman, B., Hiller, L. P., Brockman, D. E. and Martin, D. L. (1978) J. Neurochem. 30, 371-376.
(16) Tursky, T. and Lassanova, M. (1978) J. Neurochem. 30, 903-905.
(17) Sze, P. Y. and Lovell, R. A. (1970) J. Neurochem. 17, 1657-1664.
(18) Sze, P. Y., Kuriyama, K., Haber, B. and Roberts, E. (1971) Brain Res. 26, 121-130.
(19) Jung, M. J., Lippert, B., Metcalf, B. W., Böhlen, P. and Schechter, P. J. (1977) J. Neurochem. 29, 797-802.
(20) Kuriyama, K. and Yoneda, Y. (1978) Brain Res. 148, 163-179.
(21) Tzeng, S. and Ho, I. K. (1977) Biochem. Pharmacol. 26, 699-704.
(22) Blindermann, J. M., DeFeudis, F. V., Maitre, M. and Mandel, P. (1979) J. Neurochem. 32, 1357-1359.
(23) Albers, R. W. and Brady, R. O. (1959) J. Biol. Chem. 234, 926-928.
(24) Molinoff, P. B. and Kravitz, E. A. (1968) J. Neurochem. 15, 391-409.
(25) Blindermann, J. M., Maitre, M. and Mandel, P. (1979) J. Neurochem. 32, 245-246.
(26) Yamaguchi, T. and Matsumura, Y. (1977) Biochim. Biophys. Acta, 481, 706-711.
(27) Possani, L. D., Bayon, A. and Tapia, R. (1977) Neurochem. Res. 2, 51-57.
(28) Hadjian, R. A. and Stewart, J. A. (1977) J. Neurochem. 28, 1249-1257.
(29) Blindermann, J. M., Maitre, M., Ossola, L. and Mandel, P. (1977) C. R. Acad. Sc. Paris, 285, 1079-1082.
(30) Maitre, M., Blindermann, J. M., Ossola, L. and Mandel, P. (1978) Biochim. Biophys. Res. Commun. 85, 885-890.
(31) Tapia, R. and Covarrubias, M. (1978) J. Neurochem. 31, 1209-1214.
(32) Drummond, G. S. and Goldstein, M. (1978) Trans. Amer. Soc.

Neurochem. 9, 183.
(33) Ryan, L. D. and Roskoski, R., Jr. (1976) Neurochem. Res. 1, 37-45.
(34) Andrews, P. (1964) Biochem. J. (Lond.) 91, 222-233.
(35) Guion-Rain, M. C., Portemer, C. and Chatagner, F. (1975) Biochim. Biophys. Acta, 384, 265-276.
(36) Saito, K., Wu, J. Y., Matsuda, T. and Roberts, E. (1974) Brain Res. 65, 277-285.
(37) Coyle, J. T. and Enna, S. J. (1976) Brain Res. 111, 119-133.
(38) Sims, K. L. and Pitts, F. N. (1970) J. Neurochem. 17, 1607-1612.
(39) Van den Berg, C. J., Van Kempen, G. M. J., Schade, J. P. and Veldstra, H. (1965) J. Neurochem. 12, 863-869.
(40) Pitts, F. N., and Quick, C. (1967) J. Neurochem. 14, 561-570.
(41) Gaitonde, M. K. and Festing, M. F. W. (1976) Brain Res. 103, 617-621.
(42) Sze, P. Y. (1977) Brain Res. 122, 59-69.
(43) DeFeudis, F. V. (1972) Neuropharmacology, 11, 879-888.
(44) DeFeudis, F. V., Madtes, P., Ojeda, A. and DeFeudis, P. A. (1976) Exptl. Neurol. 52, 285-296.
(45) Mack, G., Simler, S. and Mandel, P. (1975) J. Physiol. (Paris), 71, 162 A.
(46) Mandel, P., Mack, G., Kempf, E., Ebel, A. and Simler, S. (1978) In: Interactions Between Putative Neurotransmitters in the Brain (S. Garattini, J. E. Pujol and R. Samanin, eds.) pp. 286-303, Raven Press, New York.

BRAIN SUCCINIC SEMIALDEHYDE DEHYDROGENASE

C. Cash, M. Maitre, L. Ciesielski and P. Mandel

Centre de Neurochimie et Institut de Chimie Biologique, Faculté de Médecine, 67085 Strasbourg Cedex, France

INTRODUCTION

Succinic semialdehyde dehydrogenase (succinate semialdehyde : NAD oxidoreductase, EC 1.2.1.24; SSADH) is the final enzyme of the GABA shunt pathway and is responsible for the oxidation of succinic semialdehyde (SSA), produced by the transamination of GABA, to succinate which then re-enters the Krebs cycle (Fig. 1). This dehydrogenase has been identified in brain tissue from several species including cat, rat and monkey (1) and man (2). SSADH has been partially purified and characterized from rat (3), monkey (4) and human brain (5). These preparations all showed a high degree of specificity for SSA and NAD. The partially purified rat brain preparation required addition of mercaptoethanol for full activity, whereas the monkey and human brain preparations retained their activity without added thiols. However, in each case, addition of mercaptoethanol markedly potentiated the inhibition of the enzymes by arsenite, but the monkey brain enzyme, previously inactivated by heat, urea or H_2O_2 treatment, could be re-activated by mercaptoethanol. Apparent K_ms of 4.7×10^{-6} M and 9.9×10^{-5} M for SSA and NAD, respectively, were reported for the rat brain enzyme, and in the case of the enzymes from monkey and human brain, apparent K_ms of 2.7×10^{-6} M and 5.3×10^{-6} M, respectively, for SSA were found.

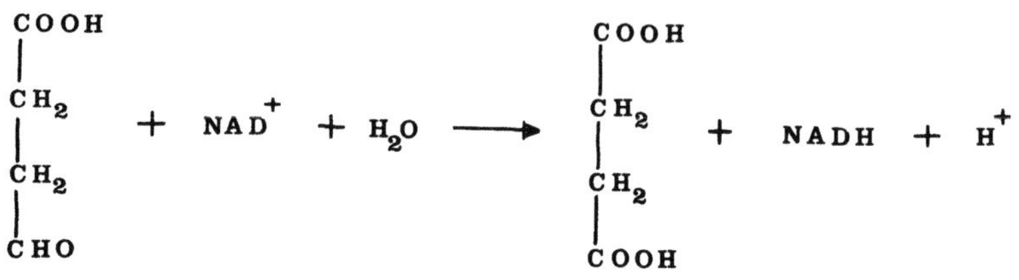

Fig. 1.

In both of the latter cases, kinetic studies using NAD as the variable substrate gave rise to biphasic double-reciprocal plots suggesting the presence of two enzyme species. Substrate inhibition occurs in all cases with SSA concentrations greater than about 10^{-4} M. Diphenylhydantoin and certain other anticonvulsants inhibited the enzyme from human brain, but the most powerful inhibitors tested were m- and p-hydroxybenzaldehydes.

SSADH is a mitochondrial enzyme (6) and its distribution in brain tissue is similar to that of GABA-transaminase (GABA-T), the enzyme which provides its substrate (7). Its activity increases by about 3-fold in rat brain during development (8). The other enzymes of the GABA shunt pathway, glutamate decarboxylase and GABA-T, also show considerable increases in activity during brain development (9,10).

Unlike that of its precursor, GABA, the concentration of SSA in brain appears to be very low. It was recently detected at a level of 0.05 – 0.1 n-mole/gram of mouse brain (11). Thus, it appears that the catabolism of SSA is not normally the rate-limiting step of the GABA shunt. This view is supported by the findings that brain SSADH activities are higher than those of GABA-T. Values of the ratio SSADH:GABA-T activity in rat brain of about 5 (12), and 1.8 (13) and in human brain of 1.5 (7) have been reported.

Table I - Purification of Succinic Semialdehyde Dehydrogenase from Rat Brain

Fraction	Volume (ml)	Units/ml	mg protein/ml	Units/mg protein	Yield	Purification
Homogenate	240	0.48	30	0.016	100	1
Pooled supernatants	306	0.3	15	0.021	80	1.3
Sephadex G200 eluate	94	0.2	2.5	0.08	16	5
DEAE cellulose eluate	116	0.08	0.1	0.87	8	55
AMP affinity column eluate	2.2	3.07	0.11	27	6	1710

1 Unit = 1 μ-mole NADH formed per minute.

Table II - Purification of Two Succinic Semialdehyde Dehydrogenases from 382 Grams of Human Brain

Fraction	Vol. (ml)	Units ml	Total units	Protein mg/ml	Units mg prot.	Yield %	Purification	Remarks
Homogenate	1500	0.097	146	34	0.003	100	1	
Supernatant	1010	0.18	185	12.5	0.015	127	5.1	
0 → 80% $(NH_4)_2SO_4$	220	0.53	115	17.8	0.03	79	10	
DE. Cellulose	248	0.22	55	1.6	0.14	38	48	
AMP-sepharose	65	0.26	17	0.06	4.3	11.5	1516	
2nd DE. cellulose	27	0.41	11	0.06	6.5	7.5	2281	DE. retained
CM. cellulose	112	0.39	44	0.64	0.61	30	215	
AMP-sepharose	128	0.085	10.8	0.01	8.5	7.4	2980	
Hydroxyapatite	11	0.26	2.9	0.017	16	1.9	5600	CM. retained

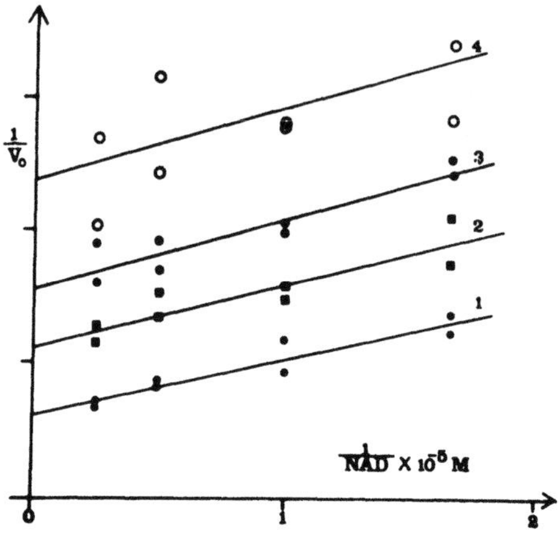

Fig. 2. Lineweaver-Burk plot of initial reaction rates at 25°C as a function of NAD concentration, and variable concentrations of the second substrate. 1. SSA, 8 x 10^{-5} M; 2. SSA, 4 x 10^{-5} M; 3. SSA, 2.5 x 10^{-5} M and 4. SSA, 1.5 x 10^{-5} M.

PROPERTIES OF HIGHLY PURIFIED SUCCINIC SEMIALDEHYDE DEHYDROGENASES ISOLATED FROM RAT AND HUMAN BRAINS

Rat Brain SSADH

Rat brain SSADH has been purified to electrophoretic homogeneity by the procedures summarized in Table I, the details of which have been published (14). Structural studies showed that the enzyme, which has a molecular weight of about 140,000, is composed of two subunits of identical molecular weights which are not attached by disulphide bridges. The subunits are probably identical, as a single N-terminal amino acid was found. By measuring the fluorescence enhancement induced in NADH when it was titrated with pure enzyme, it was found that one mole of enzyme could bind with two moles of dinucleotide. The enzyme activity was routinely assayed by measuring fluorimetrically the rate of formation of NADH at 37°C

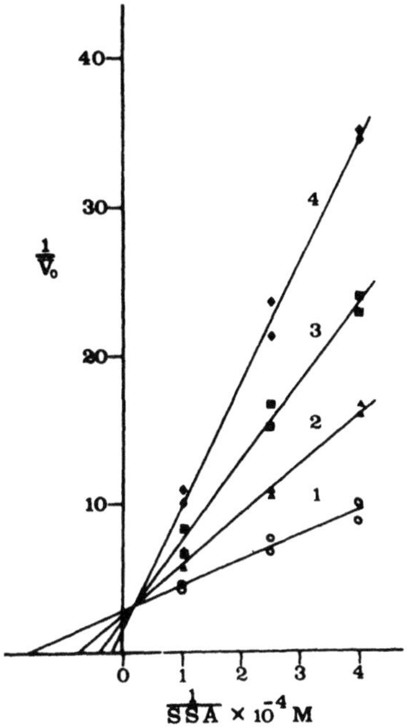

Fig. 3. Double reciprocal plot of SSADH inhibition by AMP at 3 concentrations of NAD. SSA is 1×10^{-4} M. 1. No inhibitor; 2. AMP, 0.1 mM; 3. AMP, 0.2 mM and 4. AMP, 0.5 mM.

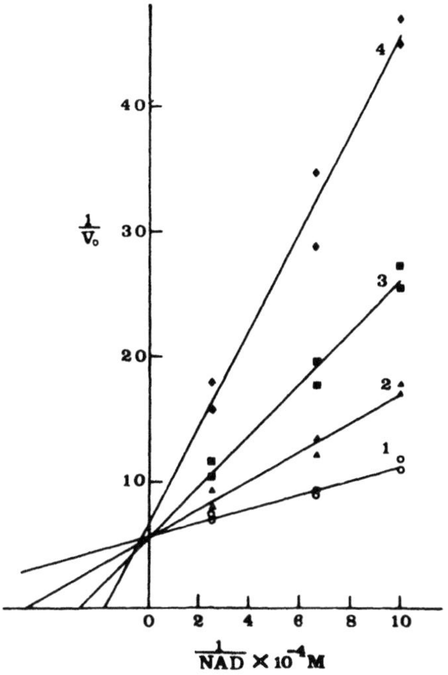

Fig. 4. Double reciprocal plot of inhibition of SSADH by 4HB at 3 concentrations of SSA. NAD concentration is 3×10^{-4} M. 1. No inhibitor; 2. 4HB at 2×10^{-6} M; 3. 4HB at 5×10^{-6} M; 4. 4HB at 1×10^{-5} M.

in the presence of 1×10^{-4} M SSA, 3×10^{-4} M NAD and 20 mM mercapto-ethanol in 100 mM Tris-HCl at the optimum pH of 8.6. To determine the Michaelis constants, one substrate was varied at various fixed concentrations of the other and the initial rates were plotted by the Lineweaver-Burk method. This revealed a series of near parallel lines which generally indicates a ping-pont catalytic mechanism.(Fig. 2). However, such a mechanism is unlikely in view of the results

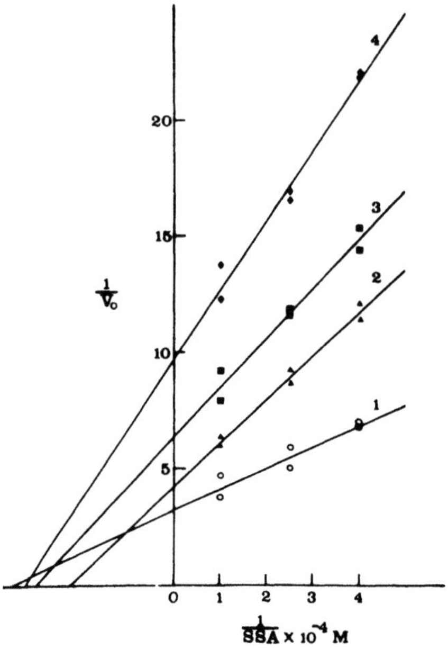

Fig. 5. Double reciprocal plot of SSADH inhibition by AMP at 3 concentrations of SSA. NAD is 3 x 10^{-4} M. 1. No inhibitor; 2. AMP, 1 mM; 3. AMP, 2 mM; 4. AMP, 5 mM.

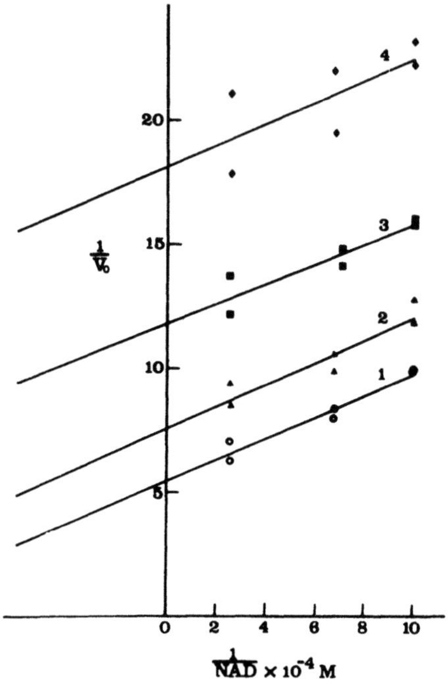

Fig. 6. Double reciprocal plot of SSADH inhibition by 4HB at 3 concentrations of NAD. SSA concentration is 1 x 10^{-4} M. 1. No inhibitor; 2. 4HB, 2 x 10^{-6} M; 3. 4HB, 5 x 10^{-6} M; 4. 4HB, 1 x 10^{-5} M.

of most of the previous investigations with NAD(P) linked dehydrogenases. Secondary plots of the intercepts versus the inverse of the substrate concentrations gave K_m (limits) of 2 x 10^{-5} M for NAD and 1 x 10^{-4} M for SSA. The value obtained with SSA is about an order of magnitude higher than that indicated when a single near saturating concentration of NAD is used, or that obtained by other investigators (8,15).

In order to further investigate the enzyme mechanism, we em-

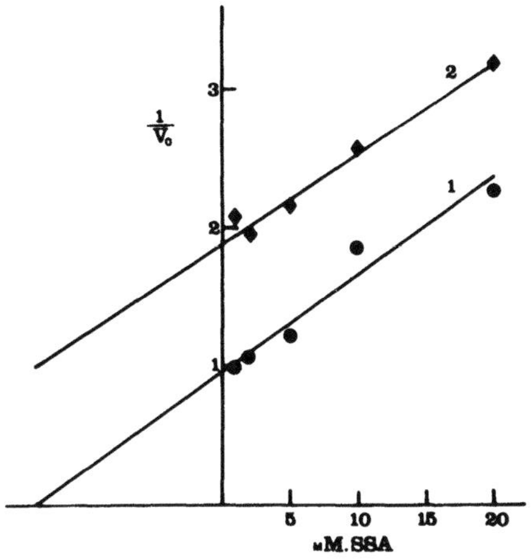

Fig. 7. Substrate inhibition by SSA. Plot of reciprocal velocity against SSA concentration. 1. NAD, 3 x 10^{-4} M; 2. NAD, 1 x 10^{-4} M.

ployed two inhibitors, 5'AMP and 4-hydroxybenzaldehyde. The former is a competitive inhibitor vis à vis NAD (Fig. 3), and the latter, a competitive inhibitor vis à vis SSA (Fig. 4). However, AMP was found to be a non-competitive inhibitor vis à vis SSA (Fig. 5), whereas 4-hydroxybenzaldehyde is an uncompetitive inhibitor vis à vis NAD (Fig. 6). This pattern of inhibitions is incompatible with a ping-pong mechanism and is indicative of an ordered mechanism whereby NAD binds first, followed by SSA, before release of the products. Thus, the uncompetitive inhibition observed with 4-hydroxybenzaldehyde shows that it binds to an enzyme-NAD complex and not to the free enzyme. At high concentrations, SSA is itself an uncompetitive inhibitor vis à vis NAD (Fig. 7). Rat brain SSADH is also inhibited by the anticonvulsant drug di-*n*-propylacetate and by similar branched-chain fatty acids (12,15).

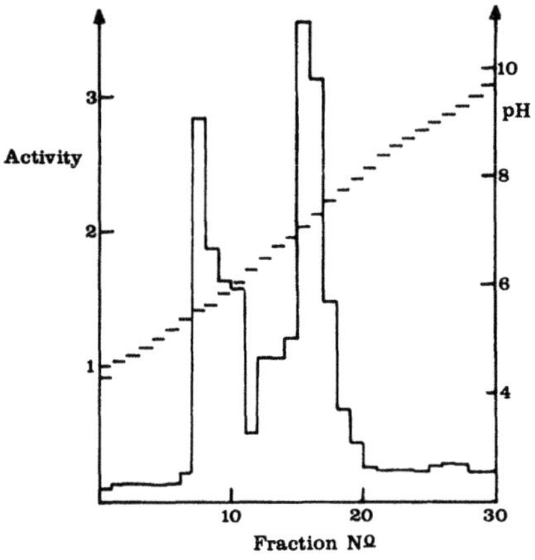

Fig. 8. Human SSADH, 0 → 80% saturation ammonium sulphate precipitate fraction. Isoelectric focussing on ultrodex gel in the presence of 3 M urea.

Human Brain SSADH

A typical purification starting with about 400 g of human brain is summarized in Table II, the details of which have been published (16). In contrast to rat brain, two SSADH activities were separated at an early stage of the purification. When a crude ammonium-sulphate precipitate fraction was applied to a column of DEAE cellulose at pH 7.2, part of the activity was retained and the unretained part was fixed by a column of CM cellulose under the same conditions. These two activities, hereafter referred to as "DE" enzyme and "CM" enzyme, were then purified separately to electrophoretic homogeneity. Structural studies revealed that both enzymes are dimers with molecular weights of about 145,000, composed in each case of similar if not identical subunits. In order to be more certain that the obtention of two enzymes was not an artifact due to protein aggregation, the crude ammonium-sulphate precipitate fraction was submitted to isoelectric focussing in the presence of 3 M urea. Two distinct bands of activity were separated at pH values of 5.6 and 7.2 (Fig.

Table III - Amino Acid Composition

Amino acid	Residues per mole	
	CM enzyme	DE enzyme
Glutamate + glutamine	81	28
Aspartate + asparagine	79	75
Threonine	171	227
Serine	95	103
Glycine	116	58
Half-cystine	3	4
Methionine	2	2
Valine	92	141
Proline	38	2
Arginine	56	43
Histidine	51	47
Lysine	74	75
Phenylalanine	48	47
Tyrosine	2	6
Leucine	109	94
Isoleucine	65	113
α-Alanine	109	98
Tryptophan	-	-
Total	1189	1173

8). There are significant differences in the amino acid compositions (Table III) and in the tryptic fingerprints (Fig. 9) of the two pure enzymes. The pH optima of both enzymes was about 8.6, but as the enzymes were not very stable at this pH, kinetic experiments were carried out at pH 7.2. The major kinetic parameters for the two enzymes were similar. The K_m (limits), determined in the same way as for the rat brain enzyme for NAD and SSA, were 1.85×10^{-5} and 2×10^{-6} M, respectively, for the CM enzyme and 1.6×10^{-5} and 3.7

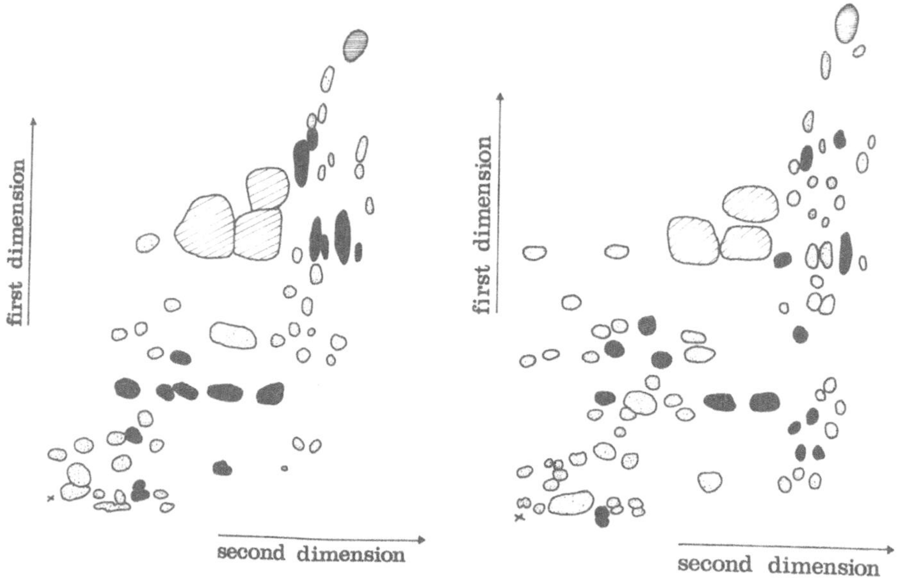

Fig. 9

x 10^{-6} M, respectively, for the DE enzyme. However, the inhibition patterns produced by 4-hydroxybenzaldehyde and AMP were different. Figure 10a and Figure 10b are, respectively, double-reciprocal and secondary plots of the inhibition produced by 4-hydroxybenzaldehyde at variable NAD concentrations for the two enzymes. Similarly, Figure 11a and Figure 11b show the double-reciprocal and secondary plots, respectively, of the inhibition produced by AMP at variable NAD concentrations. In neither case were the inhibitions of a pure type, but in each case the CM enzyme was inhibited in a linear fashion, whereas the DE enzyme was inhibited in a non-linear fashion. As found for rat brain SSADH, both human brain enzymes were inhibited to a similar extent by di-*n*-propylacetate and related branched-chain fatty acids (Table IV).

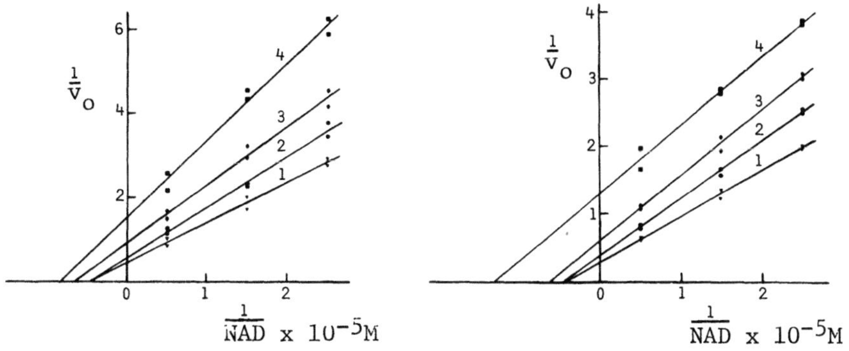

Fig. 10a. Double-reciprocal plot of inhibition by 4-hydroxybenzaldehyde (4HB) at variable concentrations. (Left) CM enzyme, (right) DE enzyme. Lines: (1) no inhibitor; (2) 0.05 mM 4HB; (3) 0.1 mM 4HB; (4) 0.2 mM 4HB.

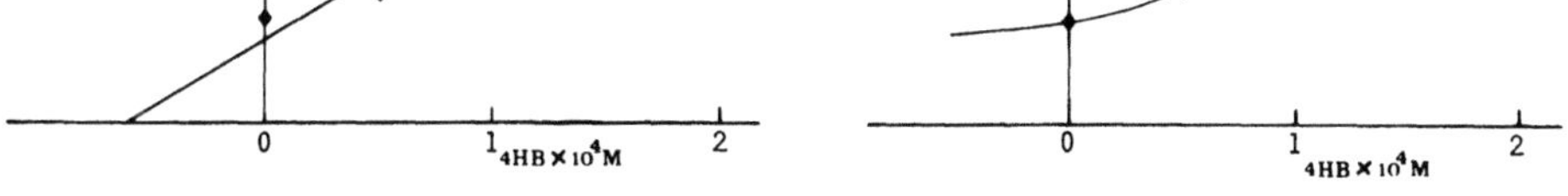

Fig. 10b. Secondary plots of slopes and intercepts versus 4-hydroxybenzaldehyde concentration at variable NAD concentrations. (a) CM enzyme, (b) DE enzyme. Lines: (1) intercepts, (2) slopes.

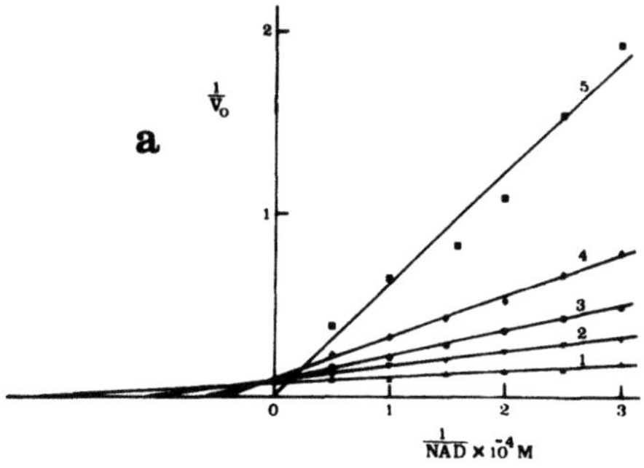

Fig. 11a. Double reciprocal plots of inhibition by AMP at variable NAD concentrations. (a) CM enzyme, (b) DE enzyme. Lines: (1) no inhibition; (2) AMP, 0.5 mM; (3) AMP, 1 mM; (4) AMP, 2 mM; (5) AMP, 5 mM.

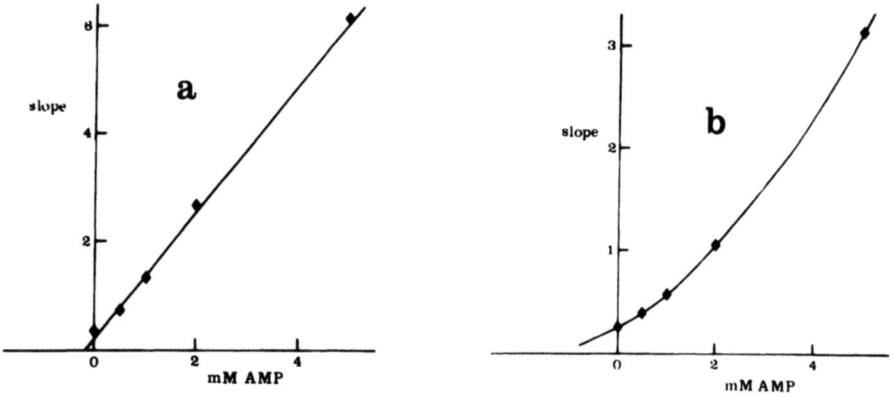

Fig. 11b. Secondary plots of slopes versus AMP concentration at variable NAD concentrations. (a) CM enzyme; (b) DE enzyme.

Table IV - Some Inhibition Constants for Rat and Human Brain SSADH

Enzyme	Substrate varied	Inhibitor	K_i (mM)
Rat	SSA	AMP	2.0 (s)
"	NAD	"	0.075 (s)
"	SSA	4-Hydroxybenzaldehyde	0.003 (s)
"	NAD	"	0.004 (i)
"	SSA	Di-_n_-propylacetate	0.8 (s)
"	NAD	"	3.0 (i)
"	SSA	Propyl-2-penten-2-oate	1.2 (s)
"	SSA	2,3,3-Tripropylpropionate	0.08 (s)
Human "CM"	NAD	AMP	0.16 (s)
"	NAD	4-Hydroxybenzaldehyde	0.07 (i)
"	SSA	Di-_n_-propylacetate	0.82 (s)
"	NAD	"	20 (s)
"	SSA	2,3,3-Tripropylpropionate	0.038 (s)
Human "DE"	NAD	AMP	N.L.
"	NAD	4-Hydroxybenzaldehyde	N.L.
"	SSA	Di-_n_-propylacetate	0.7 (s)
"	NAD	"	20 (s)
"	SSA	Propyl-2-penten-2-oate	1.2 (s)
"	SSA	2,3,3-Tripropylpropionate	0.038 (s)

The letter "s" after the K_i value indicates that this value was obtained from a secondary plot of slopes versus inhibitor concentrations of the initial Lineweaver-Burk plot. Similarly, the letter "i" indicates that the intercepts were employed. N.L. signifies that secondary plots using both slopes and intercepts were non-linear; hence K_i values were not obtained.

Some Comparisons Between Rat and Human Brain SSADH

No important structural differences were apparent in the enzymes obtained from both species, but the obtention of two SSADHs from human brain and only one from rat brain is of interest. The significance of this finding is not known at present, but two enzymes have also been obtained from mouse brain (unpublished results). Kinetically, the most important difference observed is the much higher affinity of the human brain enzymes for SSA. In this context, it is noteworthy that human brain GABA-T has a much higher affinity for GABA than does the rat brain enzyme (17). The pure uncompetitive inhibition observed with 4-hydroxybenzaldehyde when NAD was used as the variable substrate with the rat brain enzyme was not obtained with either of the human brain enzymes. Similarly, AMP was a pure competitive inhibitor with respect with the rat brain enzyme, but produced inhibition of a mixed type with the human brain enzymes.

AN ALTERNATIVE PATHWAY FOR SSA CATABOLISM?

It has long been known that brain tissue has the ability not only to oxidize SSA to succinate, but also to reduce it to 4-hydroxybutyrate (GHB) (18). GHB is of interest as it has an anesthetic-like action when administered to animals and man in large doses (19). Its natural occurrence in mammalian brain has been confirmed; it appears to be present at the low concentration of about 10^{-6} M (20). Thus, it is not known at present whether this reductive pathway is of pharmacological significance *in vivo*. However, the recent obtention of an NADPH-dependent SSA reductase from human brain which is fairly specific for SSA (21) strongly supports the view that this reaction has a significant function *in vivo*. Indeed, two distinct enzymes capable of reducing SSA to GHB were purified from human brain. One of these enzymes can reduce a wide range of aldehydes; the best substrate found was p-nitrobenzaldehyde.

Thus, its function *in vivo* might be to reduce the aldehydes produced by the actions of monoamine oxidase on the amine neurotransmitters. It appears to belong to a group of enzymes previously described for human brain (22), and is potently inhibited by diphenylhydantoin and barbiturates. Di-*n*-propylacetate inhibited this enzyme to about the same degree as the barbiturates, to which it bears no structural resemblance. However, the SSA-specific enzyme was not inhibited by barbiturates or by di-*n*-propylacetate at the same low concentrations, and thus does not appear to belong to a category of aldehyde reductases previously described for human brain.

DISCUSSION

Succinic semialdehyde is primarily oxidized to succinate by a specific dehydrogenase, the properties of which have been described. SSA can also be reduced to 4-hydroxybutyrate (GHB) by brain tissue both *in vitro* (18) and *in vivo* (23). The presence of GHB in mammalian brain has recently been confirmed (20) albeit at concentrations much lower than those attained by the brain when this substance is administered in quantities sufficient to produce anesthesia (19). Thus, it is not yet known whether this reductive pathway of SSA catabolism or the presence of endogenous GHB is of significance under normal conditions. However, the recent isolation of a fairly specific SSA reductase from human brain (21) lends support to the hypothesis that GHB might well play some natural role. It would be of great interest to know if SSADH activity can be regulated either by endogenous factors or by drugs *in vivo*, and if so, whether this regulation might affect the biosynthesis of GHB.

The anticonvulsant drug di-*n*-propylacetate (valproate) inhibits GABA-T, and its intraperitoneal administration to experimental animals induces an increase in cerebral GABA levels which parallels a protection against certain types of convulsions (24). Thus, its mode of action has been attributed directly to an inhibition of

GABA-T activity. However, this drug also inhibits SSADH (15) and non-specific aldehyde reductases that have been purified from human brain (21) and partially purified from ox brain (25). It was thus suggested that the increase in cerebral GABA levels induced by di-*n*-propylacetate administration might be due to SSADH inhibition and consequent reverse transamination of SSA to GABA (15). However, it has not been possible to increase brain GABA levels by administering dehydrogenase inhibitors (26), and *in vitro* experiments have indicated no decrease in the rate of GABA catabolism when rat brain mitochondria were incubated in the presence of di-*n*-propylacetate or the powerful SSADH inhibitor, 4-hydroxybenzaldehyde (27). Nevertheless, administration of substances which inhibit SSADH activity *in vivo* must inevitably raise the normally very low cerebral SSA levels, and thus it could be suggested that this might favor the reductive pathway of SSA catabolism and thereby increase brain GHB levels. This hypothesis is however doubtful in view of the finding that the "sleep" state induced by GHB administration resembles absence epilepsy (28) and this syndrome is antagonized by anti-petit mal drugs including di-*n*-propylacetate (29).

Further research is presently underway in order to investigate the possible modulation of SSA catabolism by endogenous and exogenous factors, and the possible physiological implications.

REFERENCES

(1) Albers, R. W. and Salvador, R. (1958) Science, 128, 359-360.
(2) Embree, L. J. and Albers, R. W. (1964) Biochem. Pharmacol. 13, 1204-1217.
(3) Kammeraat, C. and Veldstra, H. (1968) Biochim. Biophys. Acta, 151, 1-10.
(4) Albers, R. and Korval, G. (1961) Biochim. Biophys. Acta, 52, 29-35.
(5) Embree, L. J. and Albers, R. W. (1964) Biochem. Pharmacol. 13, 1209-1217.
(6) Salganicoff, L. and DeRobertis, E. (1964) J. Neurochem. 12, 287-309.
(7) Miller, A. and Pitts, F. N. (1967) J. Neurochem. 14, 579-584.
(8) Pitts, F. N. and Quick, C. (1967) J. Neurochem. 14, 561-570.

(9) Blindermann, J. M., Maitre, M. and Mandel, P. (this volume).
(10) Ossola, L., Maitre, M., Blindermann, J. M. and Mandel, P. (1977) In: Maturation of Neurotransmission. Biochemical Aspects (A Vernadakis, E. Giacobini and G. Filogamo, eds.) pp. 83-90, Karger, Basel.
(11) Matsuda, M. and Hoshino, M. (1977) Jikeikai Med. J. 24, 33-36.
(12) Cash, C., Ciesielski, L., Maitre, M. and Mandel, P. (1975) C. R. Soc. Biol. Paris, 164, 884-887.
(13) De Boer, Th. and Bruinvels, J. (1977) J. Neurochem. 28, 471-478.
(14) Cash, C., Ciesielski, L., Maitre, M. and Mandel, P. (1977) Biochimie, 59, 257-268.
(15) Harvey, P., Bradford, H. F. and Davidson, A. (1975) FEBS Lett. 52, 251-254.
(16) Cash, C., Maitre, M., Ossola, L. and Mandel, P. (1978) Biochim. Biophys. Acta, 524, 26-36.
(17) Maitre, M., Ciesielski, L., Cash, C. and Mandel, P. (1978) Biochim. Biophys. Acta, 522, 385-399.
(18) Fishbein, W. and Bessman, S. (1964) J. Biol. Chem. 239, 357-361.
(19) Laborit, H., Jouany, J. P., Gerard, J. and Fabiani, P. (1961) Neuro-psychopharmacology, 2, 490-497.
(20) Doherty, J. D., Hattox, S. E., Carter Snead, O. and Roth, R. H. (1978) J. Pharmacol. Exp. Therapeut. 207, 130-139.
(21) Cash, C., Maitre, M. and Mandel, P. (1978) C. R. Acad. Sci. Paris, 286, 1829-1832.
(22) Ris, M. and Von Wartburg, J. P. (1973) Eur. J. Biochem. 37, 69-77.
(23) Matsuda, M. and Hoshino, M. (1976) Jikeikai Med. J. 23, 189-193.
(24) Simler, S., Ciesielski, L., Maitre, M., Randrianarisoa, H. and Mandel, P. (1973) Biochem. Pharmacol. 22, 1701-1708.
(25) Whittle, S. and Turner, A. (1978) J. Neurochem. 31, 1453-1459.
(26) Baxter, C. and Roberts, E. (1961) J. Biol. Chem. 12, 3287-3294.
(27) Maitre, M., Ossola, L. and Mandel, P. (1976) FEBS Lett. 72, 53-57.
(28) Marcus, R., Winters, W., Mori, K. and Spooner, C. (1967) Int. J. Neuropharmacol. 6, 175-185.
(29) Godschalk, M., Dozoljic, M. and Bonta, I. (1976) Neuroscience Lett. 3, 145-150.

REGULATION OF GLUTAMINASE AND ITS POSSIBLE IMPLICATION FOR GABA METABOLISM

Elling Kvamme

Neurochemical Laboratory, The Oslo University Psychiatric Clinic, Vinderen, Oslo 3, Norway

INTRODUCTION

The synthesis of γ-aminobutyric acid (GABA), which occurs mainly by the glutamate decarboxylase (GAD) reaction, is commonly presumed to be controlled by the enzyme only, the activity of which is dependent on the cofactor pyridoxal phosphate. Therefore, factors influencing GAD and its cofactor have been widely discussed in the literature, whereas the possibility that the glutamate decarboxylase reaction, under certain conditions, might be limited by the concentration of its substrate, glutamate, has largely been overlooked. One reason for this is obviously the generally very high glutamate concentration in brain. However, nothing is known about the glutamate concentration in nerve endings where GABA is synthesized, and a rather low glutamate concentration is quite likely to be found in these structures.

If we consider the possibility that GABA synthesis in certain compartments and under certain conditions may be limited by the glutamate concentration, another parameter has to be taken into account when discussing GABA metabolism, namely the synthesis of glutamate and its regulation.

REGULATION OF GLUTAMATE CONCENTRATION

The most important reactions leading to glutamate synthesis are catalyzed by aspartate and alanine aminotransferases, glutamate dehydrogenase and glutaminase. In addition to phosphate-activated glutaminase (PAG), other enzymes that use glutamine as substrate have been labelled glutaminases; i.e., glutamine-oxoacid transaminase (glutaminase II) (1) and phosphate-independent glutaminase (2,3), which has been shown to be identical with γ-glutamyl transpeptidase (γ-GT) (4,5). Since the synaptosomes prepared by us are low in γ-GT, phosphate-activated glutaminase is likely to be the only glutaminase of practical importance in this tissue. We prefer to use the designation phosphate-activated glutaminase (PAG) instead of phosphate-dependent glutaminase (PDG), because the enzyme is not dependent on phosphate for its activity.

Concerning subcellular localization, the transferases are localized both in mitochondria and in cytoplasm, whereas glutamate-dehydrogenase and PAG are localized in mitochondria (6). Glutamate which is synthesized by cytoplasmic transferases does not have to cross mitochondrial membranes, and therefore should be more readily available to cytoplasmic GAD than glutamate that is produced inside the mitochondria. However, the availability of mitochondrially-produced α-oxoglutarate for cytoplasmic transferases is likely to place a severe restriction upon this reaction. The equilibrium constant of glutamate-dehydrogenase is such that it favors glutamate synthesis, provided sufficient α-oxoglutarate, reduced pyridine nucleotides and ammonia are present, but the availability of the latter two compounds may be limiting for the reaction.

PAG catalyzes a hydrolytic reaction which strongly favors the formation of glutamate and which is subjected to a very complex regulation. This reaction is well suited for transmitter synthesis because it is fast, energy-independent and independent of any other substrate or cofactor than glutamine. The reaction may be easily switched on and off by activators and inhibitors. Indeed, Hamberger et al. (7) have produced evidence that glutamine is a key precursor for transmitter glutamate in the hippocampus.

MOLECULAR PROPERTIES

PAG was first described by Krebs in 1935 (8) and it was characterized as a phosphate-activated enzyme by Errera and Greenstein in 1949 (9). Pig renal PAG was solubilized in 1954 (10) and partially purified (ca. 300-fold) in 1958 (11,12). Our group purified pig kidney PAG and pig brain PAG to apparent homogeneity (ca. 10,000-fold) in 1970 and 1973, respectively (13-15), making use of the property of the enzyme to solubilize and polymerize in a reversible manner. Rat kidney PAG was purified to apparent homogeneity in 1976 by Curthoys and coworkers (16).

Pig brain PAG, when solubilized in Tris-HCl buffer, has a molecular weight of about 120,000 - 135,000, as determined by sedimentation equilibrium centrifugation and sucrose gradient centrifugation when the protein concentration is kept below 0.1 mg/ml (Table I). Solubilized PAG has a tendency to polymerize, particularly at higher protein concentrations, a phenomenon which renders difficult the accurate determination of its molecular weight.

If the protein concentration is not too low, PAG will dimerize in the presence of phosphate, but polymers are also present (17). Insoluble polymers, with molecular weights of about 2,000,000 , are readily formed upon addition of borate to the phosphate buffer (Table I). When polymerized, the specific activity (activity per mg) of PAG increases about three-fold. For this reason we distinguish among the Tris-enzyme (monomeric form), the phosphate-enzyme (dimeric form) and the phosphate-borate-enzyme (polymeric form), all of which have different catalytic properties; e.g., the latter enzyme has a catalytic constant which is about 30-fold that of the Tris-enzyme (18,19).

Pig brain PAG is apparently composed of subunits with molecular weight of 64,000, whereas pig kidney PAG contains subunits with

Table I - Effects of Activators on the Sedimentation Coefficients and Molecular Weights of Glutaminase

Pre-incubated with	s value (S)		Molecular weight range
	Mean	Range	
None	5.8	5.7 - 6.0	120,000 - 135,000
Acetyl-CoA	6.5	6.4 - 6.8	
Phosphate	7.1		200,000 - 280,000
Phosphate-borate	53		≈ 2,000,000

Sedimentation coefficients were determined by the sucrose gradient technique and molecular weights were estimated by this method and by sedimentation equilibrium centrifugation. The Tris-enzyme was pre-incubated for 10 min at 23°C (pH 7.0) with or without activator (50 mM phosphate; 30 mM borate; 0.125 mM acetyl-CoA).

molecular weights of 64,000 and 57,000, as determined by polyacrylamide gel electrophoresis in Na^+-dodecylsulphate (15,17).

Electron micrographs (B. R. Olsen, unpublished data) of pig brain PAG did not differ from those previously reported for pig kidney PAG (17,20,21). After negative staining, the Tris-enzyme appeared to consist of cylindrical particles. In high magnification micrographs a number of molecules showed the presence of a stain-filled cleft which appeared to divide each molecule into two parts. After addition of phosphate to the Tris-enzyme, an elongated dimer composed of two Tris-enzyme molecules became a major component of this preparation. On addition of phosphate-borate buffer, large helical polymers were formed.

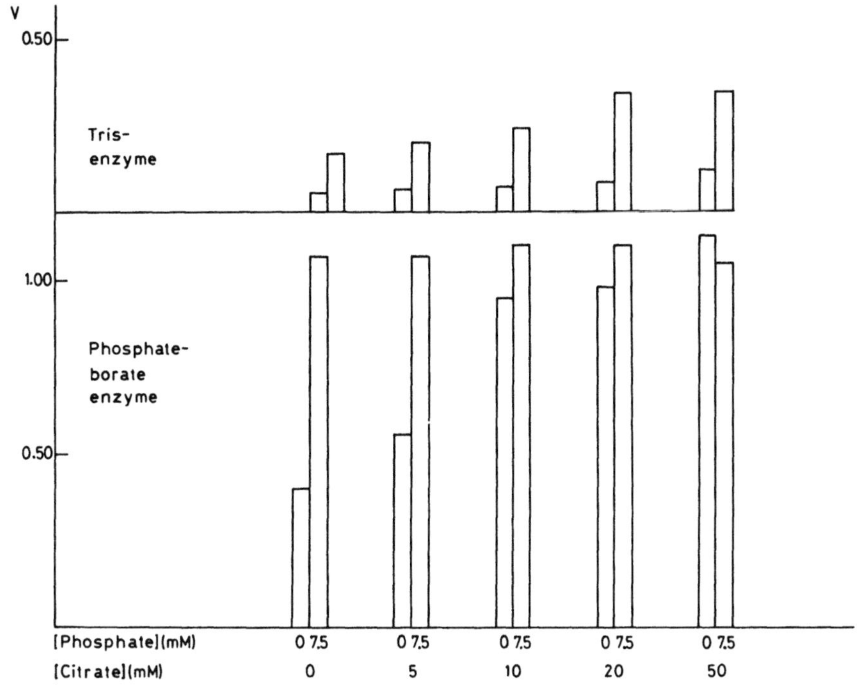

Fig. 1. Activation of glutaminase by citrate. The assay mixture contained 10 mM L-glutamine, 0.2 mM EDTA and the indicated concentrations of citrate and phosphate; v, μ-mole glutamate formed in 2 min at 23°C at pH 9.0.

REGULATORY LIGANDS

Activators

Phosphate is a potent anionic activator of brain PAG. Intermediates of the citric acid cycle, such as succinate and citrate, are less potent activators. These activators affect both the purified Tris-enzyme and the phosphate-borate-enzyme (19,22; see Fig. 1), as well as PAG in non-purified tissue preparations; e.g., in homogenates, mitochondria (23,42) and synaptosomes (Fig. 2). Anionic activation appears to be a general characteristic of this type of mammalian glutaminase; e.g., it has been shown in liver (24) and in kidney, both in purified preparations (13,16,18,25-28) and in non-purified prepa-

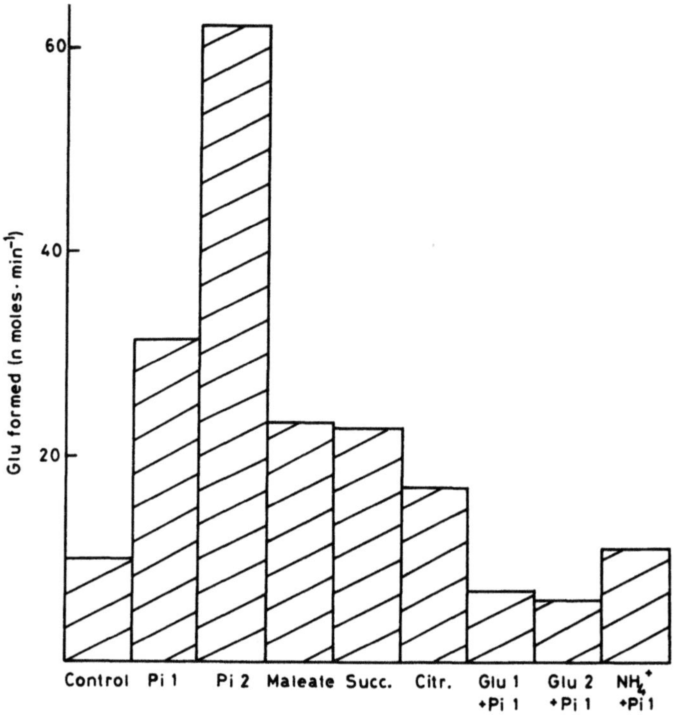

Fig. 2. Effects of activators and inhibitors on glutaminase in synaptosomes. Synaptosomes were purified according to Gray and Whittaker (83) and incubated with 2 mM [^{14}C]L-glutamine for 2 min at 25°C, pH 7.4. Glutamate and glutamine were separated by paper chromatography and counted (15). Additions: D-mannitol, 0.13 M; sucrose, 0.03 M; HEPES, 0.005 M, pH 7.4; NaCl, 0.1 M; KCl, 0.003 M; $MgCl_2$, 2 mM; bovine serum albumin, 0.5%; Antimycin A, 0.6 μg; oligomycin, 10 μg; maleate, succinate (Succ) or citrate (Citr.), 10 mM.
Phosphate: Pi 1, 5 mM; Pi 2, 10 mM. Glutamate: Glu 1, 1 mM; Glu 2, 2 mM; NH_4^+, 1 mM.

rations (29-31). Other general activators of PAG are phosphorylated compounds, such as riboflavin phosphate (32), trinucleotides (23) and acyl-CoA derivatives (33-35; Table II, Fig. 3). It has also been

Table II - Activation of Glutaminase by Acyl-CoA Derivatives

[Phosphate, mM]	Maximum activation (fold)		$10^2 \times A_{50}$ (mM)		$10^2 \times I_{50}$ (mM)	
	0	7	0	7	0	7
CoA	2.2	n.c.	1.5	n.c.	-	-
Acetyl-CoA	5.9	1.0	0.2	-	-	-
Acetoacetyl-CoA	1.3	n.c.	n.c.	n.c.	-	-
Propionyl-CoA	4.1	n.c.	0.5	n.c.	-	-
Malonyl-CoA	1.7	n.c.	0.8	n.c.	-	-
Decanoyl-CoA	1.7	3.5	1.0	1.5	-	-
Palmitoyl-CoA	1.7	1.5	0.2	0.2	1.0	1.0
Stearoyl-CoA	1.4	1.2	0.1	0.1	1.0	1.0
Oleoyl-CoA	2.5	2.0	0.4	0.4	8.0	8.0
Bromothymol-Blue	6.0	5.0	3.0	3.0	18.0	15.0

Purified brain glutaminase (Tris-enzyme) was incubated for 5 min at 23°C (pH 7.0). The reaction mixture (0.5 ml) contained 40 mM L-glutamine, 15 mM HEPES, 3 mM EDTA (as sodium salts) and 0.01 mg protein. The effectors were added in concentrations of 1 - 100 mM. Maximum activation is expressed as the ratio of maximum activity with CoA-derivative added to the activity without addition of CoA-derivative. A_{50} is the concentration of activator required to give 50% maximum activation. I_{50} is the concentration of inhibitor required to give 50% inhibition of maximum activity; n.c., not calculated.

reported that renal PAG is affected by the energy state of the mitochondria (36,37). Moreover, it has been reported that PAG of intact brain mitochondria is activated by increased oxygen pressure (38) and by acetyl aspartate, acetyl glutamate and thyroxine (39-41). Activation by anions differs for the soluble and polymerized enzyme

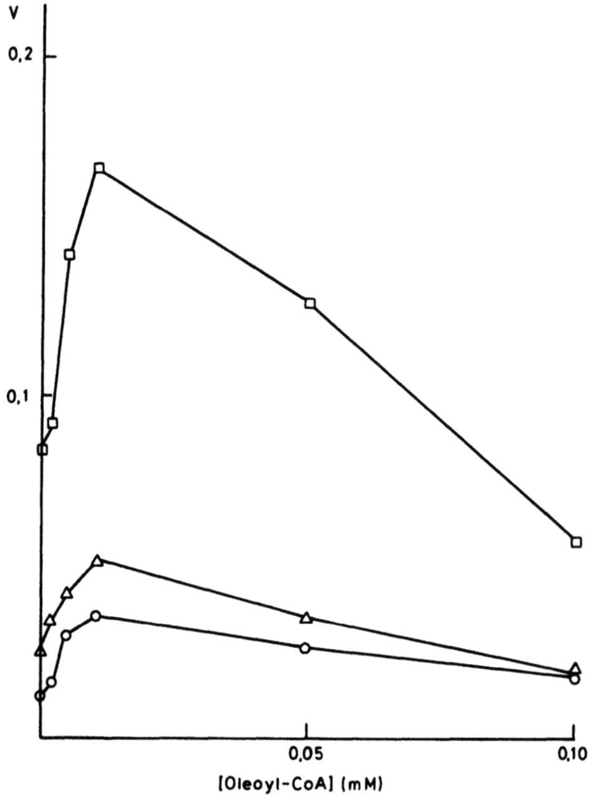

Fig. 3. Effect of oleoyl-CoA on glutaminase. The experimental conditions were the same as for data shown in Table II. o, no addition; Δ, 7.5 mM phosphate; □, 25 mM phosphate.

forms. The polymeric form of PAG is thus more susceptible to activation by low concentrations of anions than the Tris-enzyme (see, e.g., Figs. 1 and 4).

It should be noted that maleate activates the purified enzyme to the same extent as do intermediates of the citric acid cycle. This activation disappears when PAG is inactivated by heat, and therefore is not due to contamination of PAG with "maleate-dependent glutaminase" or γ-GT which is resistant to heat treatment. The same holds

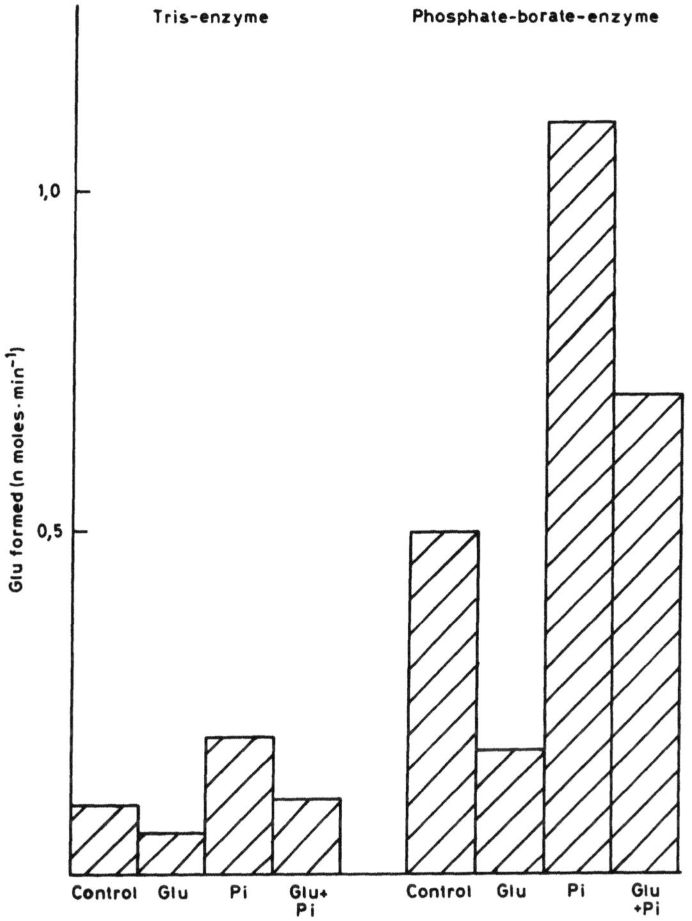

Fig. 4. Inhibition of glutaminase by glutamate. The experimental conditions were the same as those described in the legend to Fig. 1, except that ammonia formation was assayed. Glu: glutamate (10 mM) added; Pi: phosphate (7.5 mM) added.

true for maleate activation of synaptosomal PAG, which is of the same order of magnitude as that produced by intermediates of the citric acid cycle (Fig. 2).

Members of the citric acid cycle (e.g., citrate) potentiate, at low concentrations, the activation by phosphate (22,42), whereas at

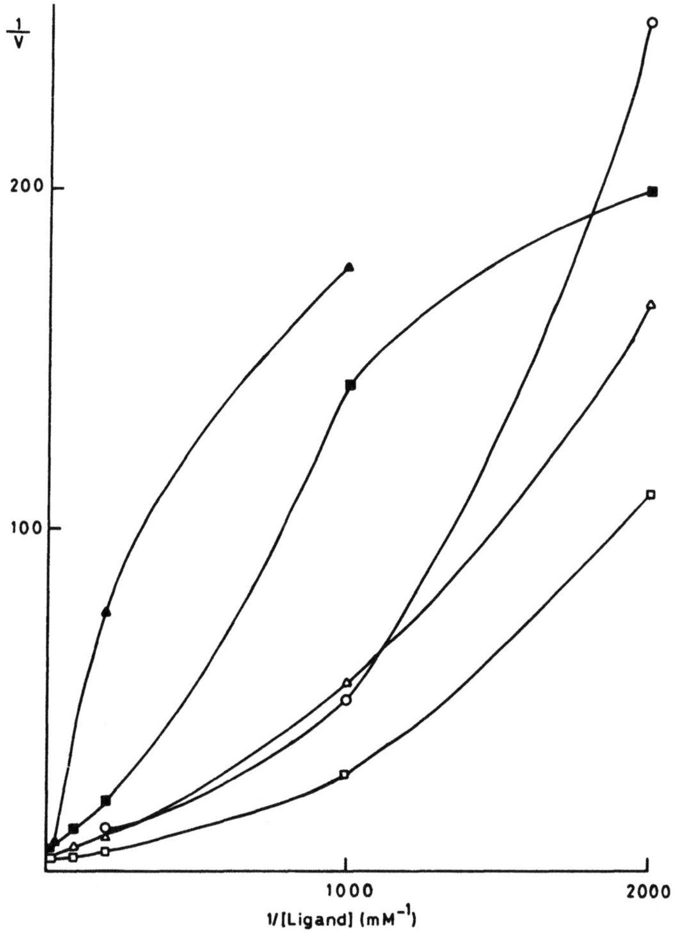

Fig. 5. Effects of decanoyl-CoA and oleoyl-CoA on double-reciprocal plots. The experimental conditions were the same as for data shown in Table II. Variable ligand oleoyl-CoA: ○, no addition; Δ, 7 mM phosphate; □, 25 mM phosphate. Variable ligand decanoyl-CoA: ▲, 7 mM phosphate; ■, 25 mM phosphate.

high concentrations these compounds produce no additional activation with regard to phosphate (Fig. 1); hence, they appear to compete for the same anionic sites. The potentiating effect appears to be a characteristic of the soluble Tris-enzyme form (22). Very interesting by Bromothymol Blue. In low concentrations, this dye strongly potentiates the activation by phosphate (Table II) by an action which increases the affinity of phosphate for anionic sites, whereas at higher concentrations it inhibits the enzyme (13,25,26). Phosphate protects PAG against this inhibition. The effects of acyl-CoA derivatives on purified soluble PAG are similar to those of Bromothymol Blue (33-35; Figs. 3 and 5; Table II); thyroxine also appears to have a similar effect on PAG in intact mitochondria (41,43). We have recently found that Ca^{2+} is an activator of synaptosomal PAG (see below).

Inhibitors

PAG is inhibited by cyclic-AMP and cyclic-GMP (23), by protons, and by long-chain acyl-CoA derivatives, the latter being required in higher concentrations than those which activate the enzyme (33,35; Fig. 3; Table II). Also, diethylstilbestrol and testosterone have been reported to counteract the activation produced by thyroxine, but these compounds appear to have little effect on the activation by phosphate, citrate or other anions (44).

Particularly interesting is the inhibition produced by the end-products of the glutaminase reaction, ammonia and glutamate. It is noteworthy that ammonia exerts different effects on non-purified and purified brain PAG. Thus, synaptosomal PAG is strongly inhibited by 1 mM ammonia (45; see also Fig. 2), whereas this concentration of ammonia has no effect on purified PAG. On the contrary, as shown in Table III, ammonia at 5 mM, or higher, has an activating effect on purified brain PAG, particularly on the soluble Tris-form. This is another example of how dangerous it may be to use uncritically results obtained with purified enzymes to explain the actions of the same enzymes operating under natural conditions inside the cell. The other end product, glutamate, is a powerful inhibitor both in non-purified (Fig. 2) and purified systems (Fig. 4). As shown in Figure 4, phosphate protects particularly the polymerized form of PAG against glutamate inhibition.

Table III - Activation of Glutaminase by Ammonium Chloride

Glutamate formed	[Ammonium chloride, mM]				
	0	5	10	20	50
Tris-enzyme	0.50	0.60	0.75	1.00	1.45
Phosphate-borate-enzyme	1.35	1.65	1.72	1.80	1.75

The experimental conditions were the same as those described in the legend to Fig. 1, except that 7.5 mM phosphate was added in the assay.

KINETIC PROPERTIES OF PURIFIED PAG

PAG is an allosteric enzyme (13,25). Double-reciprocal plots of activity against glutamine gave straight lines only at pH about 8, whereas the plots became increasingly concave-upwards at higher pH values and convex-upwards at decreasing pH values, indicating a switch from positive to negative cooperativity (15). Similarly, double-inverse plots of activity against phosphate concentration were concave-upwards (22), but such plots were generally convex-upwards against acyl-CoA derivatives with relatively short carbon chains and concave-upwards for derivatives with long carbon chains (Fig. 5; see ref. 35).

Time-dependent activation is typical for soluble PAG (18,19). Time-dependent activation occurs when the Tris-enzyme is incubated for varying lengths of time with anionic activators, such as phosphate, phosphate-borate, intermediates of the citric acid cycle and acetyl-CoA (Fig. 6). It only occurs at relatively high protein concentrations, and is competitively inhibited by glutamate. Maximal activation, about 3-fold, can be demonstrated when the Tris-enzyme is incubated with phosphate plus borate. Time-dependent activation appears to be accompanied by a conformational change of the enzyme (Table I), followed by polymerization, when the enzyme is incubated with phosphate or phosphate plus borate. This phenomenon

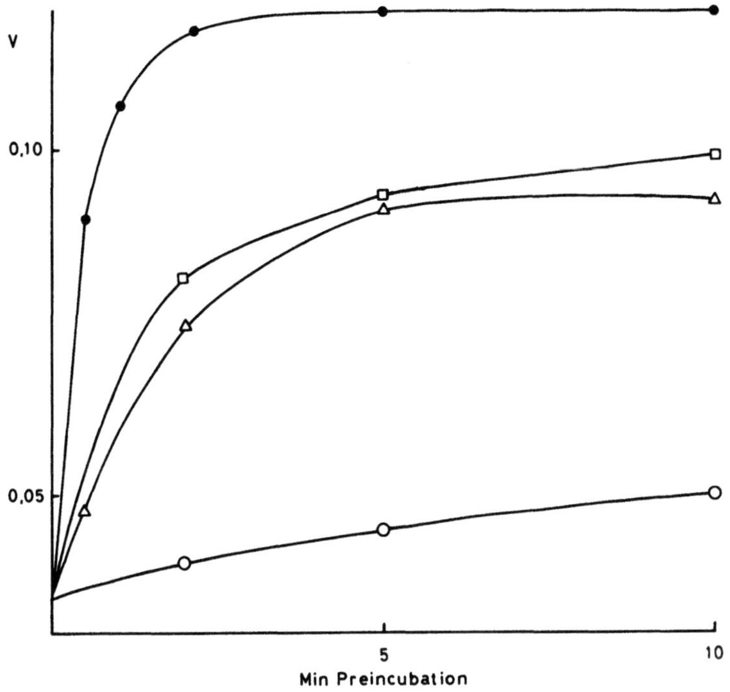

Fig. 6. Time-dependent activation of glutaminase by various ligands. Purified pig brain glutaminase (Tris-enzyme) was pre-incubated with ligand for the times indicated at 23°C (pH 8) in the presence of 5 mM Tris-HCl buffer and 3 mM EDTA before assay. A spectrophotometric assay of the rate of NH_3 formation was used, permitting the measurement of initial rates (15). ○, No addition; Δ, 0.2 M citrate; ●, 0.2 M phosphate; □, 0.125 mM acetyl-CoA. The protein concentration in the pre-incubation mixture was 0.2 mg/ml and in the assay it was 0.01 mg/ml.

does not occur at 0°C and is apparently independent of pH. Half-time of the reaction ($T_{\frac{1}{2}}$) with phosphate or phosphate plus borate was found to be 53 seconds. $T_{\frac{1}{2}}$ for pig brain brain glutaminase was similar to that for pig kidney Tris-enzyme.

PROPERTIES OF SYNAPTOSOMAL GLUTAMINASE

Rather few workers have studied synaptosomal glutaminase. Such studies certainly have many inherent difficulties when one considers the heterogeneity of brain mitochondria (46-48) and the fact that synaptosomal preparations are never pure. Thus, non-synaptic mitochondria are reported to oxidize glutamate plus malate almost twice as rapidly as synaptic mitochondria, and the pyruvate kinase activity in non-synaptic mitochondria is higher, whereas the glutamate-dehydrogenase activity is lower than in synaptic mitochondria (49). The increased glutamate oxidation in non-synaptic mitochondria appears to be due to their significantly higher specific activity of aspartate aminotransferase (50). Dienel et al. (51) reported that the activity of glutaminase and certain other enzymes is higher in non-synaptic mitochondria than in synaptic mitochondria, and that the ratio of these activities increases with age. However, Bradford and Ward (45) have found the glutaminase content of synaptosomes to be remarkably high. Thus, about 40% of the tissue glutaminase activity appears to be localized in the synaptosome fraction, which contains only about 10% of the tissue protein. Using a double-isotope technique (glutamine and glucose labelled with different isotopes), these workers have also shown that stimulus-released glutamate is derived principally (80 - 90%) from glutamine (52). This finding agreed with that of Hamberger et al. (7) which suggested that glutamine is a dominant precursor for transmitter glutamate in hippocampus. Moreover, it is of great interest that about 50% of released GABA is derived from glutamine (52). This provides evidence that glutamate which is formed by the glutaminase reaction contributes to GABA synthesis.

Concerning the properties of synaptosomal PAG, the enzyme is subject to regulation by anionic activators (Fig. 2; ref. 45) similar to the purified enzyme. Inhibition of purified and synaptosomal PAG by glutamate is also rather similar, but the effect of ammonia is qualitatively different in the two preparations. As described above, the pronounced inhibitory effect of 0.5 - 1 mM ammonia on synaptosomal PAG is completely lost during the purification procedure.

The inhibition by ammonia observed in synaptosomes might possibly be due to permeability restrictions, either on the substrate (glutamine) or on some activator. The intra-mitochondrial localization of PAG in synaptosomes is not known, but it is likely to be similar to that of kidney mitochondrial PAG. Conflicting views regarding the exact localization of kidney PAG have been expressed, but investigators in the field generally agree that PAG is localized to the inner part of the mitochondria (i.e., matrix and the inner membrane). Thus, Crompton *et al.* (53) and Kalra and Brosnan (54) have reported that PAG is localized to the matrix region, whereas Curthoys and Weiss (55) and Kovačević (56) have provided strong evidence that it is localized to the inner mitochondrial membrane.

We have approached the problem of possible permeability restrictions by using the SH reagent, N-ethylmaleimide (NEM). NEM, in concentrations up to 0.4 mM, inhibits synaptosomal PAG by about 50%, whereas higher concentrations exert no further inhibition. We found about the same proportional inhibition of PAG by NEM in various synaptosomal preparations and under different experimental conditions, and also in rather pure mitochondria isolated from pig kidney cortex. Therefore, it appears unlikely that inhibition by NEM is due to its differential effect on distinct mitochondrial species, especially when the heterogeneity of the synaptosomal preparation is taken into consideration. For this reason we feel that it is justified to distinguish between NEM-sensitive and NEM-insensitive mitochondrial PAG as being present in different compartments; i.e., NEM-accessible and NEM-inaccessible compartments. NEM-sensitive PAG generally shows a greater response to changes in external pH than NEM-insensitive PAG, indicating that the NEM-accessible compartment is easily reached by $[H^+]$ ions derived from the extra-mitochondrial space (Fig. 7). Hence, NEM-sensitive glutaminase is localized at the outer face of the inner membrane. Actually, a NEM-sensitive protein has been identified in

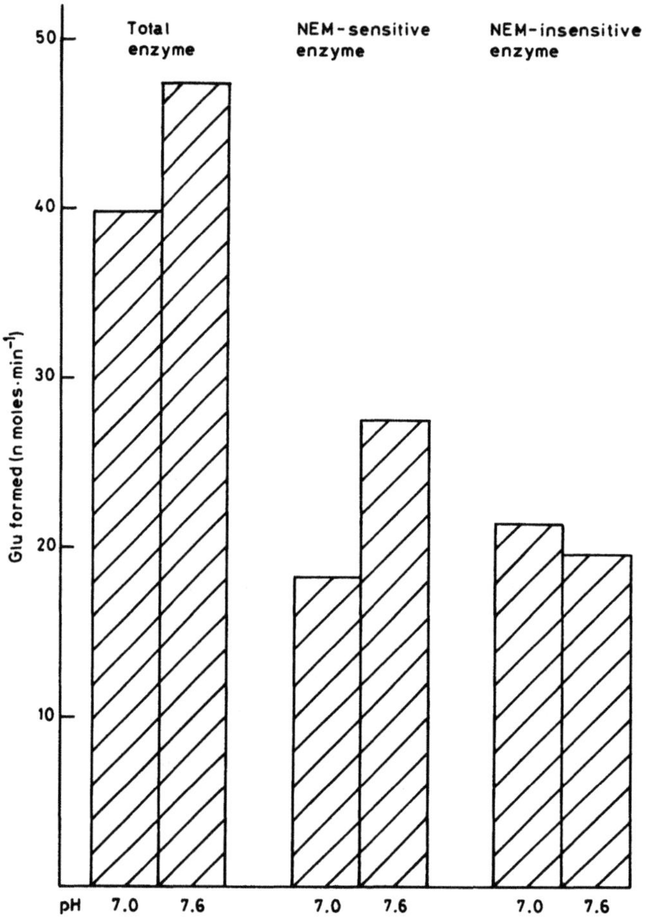

Fig. 7. Effect of N-ethylmaleimide (NEM) at pH 7.0 and pH 7.6 on glutaminase in synaptosomes. The experimental conditions were the same as those described in the legend to Fig. 2. Added NEM 0.4 mM. NEM-sensitive enzyme was calculated from total enzyme and NEM-insensitive enzyme.

this region (57) and found to be identical to the ADP-NEM-sensitive protein (58). This protein may, in fact, also be identical to the protein involved in the phosphate transport system which appears to be the major contributor to phosphate transport during oxidative

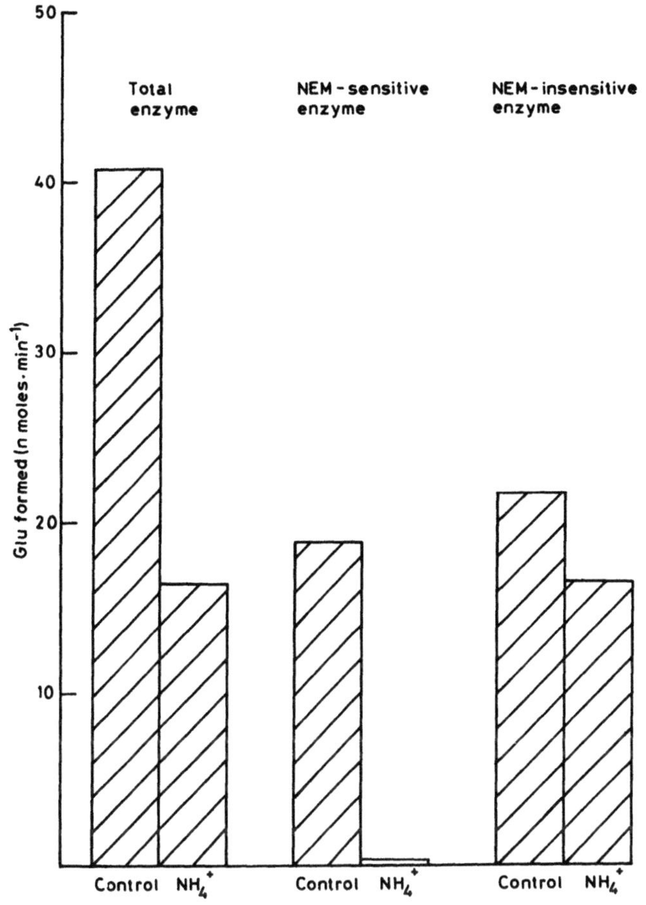

Fig. 8. Effects of ammonia on NEM-sensitive and NEM-insensitive glutaminase in synaptosomes. The experimental conditions were the same as those described in the legend to Fig. 2. Added phosphate 5 mM, NEM 0.4 mM, NH_4Cl 1 mM.

phosphorylation (59). Only the P_i/OH exchange carrier is sensitive to NEM; the other mitochondrial phosphate carrier, the P_i-dicarboxylic acid carrier, is insensitive to NEM. If ammonia inhibits glutaminase by inhibiting a membrane-bound carrier, thereby restricting the permeability of glutamine or of some activator, one should expect

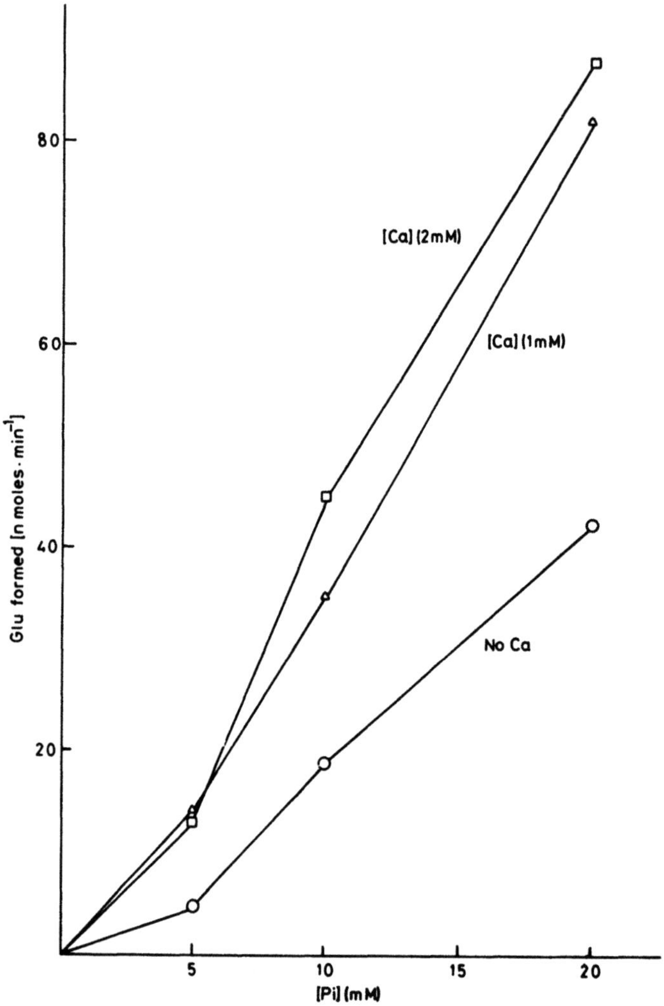

Fig. 9. Activation of glutaminase in synaptosomes by calcium. The experimental conditions were the same as those described in the legend to Fig. 2. No turbidity was observed in experiments run in the absence of synaptosomes, except in the vessel containing 20 mM phosphate and 2 mM calcium.

that ammonia would exert a stronger inhibition on NEM-insensitive PAG than on NEM-sensitive PAG. As shown in Figure 8, the opposite result was found. Ammonia is a more powerful inhibitor of NEM-sensitive PAG than of NEM-insensitive PAG. This renders the above-mentioned hypothesis less likely and points to an inhibition of the enzyme itself by ammonia; e.g. via proton inhibition. Thus, in the physiological range of pH, ammonia exists mainly in the form of NH_4^+, and less than 5% exists in the undissociated form (60).

Calcium is an activator of PAG (Fig. 9); it is the only cationic activator, to our knowledge, which potentiates phosphate activation. The concentration range for activation by calcium is rather narrow and is restricted to 0.5 - 1.0 mM It is of interest that ammonia reduces the calcium potentiation of phosphate activation by about 80% (Fig. 10). It is noteworthy that no activation by calcium of purified PAG has been found; thus ammonia inhibition as well as calcium activation are both lost during enzyme purification.

The effect of calcium on PAG remains obscure. It is well known that energy-dependent calcium uptake in mitochondria occurs concomitantly with an uptake of anions (e.g., phosphate) and a release of protons to the external medium (61). Also, Lehninger et al. (62) have shown that calcium uptake is increased when pyridine nucleotides are present in a more reduced state and vice versa. However, the exchange of calcium and protons is abolished by respiratory inhibitors, and such exchange would probably not explain the calcium activation of PAG in our synaptosomal preparations, which are incubated with oligomycin and antimycin, unless there exists a Ca^{2+}/H^+ antiport (53,63). The Ca^{2+}/H^+ antiport would have to be reversible, which has as yet to be demonstrated. Another possibility would be the existence of a calcium-dependent activator protein acting on glutaminase, analogous to the protein which activates brain cyclic nucleotide phosphodiesterase (64).

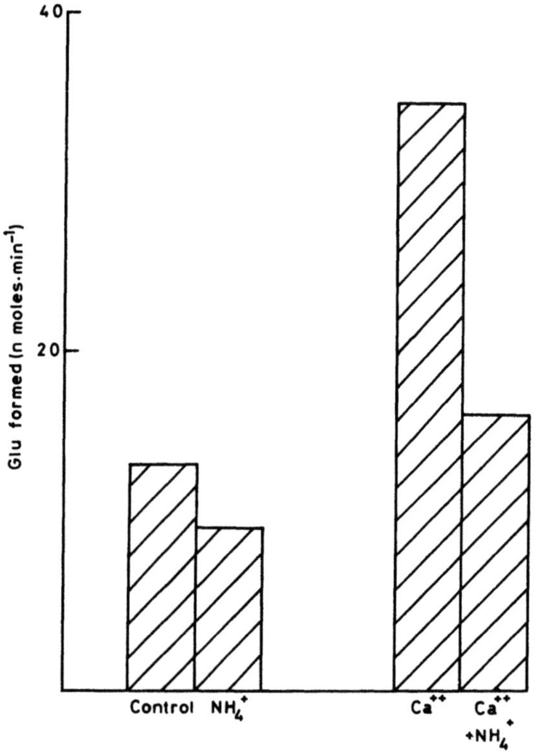

Fig. 10. Effects of calcium and ammonia on PAG activation by 10 mM phosphate in synaptosomes. The experimental conditions were the same as those described in the legend to Fig. 2. Added calcium 1 mM, ammonia 1 mM.

PHYSIOLOGICAL PERSPECTIVES

As described above, PAG of both brain and kidney is affected by an overwhelming number of activating and inhibiting ligands. However, the main activator seems to be phosphate, because many of the other highly effective activators act by potentiating phosphate activation (apparently by increasing the affinity for this ion) and are much less effective in the absence of phosphate. Thus, Bromothymol Blue, which is a model activator, affects PAG according to this principle (25) and the same holds for acyl CoA derivatives in general (33,35), in contrast to acetyl-CoA which activates only in the absence of phosphate (34). A great number of the anionic and hormone activators which have been reported in the literature are probably effective in concentrations that are too high to be of physiological interest. Sulphate, for instance, has an activating capacity similar to that of phosphate (12,45) and it also induces polymerization in cooperation with borate (25). It appears unlikely that sulphate serves as an important activator in vivo since its cellular concentration is probably not sufficiently high. However, the importance of a compound as an enzymatic regulator does not depend only upon its concentration. Of greater importance may be its range of physiological variation and its possible interactions with other variable regulatory ligands (see above).

Experiments with purified PAG have revealed that the effects of activators and inhibitors depend to a great extent upon the molecular form of the enzyme which, in turn, depends upon the ionic environment. Thus, polymerized PAG has very different kinetic properties than the soluble Tris-enzyme. The molecular form of PAG in vivo is unknown and is difficult to approach by isolation procedures since it is changed by the buffer used for isolation. However, since PAG is probably loosely bound to the inner membrane and is present in distinct mitochondrial compartments, there exists a strong possibility that it may also undergo molecular changes in vivo. In this context it is of interest that reversible movements of enzymes between the mitochondrial inner membrane and inter-membrane space have been described (65).

PAG has different functions in brain and kidney, but the molecular properties of the enzymes purified from these organs are nevertheless rather similar. However, as previously suggested (66), an enzyme with multiple regulatory possibilities may appear to be functionally different if the regulatory set of ligands that are available to the enzyme in each organ is different.

PAG of nerve endings is potentially very active, and its possible regulation by glutamate, ammonia and calcium invites interesting perspectives. Very low labelling of glutamate from precursor glutamine has been found in dorsal sensory ganglia (67), in accord with the suggestion that brain PAG is dormant in the resting state (68). This may be due to inhibition of PAG by ammonia or by glutamate. Thus, Benjamin *et al.* (60) have reported that ammonia may be loosely bound in brain slices and possibly be localized in specific brain compartments.

Concerning the role of glutamate, compartmentation of cerebral glutamine-glutamate metabolism seems to be well documented. Neurones appear to be characterized by a large glutamate pool and a small glutamine pool, whereas the opposite appears to exist for glial cells, and the concept of the glutamine cycle has been established. High activity of glutamine synthetase in glial cells and little or no such activity in neuronal perikarya, together with high-affinity uptake of glutamate localized to glial cells and high glutaminase activity in neuronal perikarya, support this concept (69,70), but it is noteworthy that the glutaminase activity in astrocytes is also remarkably high (71). Thus, the glutamine cycle implies that glutamate is actively taken up by glial cells and then rapidly converted to glutamine which may enter neurones and serve as the substrate for neuronal glutaminase (72-74; Fig. 11). Although the glutamate concentration in nerve endings is unknown, the data of Lai and Clark (49) suggest that the glutamate pool in the synaptic mitochondria is small. If this is valid, the main inhibitor of PAG in nerve endings is likely to be ammonia.

It has been reported that cerebral glutaminase activity is increased during seizures (75), by increased oxygen pressure (38), after whole-body X-irradiation (76) and by topical application of mescaline-hydrochloride to the ectosylvian gyrus (77). The mechanism of these activations is unknown, but may involve a release of calcium which activates PAG, possibly by counteracting a normal state of inhibition. The same explanation may apply to the stimulus-induced

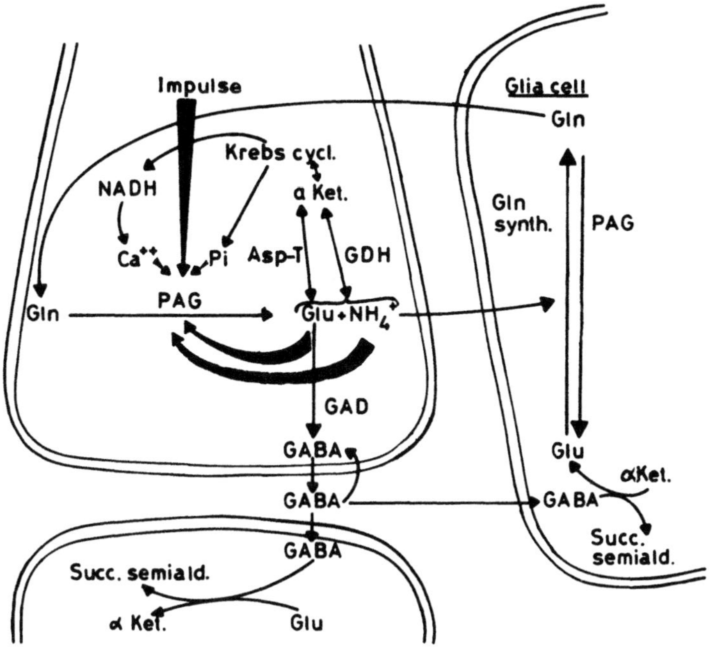

Fig. 11. Regulation of the glutaminase reaction, functioning as part of the glutamine cycle.

release of glutamate which Bradford and coworkers (52) have found to be derived from glutamine. Furthermore, calcium-dependent release of glutamate and GABA has been demonstrated from hippocampal synaptosomes where these compounds probably have a transmitter function (78), and the glutamate released by calcium has been shown to be derived mainly from glutamine (7). These findings may also be explained by the calcium activation of glutaminase which we have demonstrated. However, the effect of calcium on PAG may be distinct from its effect on the release of glutamate and GABA since the intracellular calcium ion concentration plays a general role in neurotransmitter release (79). Resting nerve cells are almost impermeable to calcium ion, but the arrival of the action potential at the nerve terminal leads to an increase in calcium conductance of the presynaptic membrane and to an influx of calcium and subsequently to a transmitter release. It has also been reported that the intracellular calcium content in the cytoplasm of a snail neurone decreases during bursting activity induced by pentylenetetrazol and accumulates near the cell membrane (80). Furthermore, there is evidence that counter-transport of three sodium ions for one calcium ion occurs in squid axons (81) and in presynaptic nerve terminals of rat brain (82). Thus, the effect of calcium on PAG may occur secondarily to its effect on transmitter release, and thereby serve to replenish depleted stores of transmitter glutamate and GABA.

In conclusion, it is suggested that the glutaminase present in synapses utilizing glutamate as a transmitter is predominantly regulated by ammonia, glutamate and calcium that are released by nervous stimulation. Calcium acts by potentiating the phosphate activation. Moreover, since about 50% of released GABA is derived from glutamine (52), this regulatory principle is also likely to play a significant role in GABA synthesis and function. This hypothetical regulatory principle, illustrated in Figure 11, suggests that the overall turnover of the glutamine cycle may be limited by the glutaminase reaction.

ACKNOWLEDGEMENTS

This research was performed in collaboration with Drs. Gerd Svenneby, Ingeborg Aa. Torgner and Borghild Tveit, as well as with Ms. Kari Bråten, Berit Randi Juel, Torun Flatebø and Bente Olsen.

REFERENCES

(1) Greenstein, J. P. and Carter, C. E. (1947) J. Nat. Cancer Inst. 7, 57-60.
(2) Katunuma, N., Huzino, A. and Tomino, I. (1967) Adv. Enzyme Regula. 5, 55-69.
(3) Katunuma, N., Katsunuma, T., Tomino, I. and Matsuda, Y. (1968) Adv. Enzyme Regula. 6, 227-242.
(4) Curthoys, N. P. and Kuhlenschmidt, T. (1975) J. Biol. Chem. 250, 2099-2103.

(5) Tate, S. S. and Meister, A. (1975) J. Biol. Chem. 250, 4619-4627.
(6) Salganicoff, L. and DeRobertis, E. (1965) J. Neurochem. 12, 287-309.
(7) Hamberger, A., Chiang, G. Nylén, E. S., Scheff, S. W. and Cotman, C. W. (1978) Brain Res. 143, 549-555.
(8) Krebs, H. A. (1935) Biochem. J. (Lond.) 29, 1951-1969.
(9) Errera, M. and Greenstein, J. P. (1949) J. Biol. Chem. 178, 495-502.
(10) Otey, M. C., Birnbaum, S. M. and Greenstein, J. P. (1954) Arch. Biochem. Biophys. 49, 245-247.
(11) Klingman, J. D. and Handler, P. (1958) J. Biol. Chem. 232, 369-380.
(12) Sayre, F. W. and Roberts, E. (1958) J. Biol. Chem. 233, 1128-1134.
(13) Kvamme, E. and Tveit, B. and Svenneby, G. (1970) J. Biol. Chem. 245, 1871-1877.
(14) Svenneby, G., Torgner, I. and Kvamme, E. (1973) J. Neurochem. 20, 1217-1224.
(15) Kvamme, E. and Svenneby, G. (1975) In: Research Methods in Neurochemistry (N. Marks and R. Rodnight, eds.) Vol. 3, pp. 277-290, Plenum Publ. Co., New York.
(16) Curthoys, N. P., Kuhlenschmidt, T. and Godfrey, S. S. (1976) Arch. Biochem. Biophys. 174, 82-89.
(17) Olsen, B. R., Torgner, I., Christensen, T. B. and Kvamme, E. (1973) J. Mol. Biol. 74, 239-251.
(18) Svenneby, G., Tveit, B. and Kvamme, E. (1970) J. Biol. Chem. 245, 1878-1882.
(19) Svenneby, G. (1972) J. Neurochem. 19, 165-174.
(20) Olsen, B. R., Svenneby, G., Kvamme, E., Tveit, B. and Eskeland, T. (1970) J. Mol. Biol. 52, 239-245.
(21) Svenneby, G. (1970) J. Neurochem. 17, 1591-1599.
(22) Svenneby, G. (1971) J. Neurochem. 18, 2201-2208.
(23) Weil-Malherbe, H. (1972) J. Neurochem. 19, 2257-2267.
(24) Blackburn, E. H., Hird, F. J. R. and Jones, I. K. (1972) Arch. Biochem. Biophys. 152, 265-271.
(25) Kvamme, E., Tveit, B. and Svenneby, G. (1965) Biochem. Biophys. Res. Commun. 20, 566-572.
(26) Kvamme, E., Svenneby, G. and Tveit, B. (1966) In: Molecular Basis of Some Aspects of Mental Activity (O. Walaas, ed.) Vol. 1, pp. 211-219, Academic Press, London.
(27) Kvamme, E., Svenneby, G. and Tveit, B. (1968) In: The Regulation of Enzyme Activity and Allosteric Interactions (E. Kvamme and A. Pihl, eds.) pp. 89-108, Universitetsforlaget, Oslo.
(28) Tveit, B., Svenneby, G. and Kvamme, E. (1970) Eur. J. Biochem. 14, 337-344.
(29) Greenstein, J. P. and Leuthardt, F. M. (1948) Arch. Biochem. Biophys. 17, 105-114.
(30) O'Donovan, J. and Lotspeich, W. D. (1966) Nature (Lond.) 212, 930-932.

(31) Katunuma, N., Tomino, I. and Nishino, H. (1966) Biochem. Biophys. Res. Commun. 22, 321-328.
(32) Weil-Malherbe, H. and Beall, G. D. (1970) J. Neurochem. 17, 1101-1103.
(33) Kvamme, E. and Torgner, I. (1974) FEBS Lett. 47, 244-247.
(34) Kvamme, E. and Torgner, I. (1974) Biochem. J. (Lond.) 137, 525-530.
(35) Kvamme, E. and Torgner, I. (1975) Biochem. J. (Lond.) 149, 83-91.
(36) Kovačević, Z., McGivan, J. D. and Chappell, J. (1970) Biochem. J. (Lond.) 118, 265-274.
(37) Kovačević, Z. (1975) Biochim. Biophys. Acta, 396, 325-334.
(38) Kritschevskaya, A. A., Gerschenovitsch, Z. and Scherbatzker, S. (1959) Biochimija, 64, 459-464.
(39) Hovhannissian, V. S., Buniatian, H. C., Badalian, L. L. and Mikirtumova, G. S. (1969) Voprosy Biochimii Mozga. Akad. Nauk Armjan. SSR, 5, 5-16.
(40) Hovhannissian, V. S., Buniatian, H. C., Ukrdumova, G. S. and Badalian, L. L. (1970) Voprosy Biochimii Mozga. Akad. Nauk Armjan. SSR, 6, 5-13.
(41) Badalian, L. L., Buniatian, H. C. and Hovhannissian, V. S. (1975) Voprosy Biochimii Mozga. Akad. Nauk Armjan. SSR, 10, 40-54.
(42) Weil-Malherbe, H. (1969) J. Neurochem. 16, 855-864.
(43) Hovhannissian, V. S., Badalian, L. L., Ukrtumian, G. S. and Sahagian, J. J. (1973) Voprosy Biochimii Mozga. Akad. Nauk Armjan. SSR, 8, 77-90.
(44) Mikirtumova, K. S., Hairapetian, H. L., Hovhannissian, V. S. and Buniatian, H. C. (1976) Voprosy Biochimii Mozga. Akad. Nauk Armjan. SSR, 11, 17-28.
(45) Bradford, H. F. and Ward, H. K. (1976) Brain Res. 110, 115-125.
(46) Balázs, R., Dahl, D. and Harwood, J. R. (1966) J. Neurochem. 13, 897-905.
(47) Blokhuis, G. G. D. (1970) Heterogeniteit van mitochondriën in hersenweefsel. Drukkerij Bronder-Offset N. V., Rotterdam.
(48) Reijnierse, G. L., Veldstra, H. and van den Berg, C. J. (1975) Biochem. J. (Lond.) 152, 469-475.
(49) Lai, J. C. K. and Clark, J. B. (1976) Biochem. J. (Lond.) 154, 423-432.
(50) Dennis, S. C., Lai, J. C. K. and Clark, J. B. (1977) Biochem. J. (Lond.) 164, 727-736.
(51) Dienel, G., Ryder, E. and Greengard, O. (1977) Biochim. Biophys. Acta, 496, 484-494.
(52) Bradford, H. F., Ward, H. K. and Thomas, A. J. (1978) J. Neurochem. 30, 1453-1459.
(53) Crompton, M., Capano, M. and Carafoli, E. (1976) Eur. J. Biochem. 69, 453-462.
(54) Kalra, J. and Brosnan, J. T. (1974) J. Biol. Chem. 249, 3255-3260.
(55) Curthoys, N. P. and Weiss, R. (1974) J. Biol. Chem. 249, 3261-3266.

(56) Kovačević, Z. (1976) Biochim. Biophys. Acta, 430, 399-412.
(57) Touraille, S., Briand, Y. and Durand, R. (1977) FEBS Lett. 84, 119-124.
(58) Klingenberg, M., Riccio, P., Aquilla, H., Schmiedt, B., Grebe, K. and Topitsch, P. (1974) In: Membrane Proteins in Transport and Phosphorylation (G. Azzone, M. Klingenberg, E. Quagliariello and N. Siliprandi, eds.) pp. 229-244, Elsevier, Amsterdam.
(59) Greville, G. D. (1969) In: Current Topics in Bioenergetics (D. R. Sanadi, ed.) Vol. 3, pp. 1-78, Academic Press, New York.
(60) Benjamin, A. M., Okamoto, K. and Quastel, J. H. (1978) J. Neurochem. 30, 131-143.
(61) Chappell, J. B. and Crofts, A. R. (1965) Biochem. J. (Lond.) 95, 378-386.
(62) Lehninger, A. L., Vercesi, A. and Bababunmi, E. A. (1978) Proc. Nat. Acad. Sci., U.S.A. 75, 1690-1694.
(63) Fiskum, G. and Cockrell, R. S. (1978) FEBS Lett. 92, 125-128.
(64) Teo, T. S., Wang, T. H. and Wang, J. H. (1973) J. Biol. Chem. 248, 588-595.
(65) Waksman, A. and Rendon, A. (1974) Biochimie, 56, 907-924.
(66) Kvamme, E., Svenneby, G. and Torgner, I. (1973) In: Central Nervous System — Studies on Metabolic Regulation and Function (E. Genazzani and H. Herken, eds.) pp. 63-69, Springer Verlag, Berlin.
(67) Johnson, J. L. (1974) Brain Res. 69, 366-369.
(68) Waelsch, H. (1951) Adv. Protein Chem. 6, 299-341.
(69) Henn, F. A., Goldstein, M. N. and Hamberger, A. (1974) Nature (Lond.) 249, 663-665.
(70) Martinez-Hernandez, A., Bell, K. P. and Norenberg, M. D. (1977) Science, 195, 1356-1358.
(71) Schousboe, A., Svenneby, G., Hertz, L. and Kvamme, K. (1979) J. Neurochem. (in press).
(72) Balázs, R., Patel, A. J. and Richter, D. (1973) In: Metabolic Compartmentation in the Brain (R. Balázs and J. E. Cremer, eds.) pp. 167-184, Macmillan, London.
(73) Van den Berg, C. J. (1973) In: Metabolic Compartmentation in the Brain (R. Balázs and J. E. Cremer, eds.) pp. 137-166, Macmillan, London.
(74) Benjamin, A. M. and Quastel, J. H. (1974) J. Neurochem. 23, 457-464.
(75) Weichert, P. and Göllnitz, G. (1969) J. Neurochem. 16, 317-322.
(76) Hăulică, A., Trandafirescu, M. and Ababei, L. (1971) J. Neurochem. 18, 2447-2450.
(77) Mison-Crighel, N., Pintilie, C. and Crighel, E. (1965) Nature (Lond.) 206, 1257-1258.
(78) Sandoval, M. E., Horch, P. and Cotman, C. W. (1978) Brain Res. 142, 285-299.
(79) Rahaminoff, R. (1976) In: Motor Innervation of Muscle (S. Thesleff, ed.) pp. 117-149, Academic Press, London.
(80) Sugaya, E. and Onozuka, M. (1978) Science, 200, 797-799.
(81) Blaustein, M. P., Russell, J. M. and De Weer, P. (1974) J.

Supramol. Struct. 2, 558-581.
(82) Blaustein, M. P. and Oborn, C. J. (1975) J. Physiol. (Lond.) 247, 657-686.
(83) Gray, E. G. and Whittaker, V. P. (1962) J. Anat. (Lond.) 96, 79-88.

ENZYMES OF GABA METABOLISM IN TISSUE CULTURE

L. Ossola, M. Maitre, J. M. Blindermann and P. Mandel

Centre de Neurochimie du CNRS, et Institut de Chimie Biologique, Faculté de Médecine, 67085 Strasbourg Cedex, France

INTRODUCTION

The development of techniques for nervous tissue culture has permitted studies on homogeneous populations of neuronal or glial cells. Such preparations represent simplified model systems which may be used to study the metabolism of neurotransmitters, providing that their morphologic and biochemical behavior *in vitro* does not differ from that which occurs *in vivo*. In this regard, many investigations have revealed that nervous tissue cultured *in vitro* maintains and/or develops morphological and functional properties similar to those observed *in vivo* (1-5). Electron-microscopic studies have shown that neuronal and glial cells grow, mature, differentiate and form synaptic complexes in culture (5,6). Neurones in culture are also able to stock and/or synthesize numerous enzymes known to be present in the nervous system *in vivo* (5,7-11). With specific regard to GABA-ergic metabolism, some works have revealed that GABA-transaminase (GABA-T) activity in primary glial cultures increases to levels which approximate its level in adult brain (12).

Dorsal root ganglion cells in primary culture have been used as a model system for studying membrane mechanisms involved in presynaptic inhibition, and the effect of GABA on these cells has been tested (13). In addition to these studies on primary cultures, others have shown that mouse neuroblastoma cells possess certain normal neuronal properties, such as neurotransmitter (e.g., GABA) synthesis (8,14). Neuroblastoma cells also have enzymes for GABA synthesis and degradation; the activities of these enzymes increase when cells differentiate (15).

Although the above-mentioned results indicate that tissue cultures, especially primary cultures, have similar morphological and biochemical properties as cells *in vivo*, we must also consider other results which have revealed striking differences between cell cultures and brain tissue. Thus, cell cultures do not always represent a convenient "model system" for studying the *in vivo* situation (7). As examples, the activities of four enzymes, choline acetyltransferase, acetylcholinesterase, lactate dehydrogenase and glutamate decarboxylase (GAD), are much lower in mouse primary culture and in C_6 glioblastoma than in a brain homogenate (16,17). Also, the activities of GAD, GABA-T and succinic semialdehyde dehydrogenase of glioblastoma C_6 and C-1300 neuroblastoma cells in culture represent less than 20% of the activities of a homogenate of mouse cerebral cortex, whereas glutamate dehydrogenase activity is equal to, or greater than *in vivo* (18).

In the present study, we have determined the activities of the enzymes of GABA metabolism in cultured cells of neuronal and glial origin and in cultured cells transformed by the SV40 virus. We also describe the characteristics of a natural inhibitor of GABA-T in astroblast primary cultures.

MATERIALS AND METHODS

Materials

The C_6 cell line was obtained from American Type Culture Collection (Rockville, Maryland). M_1 and N_{115} cell lines were isolated from C-1300 mouse neuroblastoma. The N_{115} cell line was the gift of Dr. M. Nirenberg (Laboratory of Biochemical Genetics, NIH, Bethesda, Md., USA). The M_1 cell line was developed and characterized in our laboratory (19).

Cells were grown in Falcon flasks (75 cm^3). The culture medium for all cells was composed of Eagle medium as modified by Dulbecco (DMEM medium) and contained 10% fetal calf serum (Gibco, Grand Island, N.Y.), 50 units of penicillin-G and 25 μg of streptomycin-sulfate per ml. Cultures were incubated in an isothermic oven in a circulating humid atmosphere (95% air/5% CO_2). When the cultures attained a stationary phase, a subculture was prepared after their incubation in the presence of 0.25% trypsin. Initially each flask contained approximately 1.50 x 10^5 cells. The culture medium was changed every two days. Cultures of the MT series were isolated from primary cultures of cerebral hemispheres of embryonic mice and transformed by the oncogenic SV40 virus in our laboratory. These cells remained in a rather precocious state of development; the immature character of these cells made their classification difficult. In any case, certain histochemical, morphological and ultrastructural findings favor a neuronal classification for the clones MT_{17}, MT_4 and MT_{D1} and a glial classification for MT_{16}, MT_{26} and MT_9 (20). Primary cultures of astroblasts of neonatal rat brain were prepared according to the method of Booher and Sensenbrenner (21); see also Bock et al. (22).

Preparation of Tissues

Preparation of Cell Homogenate. For measurements of enzymatic activities and for radioimmunoassays, cultures were washed with 2 x 5 ml of physiologic saline and centrifuged for 15 minutes at 1400 g. The pellet was re-extracted with 250 μl, then 100 μl, of 20 mM phosphate buffer (pH 7.4) containing 1 mM 2-aminoethylisothiouronium bromide hydrobromide, 5 mM EDTA, 0.025 mM phenylmethyl sulfonyl fluoride, 10 mg/l pyridoxal phosphate, and then homogenized at 4°C.

For GABA determination, the culture medium was discarded and cells were homogenized in 10 ml of 0.1 N hydrochloric acid, and then centrifuged for 15 minutes at 1400 g. The pellet was re-extracted with 2 ml of 0.01 N hydrochloric acid. Supernatants were pooled and placed onto anion exchange columns (15 cm x 1 cm) of the resin Dowex 50 W x 8 (H^+ form), equilibrated with 10 ml of 0.01 N hydrochloric acid. The column was washed wth 25 ml of 0.01 N hydrochloric acid. GABA was eluted with 25 ml of 2 N ammonia. The eluate was evaporated to dryness under low pressure at room temperature, and the lyophylizate was resuspended in a minimum volume of 0.01 N hydrochloric acid for GABA determination. The pellet obtained after the second GABA extraction was dissolved in 1 ml of 6 N sodium hydroxide and kept at 60°C for an hour before protein determination.

Preparation of Rat Brain Homogenate. To measure enzymatic activities, Wistar rats were killed by decapitation and their brains were rapidly removed and homogenized at 0°C in 5 mM phosphate buffer (pH 7) containing 150 mM potassium chloride, 1 mM 2-aminoethylisothiouronium bromide hydrobromide, 5 mM EDTA, 10 mg/1 pyridoxal phosphate and 0.5% Triton X-100. For GABA determinations, rats were killed in a microwave oven. The irradiation time of the head of the animal never exceeded 7 seconds; denaturation of enzymes directly involved in GABA metabolism occurred. The brains were homogenized in 10 volumes of ice cold 80% ethanol and centrifuged at 10,000 g for 30 minutes at 4°C. Pellets were re-extracted with 10 vol of ice-cold 70% ethanol and re-centrifuged. Supernatants were pooled and evaporated to dryness under low pressure at room temperature.

Protein Determination

Protein contents were determined by the method of Lowry et al. (23).

Enzyme Assays

Assay of Glutamate Decarboxylase (GAD). GAD activity was measured by trapping $^{14}CO_2$ formed from L [U-^{14}C]glutamate according to the method of Albers and Brady (24). Fifty ml of tissue homogenate (1 - 1.5 mg protein for the brain homogenate; 0.25 - 0.75 mg protein for the cell homogenate) were incubated in 6-ml conical test tubes with 50 μl incubation medium containing (final concentrations) 10 mM glutamate (0.4 μCi/μ-mole), 100 mM potassium phosphate buffer (pH 6.8), 1 mM 2-aminoethylisothiouronium bromide hydrobromide, 0.1 mM EDTA and 0.5 mM pyridoxal phosphate. The tubes were sealed with rubber stoppers which supported small cups containing 100 μl hyamine hydroxide. Each assay was performed in triplicate, incubations being carried out at 37°C for 30 minutes for brain homogenate and for 45 minutes for cell homogenates. Blanks consisted of the standard incubation mixture alone. The reaction was terminated by injection of 100 μl of 20 N sulfuric acid. One hour later, the content of each cup was washed into a scintillation vial with 10 ml scintillation fluid.

GAD activity was also measured by determining the amount of [U-^{14}C]GABA formed from L[U-^{14}C]glutamate, according to the method of Chude and Wu (25). Tissue homogenates were incubated as previously described, but the reaction was terminated by addition of 100 μl of 0.2 N sulfuric acid. The GABA formed was isolated using an anion exchange (AG 1-X8) resin column (6 cm x 0.5 cm) equilibrated with 2 N sodium acetate solution. GABA was eluted directly into scintillation vials with 3.5 ml distilled water, the radioactivity being counted after addition of 3 ml scintillation fluid.

Assay for GABA-α-Oxoglutarate Transaminase (GABA-T). GABA-T activity was determined according to the method of Dye and Taberner (26), at a constant concentration of tissue protein (1.5 mg/ml for brain homogenate; 3 mg/ml for the cell homogenate). Homogenates were incubated at 37°C in a total volume of 70 μl of 0.3 M Tris-HCl buffer (pH 8.6) containing (final concentrations) 8 mM [2,3-^{3}H]GABA (8 μCi/μ-mole), 4 mM 2-oxoglutaric acid, 1 mM 2-aminoethylisothiouronium bromide hydrobromide, 0.1 mM EDTA, 0.5 mM pyridoxal phosphate, 0.5 mM sodium succinate, 2.5 mM NAD. The reaction was started by addition of 20 μl tissue homogenate and stopped after 40 minutes for cell homogenates and after 30 minutes for brain homogenates by adding 100 μl trichloroacetic acid (10% w/v). Blanks were run under the same conditions without tissue homogenate. The product of the reaction was separated from GABA on a column (0.5 x 3 cm) of Dowex 50W X8 resin (H^+ form) by elution with 4 x 1.0 ml of water, and collected directly in scintillation vials containing 3 ml of Instagel. Each determination was run in triplicate.

Assay of GABA

GABA contents were determined by the enzymatic fluorometric assay of Graham and Aprison (27). Since some constituents of the extract, redissolved in a minimum volume of 0.01 N HCl, interfered with the fluorescence produced by the enzymatic reaction, the method of standard addition was employed (28). Assays were run under the same conditions as used for calibration curves.

Radioimmunoassay of GABA-T

The level of enzymatic protein of GABA-T was measured by radioimmunoassay using antiserum to rat GABA-T obtained by injection of purified GABA-T into an adult rabbit, and ^{125}I-labelled GABA-T was prepared by the method of Hunter (29). Radioimmunoassay for GABA-T is clearly described by Maitre et al. (this volume).
volume).

Characterization of a Natural GABA-T Inhibitor from a Four-Week-Old Primary Culture of Rat Astroblasts

Characterization of GABA-T Inhibitory Capacity of Astroblasts by Combined Incubation of Rat Brain Homogenate and Astroblast Homogenate. Ten μl of rat brain homogenate (35 μg tissue protein) were incubated with 10 μl of astroblast homogenate (0.5 - 2 mg/ml of cell protein) and GABA-T activity was measured by the method of Dye and Taberner (26). Comparison between GABA-T activity obtained by this simultaneous incubation method and by the addition of the GABA-T activities measured in the two homogenates incubated separately, allowed us to show a modulation of GABA-T activity of the rat brain homogenate by astroblast cell homogenate.

Thermal Inactivation of the Inhibitory Factor of Astroblasts. Astroblast culture homogenate (103 µg protein) was pre-incubated at 55°C for 5 to 15 minutes. Residual GABA-T activity was measured as described above to determine the GABA-T inhibitory capacity of the astroblasts after heating at 55°C.

Molecular Weight Estimation of the Inhibitory Factor. The molecular weight was determined by filtration on an Ultragel ACA 34 column (1 m x 2.5 cm). The column was previously equilibrated with 5 mM phosphate buffer containing 1 mM 2-aminoethylisothiouronium bromide hydrobromide, 0.1 mM EDTA and 0.5 mM pyridoxal phosphate. Astroblast protein (4.8 mg) was applied to the column in 1 ml of equilibrating buffer containing 3 mg bovine serum albumin, 3 mg gamma-globulin and 3 mg blue dextran. The elution was performed with equilibrating buffer at a rate of 12 ml/h, and the eluate was collected in 5-ml fractions. The proteins eluted were detected in each fraction by measuring the absorption at 280 nm. Astroblast GABA-T activity was determined by the method of Dye and Taberner (26). A simultaneous incubation of astroblasts and rat brain was performed by adding 10 µl of rat brain homogenate (35 µg protein) to a 200 µl aliquot of each fraction to determine the inhibitory factor.

Determination of the Isoelectric pH of the Inhibitory Factor. The isoelectric pH of the inhibitory factor was determined by the technique of Radola (30) using a plate of Ultrodex and Ampholines with a pH range of 3.5 - 10. The anode had 1 M phosphoric acid and the cathode had 1 M sodium hydroxide. Migration was allowed to occur for 16 hours at a constant current of 6 W, and the plate was maintained at 4°C by circulating water. The plate was divided into 30 fractions with a grid. Each fraction was removed with a spatula, diluted with 1 ml distilled water and its pH was measured. After centrifugation, the GABA-T activity of each fraction was measured (26). The inhibitory factor was determined by a simultaneous incubation by adding 300 µl of each fraction obtained by isoelectric focussing to 10 µl of rat brain homogenate (35 µg tissue protein).

RESULTS

Enzymes of GABA Synthesis and Degradation and the GABA Level of Tissue Cultures

Glutamate Decarboxylase Activity Measured as $^{14}CO_2$ and U-^{14}C GABA Production from L[U-^{14}C]Glutamate. GAD activity, measured as [U-^{14}C]GABA formed in the different cultures, is given in Table I. MT_{26} and C_6 glial cells had the lowest GAD activity; about 0.8% and 1.5%, respectively, of the GAD activity of rat brain homogenate. The highest GAD activity, present in clone MT_{19}, represented 5.3% of the GAD activity of rat brain homogenate. These results agreed with others given in the literature where GAD activity from C_6 glial cells, measured by the same method, was about 2% of that of a mouse cerebral cortex homogenate (18). Neuroblastoma M_1 and N_{115} clones and the neurone-like transformed cells MT_4, MT_{17}, MT_{D1} possessed higher GAD activity than most glial cell lines. It should be recalled that similar results have been obtained with synaptosomes or glial cell fractions (32). However, one glial type of transformed cell line possessed GAD activity similar to that of neuronal cell lines.

For brain homogenate and neuroblastoma M_1, the GAD activity determined by CO_2 production agreed with that determined by GABA production, the ratio between these two activities being about unity. For neuroblastoma N_{115}, glioblastoma C_6 and transformed cell lines, this relation no longer existed. The ratio varied from 1.8 for clone MT_4 to 5.0 for clone MT_{26} and could increase to 10 for C_6. Similar

Table I - GAD Activity of Various Tissue Cultures

Tissue	GAD activity 10^{-9} moles/min/mg protein (A)	GAD activity 10^{-9} mole/min/mg protein (B)	$\frac{A}{B}$
Rat brain	2.73 ± 0.04	2.75 ± 0.02	0.99
Cell lines			
C_6	0.396 ± 0.024	0.042 ± 0.001	9.4
N115	0.250 ± 0.006	0.102 ± 0.002	2.4
M1	0.116 ± 0.004	0.132 ± 0.007	0.8
MT26	0.116 ± 0.003	0.023 ± 0.003	5.0
MT16	0.182 ± 0.010	0.060 ± 0.001	3.0
MT9	0.311 ± 0.008	0.147 ± 0.009	2.1
MT4	0.152 ± 0.004	0.086 ± 0.002	1.8
MT17	0.242 ± 0.005	0.067 ± 0.001	3.6
MTD1	0.264 ± 0.020	0.128 ± 0.002	2.1

Glutamate decarboxylase (GAD) activity was measured by $^{14}CO_2$ formed from L [U-^{14}C]glutamate (A) and by [U-^{14}C]GABA formed from L [U-^{14}C]glutamate (B).

results were published by Wilson (16). The difference between the GAD activities determined by the two methods indicated a production of CO_2 from glutamate by metabolic pathways other than that involving GAD (18,31). This suggests that on a molar ratio the degradation of glutamate through the GABA shunt represented about 1/10th - 1/2 of the CO_2 liberated from glutamate. The GAD activity may vary for the same clonal cell line when it is determined in different batches of cultures.

GABA-α-Oxoglutarate Transaminase (GABA-T). Results concerning GABA-T activity are given in Table II. Since the specific activity of GABA-T changes with the protein concentration in the incubation medium (33,34) and remains constant for concentrations higher than

Table II - Characteristics of GABA-T and GABA Contents of Various Tissue Cultures

	GABA-T activity* 10^{-9} moles/min/mg protein	Level of GABA-T 10^{-9} moles/mg protein	Molecular activity GABA-T, sec^{-1}	GABA 10^{-9} moles/mg protein
Rat brain	0.50 ± 0.04	0.032 ± 0.006	0.26 ± 0.05	18 ± 2
Cell lines				
C_6	0.238 ± 0.046	0.0141 ± 0.0007	0.28 ± 0.02	53 ± 2
N115	0.177 ± 0.037	-	-	52 ± 1
M1	0.130 ± 0.025	0.0514 ± 0.0170	0.048 ± 0.016	75 ± 1
MT26	0.127 ± 0.001	-	-	59 ± 3
MT16	0.161 ± 0.033	-	-	48 ± 2
MT9	0.144 ± 0.037	-	-	101 ± 2
MT4	0.115 ± 0.017	-	-	61 ± 1
MT17	0.112 ± 0.013	-	-	96 ± 4
MTD1	0.042 ± 0.021	-	-	76 ± 2

*Specific activity of GABA-T was measured at constant protein concentrations of 1.5 mg/ml for rat brain and 3 mg/ml for the cell lines.

1 mg/ml in the case of rat brain homogenate and 3 mg/ml in the case of cell homogenate, it was determined at constant protein concentrations of 1.5 mg/ml for the rat brain homogenate and 3 mg/ml for the cultures. Starting with these concentrations, it was possible to compare the cellular GABA-T activities to the GABA-T activity of a rat brain homogenate. GABA-T activity of C_6 glial cells represented less than 50% of that found in brain and may reach values of about 9% for SV40 transformed neuronal type cells. Similar values were reported for a comparison between GABA-T of C_6 astrocytoma and a mouse cerebral cortex homogenate (18). Under our experimental conditions, glioblastoma C_6 GABA-T activity was higher than that of M_1 and N_{115} neuroblastoma cells. These values agreed with the higher values found for glial cells than for synaptosomes (32). In transformed cells, the presumed glial cells (MT_{26}, MT_{16}, MT_9) possessed GABA-T activities that were slightly higher than those of presumed neuronal cells (MT_4, MT_{17}, MT_{D1}). Determinations of GABA-T activity were more reproducible in different batches of cultures than those concerning GAD activity.

The level of GABA-T apoprotein in the different cell cultures was determined by radioimmunoassay (35, see also Maitre, this volume). Results provided in Table II indicate that the GABA-T level of neuroblastoma M_1 is similar to that of a rat brain homogenate and that it is reduced by about 50% for the C_6 glial cells.

Molecular activities of GABA-T, expressed as the number of molecules of GABA transformed per second at 37°C and at pH 8.6 per molecule of GABA-T, were determined for different cell cultures. The molecular activity of GABA-T for neuroblastoma M_1 represented only about 20% of that of a rat brain homogenate, whereas the level of enzymatic protein of GABA-T was similar for neuroblastoma M_1 and a rat brain homogenate (Table II). The molecular activity of GABA-T for C_6 cells was the same as that of a rat brain homogenate, whereas the level of enzymatic protein of GABA-T in C_6 cells is only about half of that of a rat brain homogenate. Thus, we can postulate that the GABA-T activity of clonal cell lines C_6 and M_1 is modulated by GABA-T effectors. A GABA-T effector has been partially characterized in a primary culture of astroblasts (36); see below.

GABA Content in Different Cell Lines. GABA contents were determined by the enzymatic fluorimetric assay of Graham and Aprison (27), corrected by the method of standard addition (28), as previously described. Results given in Table II revealed that the GABA content of a rat brain homogenate is stable and equal to 18 n-moles/mg protein. In a given tissue culture, the GABA content differed from one batch of culture to another. Thus, the differences in GABA content found after repeated determinations do not allow us to provide a definite GABA content for the tissue cultures. However, values greater than that of brain homogenate can be found in tissue cultures. Great variations for other amino acids in tissue cultures have also been described (37,38).

Characterization of a Natural GABA-T Inhibitor from a Four-Week-Old Primary Culture of Rat Astroblasts

Variation of GABA-T Activity with Astroblast Protein Concentration in the Incubation Medium. Specific GABA-T activity of astroblasts decreased with increasing protein concentration in the incubation medium (Fig. 1), when a saturating substrate concentration was used. An analogous phenomenon has been described by Waksman for a partially purified GABA-T (34) and by De Boer for GABA-T of a rat brain homogenate (33).

Characterization of GABA-T Inhibitory Capacity of Astroblasts. An homogenate of cultured astroblasts decreased the GABA-T activity of a rat brain homogenate. This inhibition was proportional to the quantity of astroblast protein added (Fig. 2) and reached a maximum of 50% under our experimental conditions.

Inactivation of GABA-T Inhibitory Capacity of Astroblast Cultures by Heating. The Inhibitory capacity of an homogenate of astroblasts decreased linearly with pre-incubation at 55°C. The

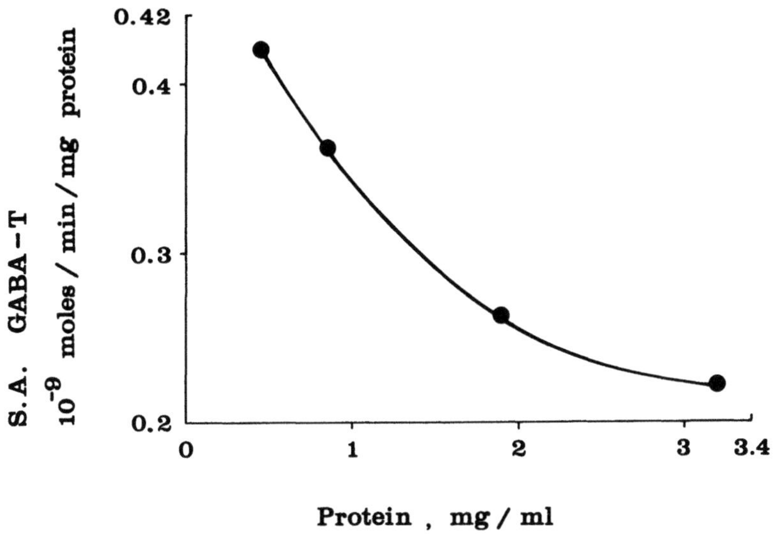

Fig. 1. Specific activity of GABA-T as a function of astroblast protein concentration.

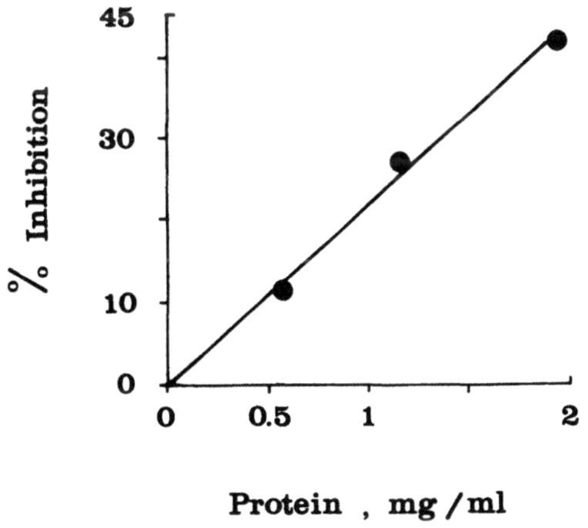

Fig. 2. Inhibition of rat brain GABA-T (homogenate) as a function of astroblast protein concentration.

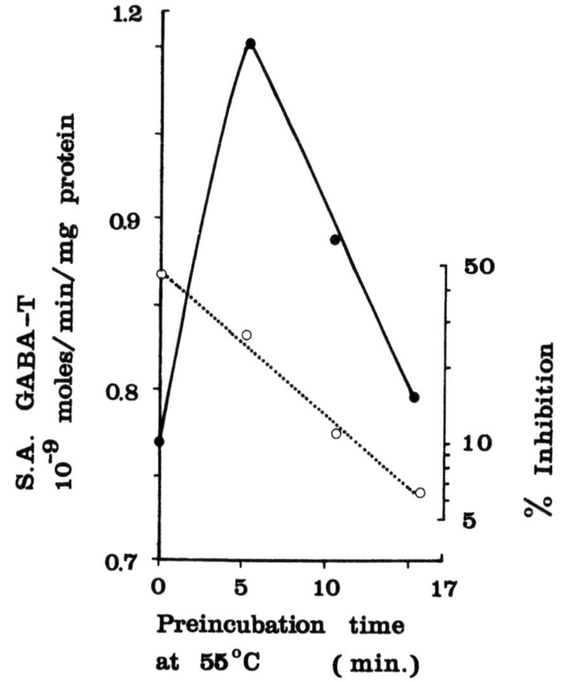

Fig. 3. Specific activity of astroblast GABA-T as a function of preincubation time (●——●), and percentage of residual inhibition of GABA-T from rat brain homogenate by preincubated astroblasts (o·········o).

period of thermal denaturation was about 6 min (Fig. 3). After 5 min at 55°C, the inhibitory effect on GABA-T disappeared and the GABA-T activity of astroblasts increased (Fig. 3). Therefore, the inhibitory factor appears to have a greater thermosensitivity than the GABA-T activity.

Dialyzable Character and Molecular Weight Estimation of the Inhibitory Factor of Astroblasts. The inhibitory factor was not dialyzable. Its elution volume on an ultrogel column suggested that it has a molecular weight of about 35,000 ± 10,000 (Fig. 4).

Determination of the Isoelectric pH of the Inhibitory Factor. The inhibitory factor and the GABA-T activity of an astroblast homogenate were separated on a pH gradient. The isoelectric pH of

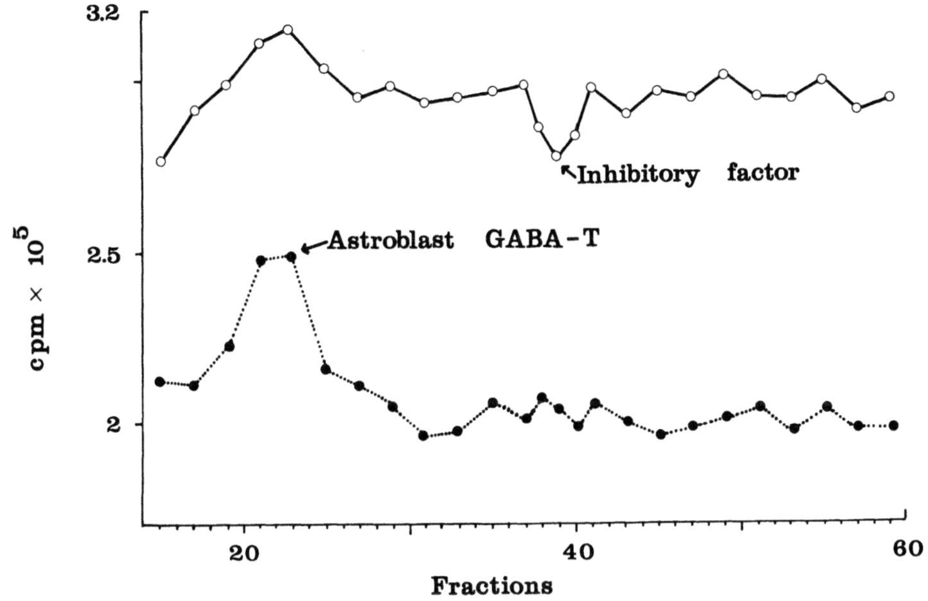

Fig. 4. Separation on ultrogel of the inhibitory factor from astroblast GABA-T. Upper plot: mixed incubation between a constant amount of rat brain homogenate and different fractions of an astroblast homogenate eluted from ultrogel column (o———o); lower plot: corresponding GABA-T activity in astroblast fractions (•·········•).

the inhibitory factor was about 5.4; it differs from the isoelectric pH of astroblast GABA-T which is about 6.4 (Fig. 5). These results suggest that a thermosensible inhibitory factor of GABA-T activity exists in primary cultures of astroblasts. The thermosensitivity, molecular weight, and isoelectric pH favor a proteinaceous nature for this inhibitory factor. As this inhibitory factor was unstable, it was technically impossible to further characterize it.

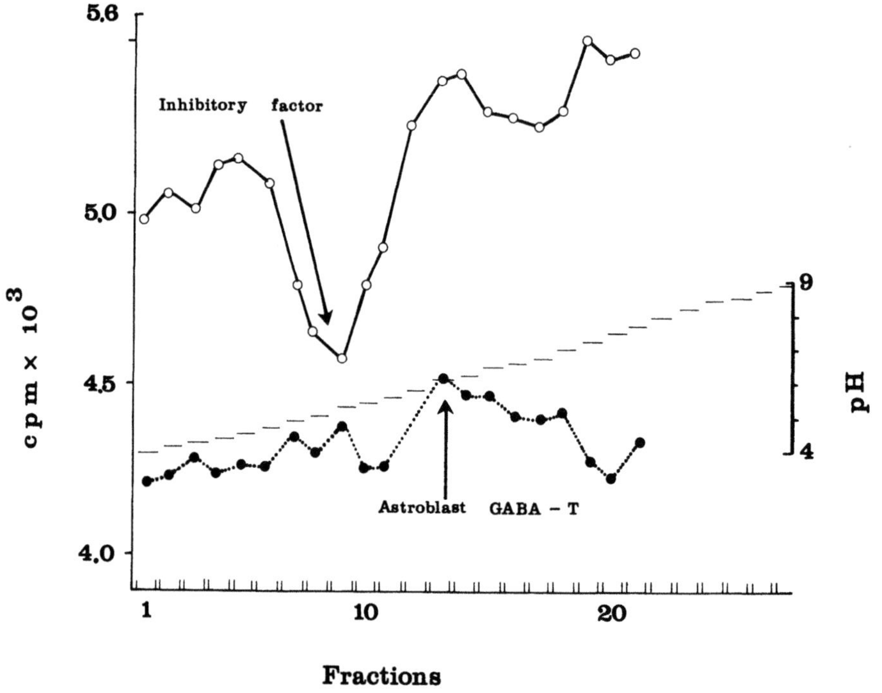

Fig. 5. Separation on pH gradient of the inhibitory factor from astroblast GABA-T. Upper plot: mixed incubation between a constant amount of rat brain homogenate and different fractions of an astroblast homogenate obtained by pH gradient (o———o); lower plot: corresponding GABA-T activity in astroblast fractions (•·········•).

DISCUSSION

Striking differences can be found when comparing the GABA-ergic system of tumoral and transformed cells in tissue culture with the GABA-ergic system *in vivo*. Such differences were observed for GAD and GABA-T. The GAD activity of tumoral or transformed cells, measured by GABA formation, varied from 0.8 % to 5% that of a rat brain homogenate. Moreover, for most cell lines the GAD activity measured by CO_2 trapping is higher than that measured as GABA formed. This indicates that the formation of some CO_2 from glutamate might occur by a metabolic pathway that is different in clonal cell lines than *in vivo*; such a pathway could involve glutamate dehydrogenase and α-oxoglutarate dehydrogenase. This metabolic pathway is also suggested by the findings that glutamate dehydrogenase activity is equal to, or greater in C_6 astrocytoma than *in vivo* (18), and that addition of NAD and CoA increases CO_2 production from glutamate in a mouse kidney homogenate (31). For a primary cell culture of the neuronal type, GAD activity, measured by the two methods (CO_2 or GABA production), gave similar values; however this enzymatic activity in culture was only about 20% of that *in vivo* (16).

GABA-T activity is very low in tissue cultures, in general, and in glial primary culture (39). However the level of this enzyme is equal to, or represents about 50% of that present *in vivo*, suggesting the possible existence of effectors of GABA-T activity in tumoral cultures. Such an effector has been partially characterized in astroblast primary cultures (36,39).

Cell cultures can serve as a useful tool for investigating the GABA-ergic system. However, we should remember that in clonal cell lines, as well as in primary cell cultures, the metabolism of GABA may differ greatly from that *in vivo*.

REFERENCES

(1) Murray, M. R. (1965) *In* : Cells and Tissues in Cultures. Methods, Biology and Physiology (E. N. Willmer, ed.) Vol. II, pp. 373-455, Academic Press, London.

(2) Crain, S. M. (1975) *In*: Metabolic Compartmentation and Neurotransmission (S. Berl, D. D. Clarke and D. Schneider, eds.) pp. 273-303, Plenum Press, New York.

(3) Nelson, P. G. (1975) Physiol. Rev. 55, 1-61.

(4) Mandel, P., Ciesielski-Treska, J. and Sensenbrenner, M. (1976) *In*: Molecular and Functional Neurobiology. (W. H. Gispen, ed.) pp. 112-157, Elsevier, Amsterdam.

(5) Hösli, L. and Hösli, E. (1978) Rev. Physiol. Biochem. Pharmacol. 81, 135-188.

(6) Sensenbrenner, M., Springer, N., Booher, J. and Mandel, P. (1972) Neurobiology, 2, 49-60.

(7) Ciesielski-Treska, J., Mandel, P., Tholey, G., and Wurtz, B. (1972) Nature (Lond.), 239, 180-181.
(8) Mandel, P., Ciesielski-Treska, J., Hermetet, J. C., Zwiller, J., Mack, G. and Goridis, C. (1973) In: Frontiers in Catecholamine Research (E. Usdin and S. Snyder, eds.) pp. 277-283, Pergamon Press, Oxford.
(9) Herschman, H. R. (1973) In: Proteins of the Nervous System. pp. 95-115, Raven Press, New York.
(10) Stefanovic, V., Ciesielski-Treska, J., Ebel, A. and Mandel, P. (1974) FEBS Lett. 49, 43-46.
(11) Hösli, L., Hösli, E. and Wolf, P. (1975) In: Cholinergic Mechanisms (P. G. Waser, ed.) pp. 309-320, Raven Press, New York.
(12) Schousboe, A., Hertz, L. and Svenneby, G. (1977) Neurochem. Res. 2, 217-229.
(13) Hösli, L., Andres, P. F. and Hösli, E. (1977) Neurosci. Lett. 6, 79-83.
(14) Hermetet, J. C., Treska, J. and Mandel, P. (1970) Histochimie, 22, 177-186.
(15) Breakefield, X. O. (1976) Life Sci. 18, 267-278.
(16) Wilson, S. H., Schrier, B. K., Farber, J. L., Thompson, E. J., Rosenberg, R. N., Blume, A. J. and Nirenberg, M. W. (1972) J. Biol. Chem. 247, 3159-3169.
(17) Ebel, A., Massarelli, R., Sensenbrenner, M. and Mandel, P. (1974) Brain Res. 76, 461-472.
(18) Passonneau, J. V., Lust, W. D. and Crites, S. K. (1977) Neurochem. Res. 2, 605-617.
(19) Ciesielski-Treska, J., Warter, S. and Mandel, P. (1975) Neurobiology, 5, 382-392.
(20) Louis, J. C. (1978) Doctoral Thesis, Université de Strasbourg.
(21) Booher, J. and Sensenbrenner, M. (1972) Neurobiology, 2, 97-105.
(22) Bock, E., Møller, M., Nissen, C. and Sensenbrenner, M. (1977) FEBS Lett. 83, 207-211.
(23) Lowry, O. H., Rosebrough, N. J., Farr, A. L. and Randall, R. J. (1951) J. Biol. Chem. 193, 265-275.
(24) Albers, R. W. and Brady, R. O. (1959) J. Biol. Chem. 234, 926.
(25) Chude, O. and Wu, J.-Y. (1976) J. Neurochem. 27, 83-86.
(26) Dye, D. J. and Taberner, P. V. (1975) J. Neurochem. 24, 997-1001.
(27) Graham, L. T. and Aprison, M. H. (1966) Anal. Biochem. 15, 487-497.
(28) Vallee, B. L. and Thiers, R. E. (1965) Anal. Chem., 6, 3463-3475.
(29) Hunter, F. W. M. (1974) Brit. Med. Bull. 30, 18-23.
(30) Radola, B. J. (1969) Biochim. Biophys. Acta, 194, 335-338.
(31) Drummond, R. J. and Phillips, A. T. (1974) J. Neurochem. 23, 1207-1213.
(32) Sellstrom, A., Sjobert, L. B. and Hamberger, A. (1975) J. Neurochem. 25, 393-398.
(33) De Boer, Th. and Bruinvels, J. (1977) J. Neurochem. 28, 471-478.
(34) Waksman, A. and Roberts, E. (1965) Biochemistry, 4, 2132-2138.
(35) Ossola, L., Maitre, M., Blindermann, J. M. and Mandel, P. (1977) In: Maturation of Neurotransmission. Biochemical Aspects. (A.

Vernadakis, E. Giacobini and G. Filogamo, eds.) pp. 83-90, Karger, Basel.
(36) Ossola, L., Maitre, M., Blindermann, J. M. and Mandel, P. (1977) J. Physiol. (Paris) 73, 52A.
(37) Lefauconnier, J. M., Partemer, C. and Chataigner, F. (1976) Brain Res. 117, 105-113.
(38) Drummond, R. J. and Phillips, A. T. (1977) J. Neurochem. 29, 101-108.
(39) Ossola, L. (1978) Thèse de 3ème Cycle, Université Louis Pasteur, Strasbourg.

SECTION II

GABA METABOLISM, UPTAKE AND RELEASE

METABOLIC COMPARTMENTATION OF THE GABA SYSTEM; RELATIONSHIP OF GABA METABOLISM TO ANESTHESIA

Sze-Chuh Cheng

Department of Anesthesia, Northwestern University
Medical School, Chicago, Illinois 60611, U.S.A.

During the last few years, we have investigated the effects of anesthetic agents on the metabolism of γ-aminobutyric acid (GABA) in rat brain. Initially, slices of cerebral cortex were used, and in later experiments synaptosomes from forebrain were used. The synaptosomes were prepared using sucrose density centrifugation (1) and had approximately 5% impurity in vehicle count. The surface brain slices were about 0.35 mm thick.

First we studied the effects of halothane on the synthesis of acetylcholine (ACh) and GABA. Brain slices were incubated in bicarbonate-buffered Krebs solution containing three different labelled substrates: [1-^{14}C]acetate, [5-^{14}C]citrate and [2-^{14}C]pyruvate. After incubation, metabolites were isolated using ion-exchange chromatography (2) and their specific radioactivities were determined. The overall incorporation of these substrates in the presence of halothane (3%) differed from the controls according to the particular labelled substrate employed. After one hr of incubation at 30°C, halothane caused a decrease in pyruvate incorporation (from a control value of 991 ± 35 to 645 ± 54 p-moles/mg protein; $p < 0.015$), an increase in citrate incorporation (from a control value of 439 ± 17 to 560 ± 45 p-moles/mg protein; $p < 0.05$) and no change in acetate incorporation (147 ± 12 to 147 ± 9 p-moles/mg protein). Since it is known that pyruvate enters primarily the large metabolic compartment (MC) and that acetate and citrate enter primarily the small MC (3), it can be concluded that: (1) halothane inhibited the large MC; and, (2) there occurred a subdivision of the small MC, one part of which was stimulated and the other of which was not (4). Inhibition of the large MC was reflected in decreased specific radioactivities of several other metabolites derived from pyruvate. The stimulation of the citrate-related small MC was caused primarily by an unusually large increase in the specific radioactivity of glutamine (from a control value of 726 ± 28 to 1066 ± 12 cpm/n-mole; $p < 0.001$) although an increase of radioactivity in the lipids extracted from the trichloroacetic acid precipitate (from a control value of 281 ± 13 to 531 ± 46 cpm/mg protein; $p < 0.001$) also occurred. In contrast, the transfer of radioactivity from acetate to glutamine was decreased (from a control value of 221 ± 20 to 142 ± 12 cpm/n-mole; $p < 0.015$), although that to lipids increased (from a control value of 160 ± 8 to 295 ± 18 cpm/mg protein; $p < 0.001$). These differences further accentuate the distinction between acetate-utilizing and the citrate-utilizing subdivisions of the small MC. The stimulation of a small MC is further supported by our observation that thiopental stimulated the incorporation of CO_2 into glutamine (5). Whether this is the same citrate-utilizing small MC is not known, but it cannot be associated with the acetate-utilizing small MC. Ammonium ion also increased the incorporation of CO_2 into glutamine (6,7), suggesting that it also exerted its effect on the citrate-utilizing small MC. The morphological correlate of this citrate-related small MC is not known, but that of the large MC appears to include synaptic components (2). The inhibitory effect of halothane on the large MC can, therefore, be construed to mean an inhibition of synaptic metabolism.

Regarding the neurotransmitters, ACh and GABA (Table I), their syntheses from pyruvate were inhibited by halothane (4). Using labelled acetate, a possible stimulation of ACh synthesis and a definite inhibition of GABA synthesis by halothane were observed. An inhibition by halothane of GABA synthesis from citrate was seen, but

Table I - Effects of Halothane (3%) on the Contents and Specific Radioactivities of ACh and GABA in Rat Brain Cortex Slices

Condition	n-moles/mg protein	[2-^{14}C]Pyruvate	[1-^{14}C]Acetate	[5-^{14}C]Citrate
		(cpm/n-mole)		
ACh				
Control	0.339 ± 0.017	92.7 ± 5.1	39.4 ± 2.9	3.0 ± 1.5
Halothane	0.253 ± 0.027	46.1 ± 3.3	47.1 ± 1.0	5.2 ± 2.7
p	< 0.02	< 0.01	< 0.07	n.s.
GABA				
Control	19.0 ± 0.9	68.5 ± 3.4	29.1 ± 2.4	10.5 ± 3.3
Halothane	25.2 ± 1.7	31.5 ± 3.6	21.0 ± 1.6	6.7 ± 1.0
p	< 0.01	< 0.01	< 0.05	n.s.
N	21	6	8	7

One-hr incubations were carried out at 30°C in Krebs-bicarbonate solution. ACh and GABA were isolated from trichloroacetic acid extracts of the brain slices.

an effect of this agent on ACh synthesis cannot be assessed due to a lack of citrate entry into the ACh pathway. Since the ACh content of halothane-treated slices was lower than that of control slices, there obviously occurred a reduced turnover of ACh. This has been substantiated recently in other studies (8). The GABA content, however, increased above the control value. Such data led us to postulate that a reduction of GABA catabolism is the primary cause of all of these metabolic changes. Our hypothesis on anesthetic action (9, 10) is outlined below:

GABA-accumulation
↓
synaptic inhibition
↓
anesthesia
(reduced functional activity)

GABA accumulation may originate from inhibition of Complex I of the electron transport chain by anesthetic agents (11-18). This would lead to stockpiling of NADH, and then GABA. Other substrates for NADH oxidation might also accumulate, but these have no transmitter function. An increase in NADH has not been observed in the brain, but has been seen in kidney (19). GABA, which is present in high concentration only in nervous tissue, is a putative central inhibitory neurotransmitter. An induced inhibition of transmission at brain synapses causes reduced functional activity which can be recognized as anesthesia and which is expressed as reduced energy utilization. As a consequence, O_2 utilization by the brain is reduced (20,21) and cerebral energy stores are elevated (12).

In order to determine whether the reduction of GABA catabolism is the only synaptic action of anesthetic drugs, we studied other synaptic processes in brain slices. Uptake, release and catabolism of labelled GABA (13) were investigated. Data on GABA synthesis (Table II) were cited only for [2-^{14}C]pyruvate (see also Table I).

Table II - Effects of Halothane (3%) on GABA Metabolism in Rat Brain Cortex Slices

Parameter	% Change from control
Synthesis from [2-^{14}C]pyruvate	-46**
Degradation of [1-^{14}C]GABA	-41**
Retention of [1-^{14}C]GABA	+19*
Uptake of [1-^{14}C]GABA	+25
Release of [1-^{14}C]GABA	+15

Incubations were carried out for 15 minutes at 30°C in Krebs-bicarbonate solution; * $p < 0.05$; ** $p < 0.01$.

Halothane (3%) had no effect on the uptake of [1-^{14}C]GABA or on the release of [1-^{14}C]GABA that had been pre-loaded into brain slices. The metabolism of [1-^{14}C]GABA involves its entry into the tricarboxylic acid cycle as 1-labelled succinate which then randomizes into 1,4-labelled succinate and this is followed by the liberation of both these labelled carboxyl carbons as $^{14}CO_2$ during the next cycle of metabolism. An unequivocal inhibition of GABA catabolism by halothane can be seen from either the amounts of pre-loaded [1-^{14}C]GABA remaining in the slices after incubation or from the amounts of metabolites formed from pre-loaded [1-^{14}C]GABA. $^{14}CO_2$ was a major metabolite, and its formation from pre-loaded [1-^{14}C]GABA was inhibited in a dose-related fashion by halothane (regression coefficient = 7.64 ± 4.19 = mean ± 95% C.L. in % inhibition/% halothane; $p < 0.04$). These observations further support our hypothesis (9,10) that accumulation of GABA might be the "final common pathway" by which a variety of anesthetic drugs produce the state of anesthesia.

If our hypothesis for anesthesia is proven to be valid, then, the results obtained with halothane in brain slices should remain valid in experiments with brain synaptosomes and with other anesthetic agents. Synaptosomes prepared from rat forebrains using sucrose density gradients were suspended and incubated in a medium containing

Table III - Comparison Between Inhibition of Rat Forebrain Synaptosomal Liberation of CO_2 from GABA and Pharmacological Potency of Neurotropic Drugs

Drug	I_{10}	ED_{50} (ref.)
Halothane	0.8%	1.0% (33)
Enflurane	1.2%	1.7% (33)
Ether	4.5%	3.2% (33)
Thiopental	0.29 mM	0.24 mM (34), 2.8 mM (35)
Pentobarbital	0.45 mM	0.071 mM (36)
Phenobarbital	4.0 mM	0.11 mM (37)
Paraldehyde	17 mM	4.5 mM (38)
Ketamine	0.75 mM	0.13 mM (39)
Ethanol	78 mM	28 mM (40)
Morphine	2.6 mM	0.15 mM (41)

I_{10} = 10% inhibition of synaptosomal conversion of GABA to CO_2; calculated from regression lines relating % inhibition to drug concentration; ED_{50} = pharmacological potency as the concentration required for 50% effectiveness of each drug, estimated from data provided in the references.

mannitol, sucrose, phosphate buffer and succinate (24). This medium was modified slightly from that used in a study of mitochondrial metabolism (25). Flasks with a side-arm for substrate and a center well for NaOH were employed (26). Because of the time required for equilibration with the volatile anesthetics, these experiments involved an initial equilibration period, followed by addition of [1-^{14}C]GABA (10 μM) from the side-arm into a vessel which remained tightly sealed to maintain a constant partial pressure of the anesthetic. $^{14}CO_2$ liberated from GABA metabolism was trapped by the NaOH in the center well at the end of incubation by introducing a strong acid to the incubation mixture before breaking the atmospheric seal.

The $^{14}CO_2$ trapped in the NaOH was counted. The data thus collected concerned not only the catabolism of GABA but also its uptake and release, since both of these latter processes can effectively alter intra-synaptosomal GABA concentration. However, this simple CO_2 trapping experiment can provide an overall assessment of all these processes prior to identifying the exact process that is affected. Therefore, it is a valuable tool for surveying the effects of various anesthetic agents.

Results (26) summarized in Table III are based on regression lines relating % inhibition to the concentration of an anesthetic or neurotropic agent. I_{10} represents the concentrations of such agents which caused a 10% inhibition of the GABA → CO_2 reaction. This value was chosen because of its similarity to the minimum alveolar concentration (MAC) values for anesthesia with diethyl ether, halothane and enflurane. Thiopental, an anesthetic commonly used for induction, has an I_{10} value close to its ED_{50} value for anesthesia. Pentobarbital, paraldehyde, ketamine and ethanol have I_{10} values within 5 times their respective ED_{50} values, whereas phenobarbital, morphine and d-tubocurarine have I_{10} values that are much greater than their corresponding ED_{50} values. A positive correlation can be established between anesthetic potency of a compound and its ability to inhibit the GABA → CO_2 process.

This method involving CO_2 production determines an overall process which includes GABA uptake and release, as well as GABA catabolism. The next step is to dissociate these three processes and to define the specific process that is sensitive to anesthetic agents. GABA-transaminase was studied first because it catalyzes the rate-limiting step of GABA catabolism in synaptosomes (17). Saturated aqueous solutions of volatile anesthetics were added to an incubation mixture containing Triton X-100-solubilized synaptosomes after temperature equilibration. Then, GABA was added to initiate the reaction in a closed cuvette. Absorbancy changes at 340 nm were followed in an aldehyde dehydrogenase coupled system (28,29). Determinations

Table IV - Comparison Between Inhibition of Rat Synaptosomal GABA-Transaminase Activity and Pharmacological Potency of Neurotropic Drugs

Drug	I^*_{10}	ED_{50}
Halothane	10.4%	1.0%
Enflurane	9.2%	1.7%
Thiopental	0.17 mM	0.24 mM
Pentobarbital	0.75 mM	0.071 mM
Phenobarbital	0.8 mM	0.11 mM
Ethanol	7.9%	0.125%
Morphine	5.5 mM	0.15 mM

I^*_{10} = 10% inhibition of GABA-transaminase activity in solubilized rat forebrain synaptosomes, calculated from regression lines relating % inhibition to drug concentration; ED_{50} = pharmacological potency (see Table III).

with non-volatile neurotropic agents did not require closed cuvettes. Reaction rates were calculated and compared with those of simultaneously-run controls. The percentage inhibition produced by any agent was then compared to its concentration. Again, I^*_{10} values (see Table IV) were calculated from the regression lines.

Results (30) shown in Table IV revealed that volatile anesthetics did not inhibit GABA-transaminase; a slight inhibitory action was observed only at very high concentrations. On the other hand, barbiturates, especially thiopental, did inhibit GABA-transaminase, the I^*_{10} concentrations being of the same order of magnitude as those concentrations required to inhibit synaptosomal GABA catabolism. For thiopental, this concentration was of the same order of magnitude as its ED_{50} for anesthesia, suggesting that thiopental-induced inhibition of GABA catabolism might be of importance in the mechanism of its anesthetic action (24). Both ethanol and morphine had little inhibitory effect.

Thiopental inhibition of GABA-transaminase from solubilized synaptosomes was further investigated using kinetic methods. This inhibition was found to be non-competitive (K_I = 1.1 mM) when either GABA or 2-oxoglutarate was the limiting substrate. Maximal inhibition was approximately two-thirds of the total GABA-transaminase activity (24).

In order to study GABA uptake by synaptosomes, the magnitudes of GABA release and of the catabolism of accumulated GABA must be known. The effects of anesthetic agents on GABA catabolism are discussed above. The release of GABA from GABA-loaded synaptosomes was also studied. Washout experiments with synaptosomal suspensions were difficult to perform, especially with volatile agents, so synaptosomal beds were made and treated as slices (31). Synaptosomes were packed on a piece of white polyester fabric (1 cm, diameter). This was covered with another piece of cloth, and the "sandwich" was clamped in silverwire frames. This assembly was easily transferred through a series of vessels containing 2 ml Krebs-phosphate solution at 2.5-minute intervals in order to study the washout of pre-loaded GABA. At the end of each experiment, the assembly was dropped into hot NaOH on a steam bath to dissolve the synaptosomes for protein determination. KCl (55 mM) depolarization, which was used to test the system, produced a sharp peak of released radioactivity (Fig. 1). The baseline amounts of radioactivity in each wash represent the release of both un-metabolized GABA and its metabolic products (e.g., CO_2). The induced release is probably all GABA, but this requires verification.

The effects of water-soluble neurotropic agents on the release of GABA were tested before studying volatile anesthetics. Thiopental caused inhibition of this release, but it was not clear if the inhibition was mediated through a removal of Ca^{2+} which was precipitated by thiopental. Ethanol and paraldehyde were required in relatively high concentrations to produce release. Since diazepam might reduce GABA turnover (32) and since it has been used clinically to induce anesthesia, we studied the water-soluble benzodiazepine,

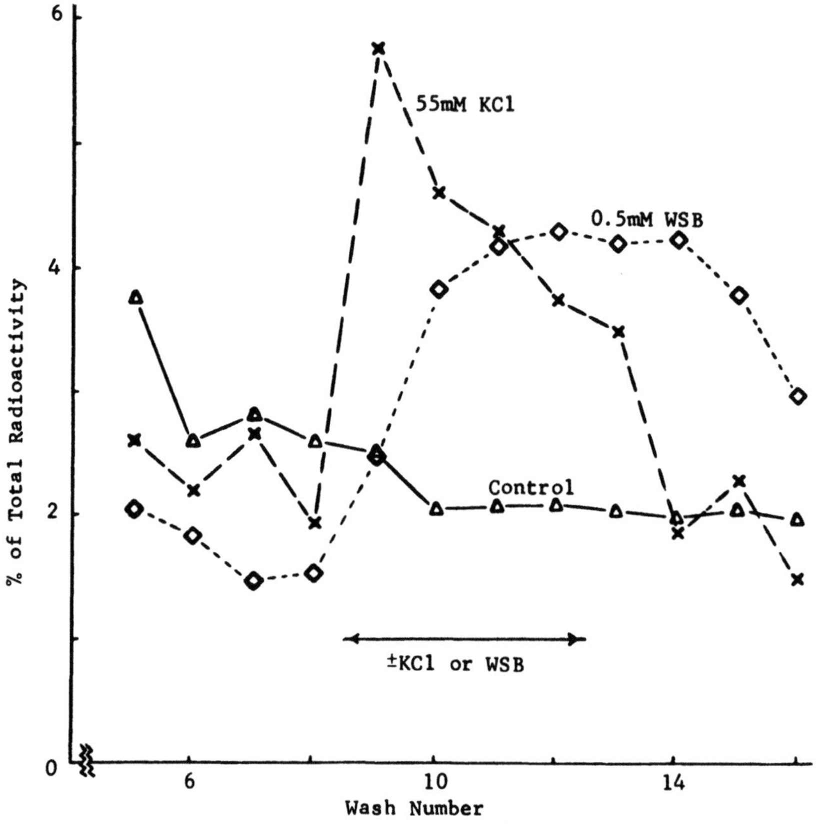

Fig. 1. Release of radioactivity from rat brain synaptosomes pre-loaded with [1-^{14}C]GABA. Rat forebrain synaptosomes were made into "synaptosomal bed assemblies" (see text). The first four washes of the beds in Krebs-phosphate solution showed a rapid decline in radioactivity. These were not used in the calculation of total radioactivity because they represented a large portion of radioactivity which was associated with non-synaptosomal material of the beds. The radioactivity found in washes Nos. 5 — 16 (later to No. 20) and the radioactivity that remained in the synaptosomes were taken as 100%. WSB = RO-21-3981, a water-soluble benzodiazepine. Both WSB and KCl were added individually to washes Nos. 9 — 12 only.

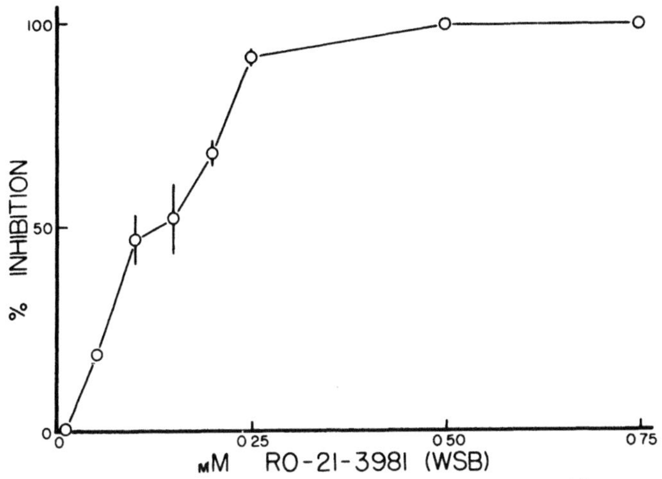

Fig. 2. Effects of WSB on the production of $^{14}CO_2$ from [1-^{14}C]GABA by synaptosomes. Rat forebrain synaptosomes were incubated for 1 hr at 30°C. Inhibition was calculated using concurrently-run controls and is given as means ± S.E.M.; (4 — 8 samples per point).

8-chloro-6-(2-fluorophenyl)-1-methyl-4H-imidazo-(1,5-a)(1,4)benzodiazepine maleate (RO-21-3981; WSB). It inhibited synaptosomal liberation of $^{14}CO_2$ from [1-^{14}C]GABA at less than 1 mM (Fig. 2), but failed to inhibit GABA-transaminase. Stimulation of GABA release by this WSB (measured as radioactivity) is shown in Figure 3. The time course of its action was much slower than that of KCl (Fig. 1). Therefore, it probably did not depolarize the synaptosomal membrane by its own action, but rather, made the membrane more leaky. The regaining of membrane intactness was also a slow process. These observations suggested a slow rate constant for the binding of WSB to a receptor. The bound form somehow appeared to interfere with the normal process of keeping GABA within the synaptosomes. The increased release of GABA produced by this WSB was dose-related (Fig. 3) and the effective dose range was very narrow. At 0.25 mM, it exerted no effect on

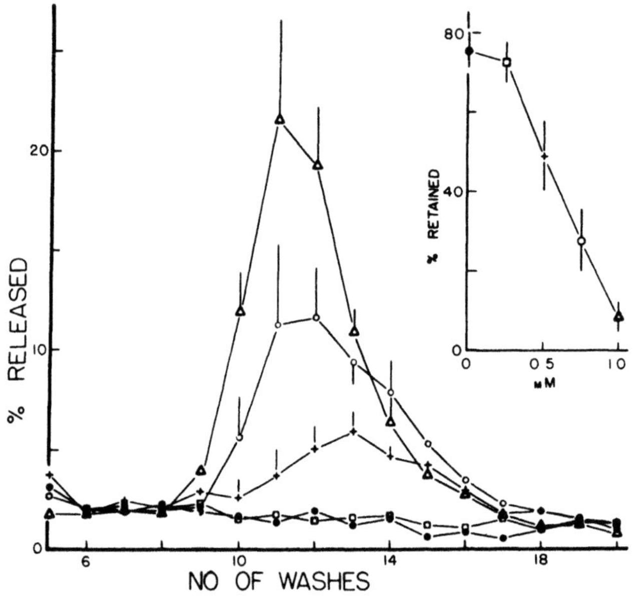

Fig. 3. Effects of WSB on the release of radioactivity from rat striatal synaptosomes pre-loaded with [1-^{14}C]GABA (see legend to Fig. 1). In addition, the S.E.M. values (n = 5) are given only for the induced release for clarity. WSB was added only to washes Nos. 9 – 12 at 0.25 mM (□), 0.50 mM (+), 0.75 mM (○) and 1.0 mM (△). Controls (●) had no WSB. Insert shows the amounts of radioactivity retained by the synaptosomes after 20 washes.

release, but at 1 mM essentially no GABA was left in the synaptosomes after induced release (Fig. 3). These observations indicated an allosteric mechanism for WSB action.

Increased release of GABA by this WSB suggests a relationship between its depressant action and GABA release. The required concentration of WSB was high, but this might be due to its increased water solubility and decreased lipid solubility, as compared with other benzodiazepines, which would reduce its pharmacological action. We are currently testing the possibility of studying the effects of volatile anesthetics on GABA release. Enhanced release could account for reduced catabolism by reducing the level of intracellular GABA available for enzymatic degradation. Experiments on uptake of GABA are still in the planning stage.

In summary, we have provided some evidence for an involvement of GABA in the mechanism of anesthetic action (9,10). Although much more data are still needed, this report supports the hypothesis that anesthesia is mediated by an increase in GABA-ergic synaptic inhibition.

ACKNOWLEDGEMENTS

I wish to thank Dr. E. A. Brunner of the Department of Anesthesia for his generous support and for his many helpful discussions. Technical help was rendered by several assistants, among whom were Ms. Elena Molina and Ms. Janice Miller. The water soluble benzodiazepine (WSB; RO-21-3981) was a gift from Hoffmann-La Roche Company, which partially financed the study with this drug.

REFERENCES

(1) Whittaker, V. P. (1965) Progr. Biophys. Mol. Biol. 15, 39.
(2) Nakamura, R., Cheng, S.-C. and Naruse, H. (1970) Biochem. J. (Lond.) 118, 443.
(3) Cheng, S.-C. (1973) In: Metabolic Compartmentation in the Brain (R. Balázs and J. Cremer, eds.) p. 107, Macmillan, London.
(4) Cheng, S.-C. and Brunner, E. A. (1978) J. Neurochem. 30, 1421.
(5) Cheng, S.-C., Naruse, H. and Brunner, E. A. (1978) J. Neurochem. 30, 1591.
(6) Berl, S., Takagaki, G., Clarke, D. D. and Waelsch, H. (1962) J. Biol. Chem. 237, 2570.
(7) Waelsch, H., Berl, S., Rossi, C. A., Clarke, D. D. and Purpura, D. P. (1964) J. Neurochem. 11, 717.
(8) Ngai, S. H., Cheney, D. L. and Finck, A. D. (1978) Anesthesiology, 48, 4.
(9) Cheng, S.-C. and Brunner, E. A. (1975) Anesth. Analg. 54, 242.
(10) Cheng, S.-C. and Brunner, E. A. (1975) In: Molecular Mechanics of Anesthesia (B. R. Fink, ed.) p. 509, Raven Press, New York.

(11) Chance, B. and Hollunger, G. (1963) J. Biol. Chem. 238, 418.
(12) Miller, R. N. and Hunter, F. E. (1970) Mol. Pharmacol. 6, 67.
(13) Harris, R. A., Munroe, J., Farmer, B., Kim, K. C. and Jenkins, P. (1971) Archs. Biochem. Biophys. 142, 435.
(14) Britt, B. A., Kalow, W. and Endrenyi, L. (1972) Biochem. Pharmacol. 21, 1159.
(15) Cohen, P. J. and MacIntyre, R. (1972) In: Cellular Biology and Toxicity of Anesthetics (B. R. Fink, ed.) p. 109, Williams & Wilkins, Baltimore.
(16) Rosenberg, H. and Haugaard, N. (1973) Anesthesiology, 39, 44.
(17) Smith, H. (1973) Biochem. Pharmacol. 22, 773.
(18) Miller, R. N., Miller, E., Caine, J. and Stabler, M. (1974) Anesth. Analg. 53, 132.
(19) Chance, B., Cohen, P., Jöbsis, F. and Schoener, B. (1962) Science, 137, 499.
(20) Matteo, R. S., Hoech, G. P. and Hoskin, F. G. (1969) Anesthesiology, 30, 156.
(21) Weiss, G. B., Hertz, L. and Goodman, F. R. (1972) Biochem. Pharmacol. 31, 625.
(22) Brunner, E. A., Passonneau, J. V. and Molstad, C. (1971) J. Neurochem. 18, 2301.
(23) Cheng, S.-C. and Brunner, E. A. (1974) Trans. Amer. Soc. Neurochem. 5, 116.
(24) Cheng, S.-C. and Brunner, E. A. (1979) Biochem. Pharmacol. 28, 105.
(25) Clark, J. B. and Nicklas, W. J. (1970) J. Biol. Chem. 245, 4724.
(26) Cheng, S.-C. and Brunner, E. A. (1976) Trans. Amer. Soc. Neurochem. 7, 194.
(27) Roberts, E. and Hammerschlag, R. (1976) In: Basic Neurochemistry (G. J. Siegel, R. W. Albers, R. Katzman and B. W. Agranoff, eds.) p. 218, Little, Brown & Co., Boston.
(28) Baxter, C. F. (1972) Meth. Neurochem. 3, 1.
(29) Jakoby, W. B. (1962) Meth. Enzymol. 5, 765.
(30) Cheng, S.-C. and Brunner, E. A. (1977) Trans. Amer. Soc. Neurochem. 8, 248.
(31) DeBelleroche, J. S. and Bradford, H. F. (1972) J. Neurochem. 19, 585.
(32) Costa, E., Guidotti, A. and Mao, C. C. (1976) In: GABA in Nervous System Function (E. Roberts, T. N. Chase and D. B. Tower, eds.) p. 413, Raven Press, New York.
(33) Eger, E. I., II. (1974) Anesthetic Uptake and Action. Williams & Wilkins, Baltimore.
(34) Bollman, J. L., Brooks, L. M., Flock, E. V. and Lundy, J. S. (1950) Anesthesiology, 11, 1.
(35) Patel, V. K. and Johnson, G. E. (1975) Can. J. Physiol. Pharmacol. 53, 669.
(36) Kalant, H., Khanna, J. M. and Marshman, J. (1970) J. Pharmacol. Exp. Therapeut. 175, 318.
(37) Rapport, R. L., II, and Kupferberg, H. J. (1973) J. Med. Chem. 16, 599.

(38) Figot, P. P., Hines, C. H. and Way, E. L. (1952) Acta Pharmacol. Toxicol. 8, 290.
(39) Cohen, M. L., Chan, S.-L., Way, W. L. and Trevor, A. J. (1973) Anesthesiology, 39, 370.
(40) Harger, R. N. and Hulpieu, H. R. (1956) In: Alcoholism (G. N. Thompson, ed.) p. 103. C. C. Thomas, Springfield, Ill.
(41) Dahlström, B. and Paalzow, L. (1975) J. Pharm. Pharmacol. 27, 172.

GABA METABOLISM IN CULTURED GLIAL CELLS

M. Tardy, J. Bardakdjian and P. Gonnard

Département de Biochimie, C.H.U. Henri Mondor,
94010 Créteil, France

The necessity for a precise, sensitive and rapid method to measure GABA-transaminase (GABA-T) activity in all tissues, including cell cultures, led us to develop a radioactive procedure. This was possible when the suitable substrate, [1-^{14}C]α-ketoglutarate, became commercially available.

This transamination occurs between the radioactive keto acid and unlabelled GABA at pH 8.9. The products of the reaction are [1-^{14}C]glutamate and succinic semialdehyde. After incubation for one hour at 37°C, transamination is terminated by acidification of the medium and a bacterial glutamate decarboxylase, active at pH 5, is then added. $^{14}CO_2$ evolved from [1-^{14}C]glutamate is trapped in hyamine and its radioactivity is determined. The $^{14}CO_2$ evolved corresponds stoichiometrically to the amount of [1-^{14}C]glutamate formed; i.e., to the transaminase activity. This method was compared with that of Waksman and Roberts (1) as a reference and the results were similar.

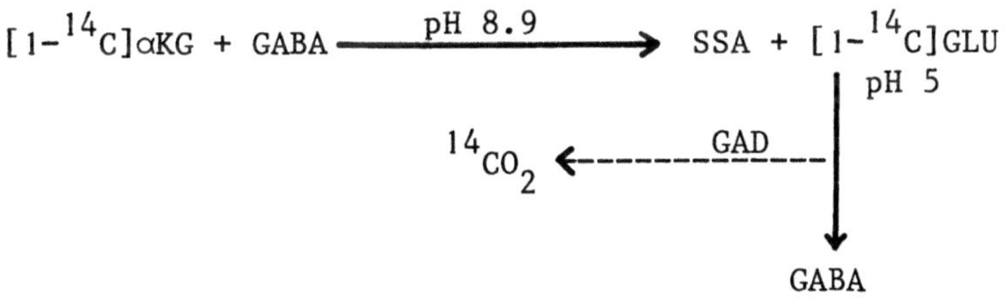

GABA-T is mainly a mitochondrial enzyme which is localized in the inner matrix and is loosely bound to the inner membrane. Cerebral GABA-T activity has been found in nerve endings as well as in the larger mitochondria of cell perikarya. In order to study the regulation of the different GABA pools, GABA-T has been determined in these compartments.

Our first studies concerned pig brain GABA-T. Nerve terminals (synaptosomes) were separated from postsynaptic mitochondria by the method of Gray and Whittaker (2), as modified by Israel and Frachon Mastour (3). Synaptosomes and postsynaptic mitochondria were ruptured by osmotic shock, and GABA-T activity was measured in dialyzed fractions of the two compartments. Table I indicates the characteristics of synaptosomal GABA-T (GABA-T S) and postsynaptic mitochondrial GABA-GABA-T (GABA-T M). The most important differences are in their specific activities and their affinities for α-ketoglutarate.

GABA-T S and GABA-T M were compared to the so-called "Forms I and II" of purified pig brain enzyme (4) and it was concluded that similarities existed between GABA-T S and Form II. The GABA content of glial cells is very low and is associated with a low glutamate decarboxylase (GAD) activity. However, the affinity of GABA-T for GABA is much higher in glial cells than in the two above-mentioned compartments. The specific activity of GABA-T represents only 45% that of GABA-T S. Glial GABA-T is a mitochondrial enzyme, and is identical with GABA-T S and GABA-T M in terms of charge, stability, and molecular weight, or size.

Reuptake of released neurotransmitters into nerve terminals is a possible conservation mechanism, but some doubt exists whether this is the major method for removal of neurotransmitters from the extracellular fluid. Henn and Hamberger (5) showed that glial cells were able to concentrate 20 times more GABA than the neuronal fraction. The question is: Could this concentrated GABA be released directly into nerve endings, or is there a glial GABA metabolism which is able to transform GABA and to synthesize new GABA by an

Table I - Comparison of the Properties of GABA-T M and GABA-T S with Those of GABA-T I and II of Purified Pig Brain

Property	GABA-T M	GABA-T I	GABA-T S	GABA-T II
K_m αKG (mM)	1.2	1.3	0.4	0.4
K_m GABA (mM)	2.6	2.4	2	2.3
pH_O	8.5 - 8.7	8.6 - 8.7	8.6 - 8.8	8.7 - 8.8
$S_{20}W$	5.7	5.5	5.5	6.5
Specific Activity (μ-mole/hr/mg protein)	2.6	500	1	100

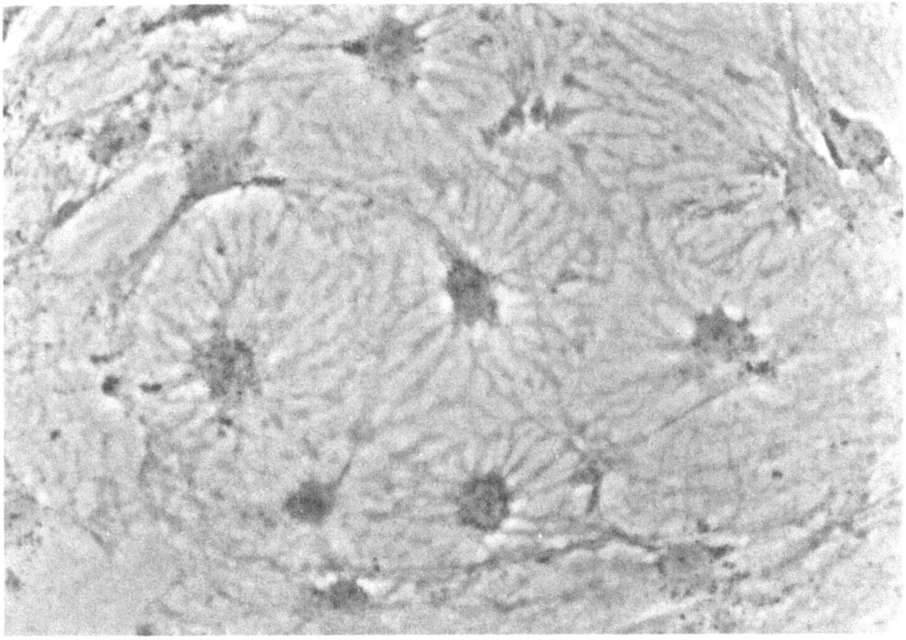

Fig. 1. Morphology of dB-c-AMP-differentiated glial cells after a three-week growth period in culture.

Table II - Specific Activities of GABA-T in Neonatal Mouse Brain, in Astrocytes Cultured from Neonatal Mouse Brain, and in Adult Mouse Brain

Material and condition	Specific Activity of GABA-T (n-mole/min/mg protein)
Neonatal brain	3.00 ± 0.02
1 week in culture	0.62 ± 0.08
2 weeks in culture	2.91 ± 0.30
3 weeks in culture	4.96 ± 0.20
4 weeks in culture + dB-c-AMP	6.15 ± 0.1
Adult brain	6.23 ± 0.05

indirect route, perhaps via the formation of a suitable metabolite such as glutamine? The existence of GABA metabolism in glial cells was suggested in 1971 by Hokfelt et al. (6). A measurable GABA-T activity has been shown in astrocytes by Schousboe et al. (7).

In order to investigate the relationship between the GABA concentration of glial cells and GABA metabolism, we chose to use the astroglial culture developed by Booher and Sensenbrenner (8) as adapted by Schousboe et al. (9). In our experiments, cultures obtained from dissociated cerebral hemispheres of newborn mice, under the conditions described by Schousboe, grew for about 15 - 20 days before confluence. Addition of 0.1 mM dibutyryl-cyclic-AMP (dB-c-AMP) to the culture medium, after two weeks of growth, induced cell differentiation (see Fig. 1). These cells have the morphological aspect of astrocytes. At about the 12th day of growth, new cell types appeared which were much smaller and darker, and which resembled oligodendrocytes. These cells disappeared at about the 20th day. We have not found any explanation for this observation.

GABA-T activity in our astrocytes increased for three weeks, reaching a maximum of 4.7 – 4.9 n-moles GABA metabolized/min/mg cell protein. Table II shows that after three weeks of growth, cells

Table III - Comparison of the Properties of GABA-T in Glial Cells with Those of the Synaptosomal and Post-synaptosomal Enzyme of Mouse Brain

Property	Type of GABA-T		
	S	M	G
K_m GABA (mM)	3	3.4	1.8
K_m αKG (mM)	3.1	0.7	1.2
I_{50}	45°	45°	45°
pH_0	8.85	8.8	8.85
$S_{20}W$	5.8	5.1	5
pH_I	6	6	6 - 6.5
Specific activity (n-mole/min/mg protein)	11	48	5

possessed a GABA-T activity which represented about 60% of the activity of this enzyme in mouse brain homogenates. The addition of dB-c-AMP to the culture medium, in the presence of serum, led to a value which approximated that of adult brain. The properties of this glial GABA-T have been compared with those of the synaptosomal and extra-synaptosomal compartments of adult mouse brain.

Table III indicates that the K_m with respect to α-ketoglutarate and GABA differed in the three compartments. As the specific activity of GABA-T M was much higher than that of GABA-T S, this could be due to a rapid metabolism of the GABA taken up by the neuronal perikarya. In nerve endings, however, steady-state levels of GABA might be maintained by re-uptake and by a lower GABA-T activity. These results differ from another observation made with C_6 glioma cells cultured under the same conditions as the primary culture of astrocytes. Table IV shows that C_6 cells can contain about 20 n-moles GABA/mg protein, which is much higher than the GABA concentration of normal cells. A much higher GAD activity is associated with these high GABA levels; GABA-T is much lower.

Table IV - GABA Concentrations and GAD and GABA-T Activities in Normal and Neoplastic Glial Cells

Preparation	GABA	GAD	GABA-T
C_6 Glioma	20 ± 1.8	0.5	1
Glial cells (astrocytes; 2 weeks)	3.5 ± 0.3	0.15	3.0
Adult mouse brain	10 ± 1.8	1.4	6.23

GABA is expressed in n-moles/mg protein; GAD and GABA-T specific activities are expressed in n-moles/mg protein/min.

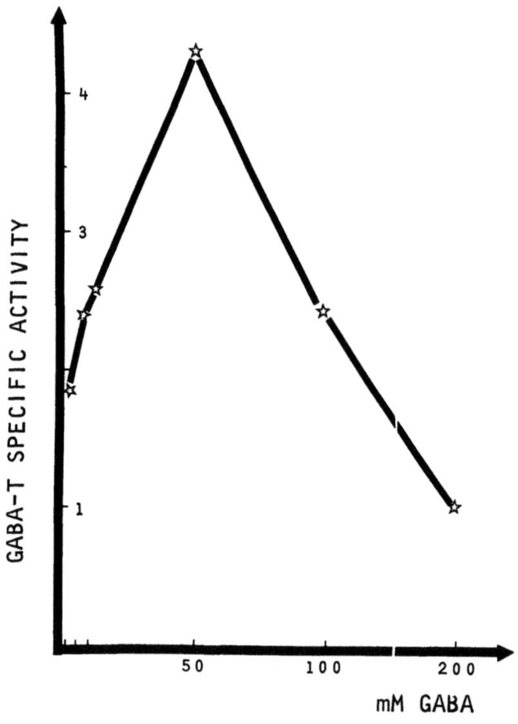

Fig. 2. Effect of various concentrations of GABA (in the culture medium) on glial GABA-T.

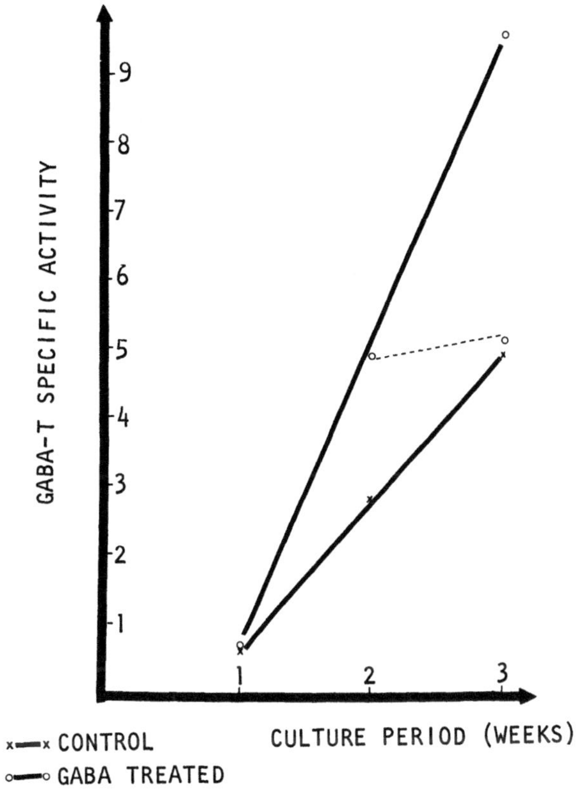

Fig. 3

The cells that we studied were grown without addition of GABA to the growth medium. When different concentrations of GABA were added to the medium from the beginning of the culture period, GABA-T activity of the cells increased during one to three weeks in culture. At three weeks, this increase was maximal. Fig. 2. shows the responses of GABA-T activity to different concentrations of GABA added to the culture medium. This curve corresponds to a confluent culture. Maximum increase in GABA-T was observed with 50 mM GABA, but the increase was also significant at 5 mM (> 20%). When GABA was removed from the medium (by changing the medium) at about two weeks of the growth period, this GABA-T increase abruptly ceased (Fig. 3).

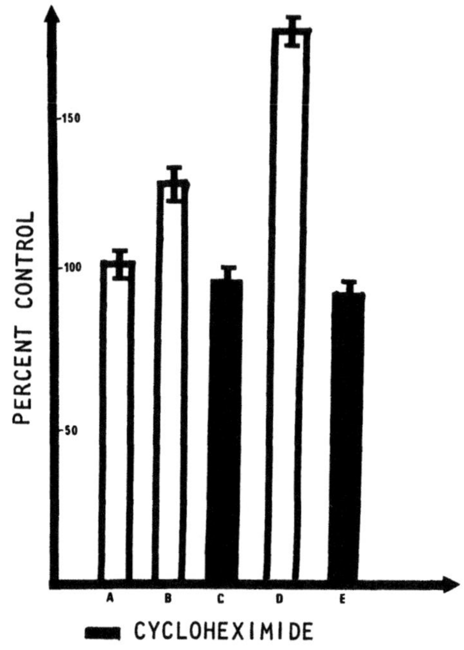

Fig. 4. Mean GABA level as percentage of control GABA-T activity (+ S.E.M.). A, Glia under normal conditions; B, with 5 mM GABA; C, with 5 mM GABA + 0.33 μg/ml cycloheximide; D, with 50 mM GABA; E, with 50 mM GABA + 0.33 μg/ml cycloheximide (48 hrs).

As 50 mM GABA is a high concentration, we controlled the viability of the cells with Trypan Blue coloration in the presence or absence of GABA. Viability was 85% for control cells and 82% for cells kept in the presence of 50 mM GABA at three weeks. Protein content per dish showed no significant difference during the whole growth period and the number of cells per mg protein was of the same order under the two conditions. What we did observe was a particular morphology of the cells kept in the presence of GABA. These cells possessed shorter prolongations and a more flattened morphology. Considering physico-chemical and functional properties of the stimulated enzyme, no differences between control and GABA-exposed cells could be observed in terms of charge, molecular weight and K_m of the enzyme. The only difference was an increased V_{max}. GABA has no direct action *in vitro* on GABA-T activity in glial homogenates.

The above results are in favor of an enzyme induction. This type of induction could be due to two mechanisms: The first one involves a *de novo* protein synthesis. If this is in play, then, GABA might be expected to induce other enzymes involved in neuronal transmission in these cells. The second is that GABA acts as an inducer by interacting with glial GABA "receptor" sites; then agonists and antagonists should mimic and interfere, respectively, with this process. From our results obtained in the presence of GABA, aspartate aminotransferase (AAT) was increased in the same way as GABA-T, but GAD decreased to about 50% of its value and glutamate dehydrogenase (GDH) did not vary. Secondly, when cycloheximide (0.33 μg/ml) was added to the culture medium and kept in contact with the cells for 48 hours it blocked completely the GABA-T increase (Fig. 4). Therefore, it can be concluded that the increased GABA-T corresponds to a *de novo* protein synthesis. This is the first demonstration that GABA-T is an inducible enzyme. Induction of GABA-T by high concentrations of GABA revealed that glial cells have a potential for catabolizing large amounts of GABA to produce high levels of glutamate. As glutamine synthetase is present in high concentration in glial cells, it is not improbable to consider that a high concentration of glutamine could be attained in glia. This increased glutamine could serve as a precursor for glutamate and GABA in the synapse, and therefore might be associated with their release from nerve terminals.

In sum, GABA-T has been shown to exist in high concentration in glial cells. This enzyme possesses a higher affinity for GABA than does GABA-T of synaptic or postsynaptic compartments. Induction of GABA-T by high concentrations of GABA shows that glial cells can catabolize large quantities of GABA. In this way glutamate is formed

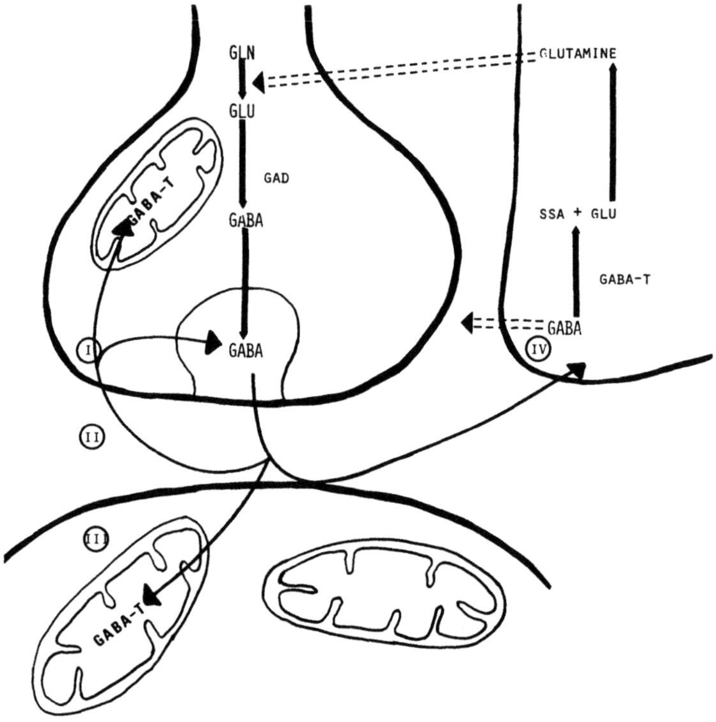

Fig. 5

and can be transformed into glutamine since glia contain a high glutamine synthetase activity. The glutamine thus formed might be taken up by nerve endings and used as a GABA precursor in this compartment (Fig. 5).

REFERENCES

(1) Gonnard, P., Wicker, A., Kouyoumdjian, J. C. and Bloch-Tardy, M. (1973) Biochimie, 55, 509.
(2) Gray, E. G. and Whittaker, V. P. (1962) J. Anat. (Lond.) 96, 79.
(3) Israel, M. and Frachon Mastour, P. (1970) Arch. Anat. Micros. 59, 383.
(4) Tardy, M., Rolland, B., Fages, C. and Gonnard, P. (1976) Experientia, 32, 434.
(5) Henn, F. A. and Hamberger, A. (1971) Proc. Nat. Acad. Sci., U.S.A., 68, 2686.
(6) Hökfelt, T. and Ljungdahl, A. (1970) Brain Res. 22, 391.
(7) Schousboe, A., Hertz, L. and Svenneby, G. (1977) Neurochem. Res. 2, 217.
(8) Booher, J. and Sensenbrenner, M. (1972) Neurobiology, 2, 97.

GABA TURNOVER AS A TOOL TO EXPLORE THE FUNCTION OF GABA-ERGIC SYNAPSES: PHYSIOLOGICAL AND PHARMACOLOGICAL STUDIES

F. Moroni

Department of Pharmacology, University of Florence, Florence, Italy

INTRODUCTION

Since the first proposal that γ-aminobutyric acid (GABA) might be an inhibitory neurotransmitter in the central and peripheral nervous systems of various animal species (1), many efforts have been made to understand how GABA-ergic synapses are regulated.

Numerous investigators have studied the postsynaptic events of GABA-ergic neurotransmission, focussing their attention on GABA-receptors (e.g., 2,3). Unfortunately, technical limitations have made the biochemical events taking place at a presynaptic site of GABA-ergic neurotransmission difficult to understand. In order to clarify the function of GABA-ergic terminals, we have resorted to a specific mass fragmentographic method which allows us to measure not only glutamate and GABA in discrete nuclei of rat brain, but also the turnover rate of GABA (TR_{GABA}) (4,5). This method appears helpful for studying the presynaptic events that regulate GABA-ergic synapses *in vivo*. In fact, it is widely accepted that by measuring the turnover rate of a putative neurotransmitter in a brain area that contains axon terminals, inferences can be made on the functional activity of a biochemically defined population of such terminals. This belief has been developed mainly from studies on catecholaminergic, serotonergic and cholinergic neurones (6-9). Previous work suggests that this concept might be extended to GABA-ergic neurones in certain brain nuclei (4,5). In this paper the research potentials and the limitations of this technique will be discussed. Moreover, using this approach, evidence will be presented for a regulation of GABA-ergic neurones in the basal ganglia by opiate receptor activation.

ESTIMATION OF TR_{GABA}

TR_{GABA} was estimated in certain nuclei of the rat brain by monitoring the ^{13}C-enrichment of GABA and glutamate during constant rate infusion of [^{13}C]glucose (50 μ-moles/kg/min). Concentrations of GABA, glutamate and their [^{13}C]analogues were measured by a sensitive and specific mass fragmentographic technique (4,10).

Briefly, the rats were killed by microwave radiation (11) after infusion of [^{13}C]glucose (4), and the brains were removed and frozen on dry ice. The frozen brains were mounted on a microtome stage (-4°C) and sliced into coronal sections of 400 microns. Various nuclei were punched out (12) and homogenized in ethanol-water solutions containing the appropriate amount of internal standards (4). After centrifugation, the supernatant was evaporated to dryness under nitrogen and the carboxylic groups of GABA, glutamate and their [^{13}C]analogues were esterified with 1,1,1,3,3,3-hexafluoroisopropanol (HFIP), while the amino groups were acylated with pentafluoropropionic anhydride (PFPA). After evaporation, the residue was dissolved in ethyl acetate and then a small aliquot was injected into a gas chromatograph-mass spectrometer (LKB 9000).

Like all methods used to measure the turnover rates of catecholamines (6) or acetylcholine (7) _in vivo_, this method is based on a number of assumptions and involves several approximations. GABA is an intermediary metabolite of glucose, and therefore, a certain portion of TR_{GABA} must be ascribed to non-specific metabolic activity of the tissue. Hence, our estimations of TR_{GABA} include such background and do not measure the absolute value of TR_{GABA} occurring in neurones. The method, in fact, implies that in the brain newly-formed [^{13}C]glutamate serves only as a GABA precursor. Naturally this is not valid. However, errors deriving from this assumption can probably be minimized by selecting areas such as the globus pallidus, substantia nigra or caudate nucleus, in which the pool size of glutamate that functions as a GABA precursor is greater than other glutamate pools.

Another problem that besets these measurements is the recycling of variant ^{13}C atoms. GABA is known to be metabolized to succinic semialdehyde which is a substrate for the tricarboxylic acid cycle. By this route, the ^{13}C incorporated into GABA can be reincorporated into glutamate. When the ^{13}C label is administered by pulse injections, the problem created by such recycling may not have a practical solution. Conversely, when the label increases at a consistent rate (for instance, during constant rate infusion), one can calculate the efflux rate of ^{13}C from the precursor-product relationship at a time when the deviations caused by ^{13}C recycling are minimal.

EFFECTS OF ELECTRICAL STIMULATION AND LESIONING OF THE STRIATONIGRAL PATHWAY

The existence of a GABA-ergic pathway connecting the striatal formation with the midbrain is well documented biochemically (13,14), anatomically (15) and electrophysiologically (16). We took advantage of this well-identified pathway in order to ascertain whether or not our measurements of TR_{GABA} reflect the activity of GABA-ergic terminals. To answer this question, we electrically stimulated or lesioned this pathway. Table I shows the effects of electrical stimulation of the caudate nucleus or of hemi-transection of the brain on TR_{GABA} of substantia nigra. These results demonstrate

Table I - Effects of Electrical Stimulation or Lesioning of the Strionigral Pathways on TR_{GABA} in Substantia Nigra

Treated side	TR_{GABA} percent of control side
Electrical stimulation	400
Hemi-transection	30

Electrical stimulation of the caudate nucleus was performed in unanesthetized rats with monopolar electrodes (Rhodes Medical Instruments), 0.25 mm of shaft diameter being inserted through chronically-implanted cannulae. The following parameters were used: 0.5 msec stimulus duration; 20 Hz frequency; 10 volts. The total stimulation time was 20 minutes. The hemi-transection was performed one week before the experiments using a 3 mm-wide knife lowered with an angle of 60° to the level of the coronary suture. The knife was inserted near the midline and then moved outwards to achieve complete hemi-transection. TR_{GABA} on the control side was 380 ± 40 n-moles/mg protein/hr.

clearly that in the substantia nigra TR_{GABA} increases or decreases when the activity of the GABA-ergic striatonigral pathway is enhanced or suppressed. These data, therefore, support the idea that our measurements of TR_{GABA} provide an acceptable index of the functional activity of GABA-ergic neurones.

INTERACTIONS BETWEEN GABA-ERGIC NEURONES AND OPIOID RECEPTOR AGONISTS AND ANTAGONISTS IN THE CAUDATAL-PALLIDAL-NIGRAL SYSTEM

Various types of GABA-ergic neurones exist in the caudate nucleus; e.g., short-axoned neurones which are intrinsic to the neostriatum (17) and long-axoned neurones which project to the globus pallidus and substantia nigra (13,14). The globus pallidus and

Table II - Steady-State Contents of Some Neuroregulators in the Caudate Nucleus, Globus Pallidus and Substantia Nigra

	Caudate nucleus	Globus pallidus	Substantia nigra
		n-moles/mg protein	
Acetylcholine[a]	0.73	0.45	0.11
Dopamine[b]	0.62	-	0.14
GABA[c]	23	58	88
Glutamate[d]	100	88	92
MET-enkephalin[e]	20	150	1.2
LEU-enkephalin[f]	3	-	-
Substance P[g]	1.2	2.0	13

[a] Moroni et al. (32); [b] Costa et al. (33); [c,d] Moroni et al. (8); [e,f] Hong et al. (19); [g]Hong et al. (34).

substantia nigra are, therefore, very rich in both GABA (Table II) and GABA-ergic nerve terminals (18). Several other putative neurotransmitters are present in particularly high concentrations in the caudate nucleus, globus pallidus and substantia nigra (Table II). For example, in the caudate nucleus and globus pallidus the concentrations of MET- and LEU-enkephalin are higher than in other regions of the CNS (19,20). The simultaneous presence in the same brain nuclei of very high concentrations of these endogenous opiate receptor ligands and of the previously described GABA-ergic neuronal populations prompted us to study how GABA-ergic and enkephalinergic neurones might interact. Since the enkephalins cross the blood-brain barrier very slowly and are rapidly metabolized when administered intra-cerebrally (21), we performed our studies using morphine or β-endorphin as opiate receptor agonists and naltrexone as an antagonist.

Table III shows the effects of subcutaneous administration of various doses of morphine and intra-ventricular (ivt) administration

Table III - Actions of Morphine and β-Endorphin on TR_{GABA} in Various Telencephalic Brain Nuclei

	Caudate nucleus	Substantia nigra	Globus pallidus
	TR_{GABA} (n-moles/mg protein/hr)		
Saline	690 ± 100	680 ± 58	520 ± 93
Morphine 35 μ-moles/kg	490 ± 20*	1200 ± 180**	750 ± 50**
Morphine 70 μ-moles/kg	480 ± 18*	1500 ± 225**	730 ± 45**
Morphine 140 μ-moles/kg	300 ± 15**	1900 ± 230**	1480 ± 90**
β-Endorphin 0.29 n-mole	480 ± 30*	1580 ± 250**	1100 ± 290*
β-Endorphin 2.9 n-moles	390 ± 20**	1600 ± 220**	1400 ± 250**

Each value represents the mean ± S.E.M. of at least 6 determinations; * $p < 0.05$; ** $p < 0.01$.

Morphine and β-endorphin were administered 30 minutes before sacrifice. Morphine was dissolved in 0.5 ml of 150 mM NaCl and injected subcutaneously. β-Endorphin was dissolved in 10 μl of 150 mM NaCl and administered intra-ventricularly.

of β-endorphin on TR_{GABA} in the substantia nigra, globus pallidus and cuadate nucleus. TR_{GABA} decreased in caudate nucleus but increased in globus pallidus and substantia nigra. These changes were dose related, and could be antagonized by previous administration of naltrexone (15 μ-moles/kg, i.p.) (Table IV). In order to ascertain the anatomical localization of the actions of morphine or β-endorphin, we injected both of these agents directly into the caudate nucleus, globus pallidus and substantia nigra. These injections were performed through chronically-implanted stainless steel cannulae, stereotaxically positioned (22) in the respective nuclei. Figure 1

Table IV - Antagonism by Naltrexone of the Effects Induced by Intra-Ventricular Administration of β-Endorphin

	Caudate nucleus	Globus pallidus	Substantia nigra
	TR_{GABA} (n-moles/mg protein/hr)		
Saline	690 ± 100	520 ± 93	680 ± 58
β-Endorphin 2.9 n-moles/rat	380 ± 50*	1400 ± 250**	1600 ± 220**
β-Endorphin 2.9 n-moles/rat + Naltrexone 15 μ-moles/kg	850 ± 95	610 ± 140	750 ± 90

Each value is the mean ± S.E.M. of at least 6 determinations. Naltrexone was administered i.p. 45 minutes before sacrifice; β-endorphin was administered intra-ventricularly 30 minutes before sacrifice. * $p \leqslant 0.05$; ** $p \leqslant 0.01$.

shows that intra-caudatal administration of both morphine and β-endorphin decreased TR_{GABA} in the caudate nucleus but increased it in the globus pallidus. The steady-state contents of glutamate and GABA in the caudate nucleus and globus pallidus were not changed after general or local administration of morphine or β-endorphin.

Figure 2 shows the antagonism by intra-caudatal injections of naltrexone of the modification in TR_{GABA} induced by subcutaneously-administered morphine (70 μ-moles/kg). It is clear that naltrexone completely antagonized both the morphine-induced decrease of TR_{GABA} in the caudate nucleus and the morphine-induced increase of TR_{GABA} in the globus pallidus. When morphine (130 n-moles) or β-endorphin (2 n-moles) were injected into the globus pallidus, TR_{GABA} and the steady-state of glutamate and GABA were not modified in the caudate nucleus or in the substantia nigra (data not shown). When morphine or β-endorphin were injected into the substantia nigra, TR_{GABA} and the steady-state of GABA and glutamate in the globus pallidus or in

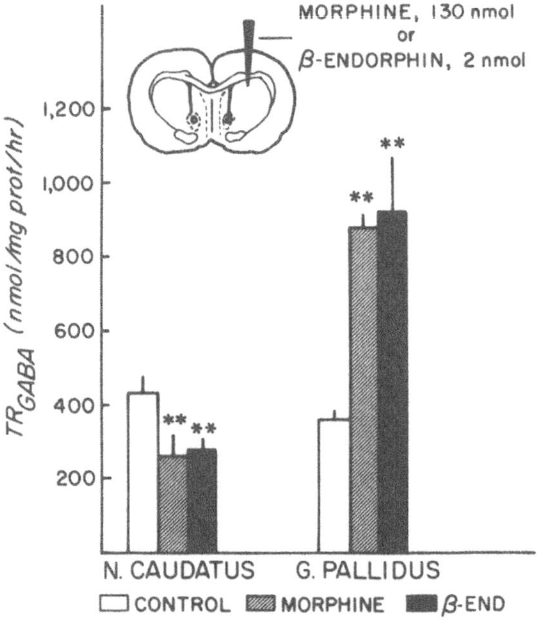

Fig. 1. Effect of intra-caudatal administration of morphine and β-endorphin on TR_{GABA} of the caudate nucleus and globus pallidus. Morphine and β-endorphin were administered in a volume of 2 μl over a period of 30 - 50 seconds, and animals were killed 30 minutes later. Each bar is the mean ± S.E.M. of results from at least 6 animals. ** $p \leqslant 0.01$.

the caudate nucleus were not changed (data not shown).

In order to investigate whether or not the increases of TR_{GABA} in globus pallidus and substantia nigra elicited by morphine or β-endorphin are operative during some of the pharmacological effects of opiate receptor agonists (9,23) we have studied the effects of injection of a specific agonist of GABA-receptors, muscimol, into the globus pallidus or substantia nigra on some of the better-studied actions of the opiates: analgesia and catalepsy. When bilateral cannulae were implanted in the globus pallidus and substantia nigra,

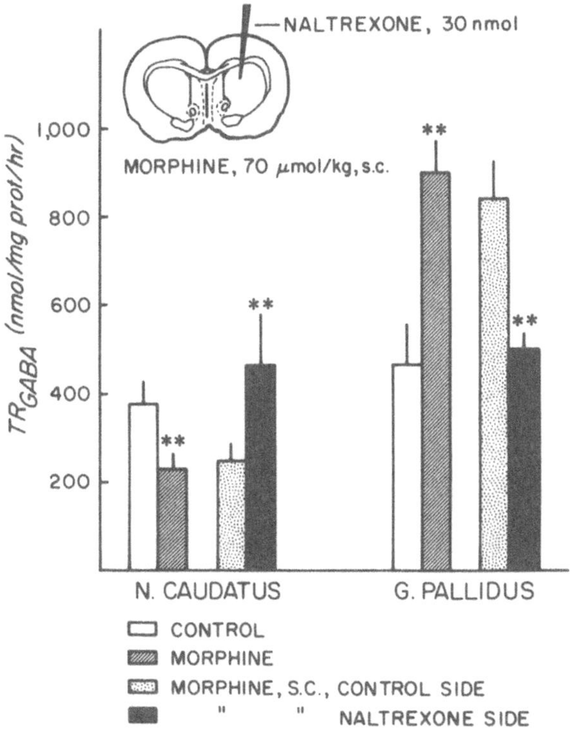

Fig. 2. Effects of subcutaneously-injected morphine and intra-caudatal administration of naltrexone on TR_{GABA} of caudate nucleus and globus pallidus. Naltrexone was administered in a volume of 2 μl; morphine was administered in a volume of 0.5 ml immediately before naltrexone (see legend to Fig. 1).

bilateral injections into the substantia nigra of up to 1 n-mole of muscimol failed to induce analgesia or catalepsy. However, when muscimol was injected bilaterally into the globus pallidus, a clear-cut cataleptogenic state was induced, and the intensity of catalepsy was dose dependent. The catalepsy began 10 - 15 minutes after the injection and lasted up to 3 hours. Thus: (1) activation of opiate-receptors by morphine or β-endorphin decreased GABA utilization in the caudate nucleus, while increasing it in the globus pallidus and substantia nigra; (2) the effects on GABA-ergic neurones in the caudate nucleus and globus pallidus are probably due to a stimulation of opiate-receptors localized in the head of the caudate nucleus; (3) morphine- or β-endorphin-induced catalepsy is probably mediated through an action on caudatal GABA-ergic neurones projecting to the globus pallidus.

ARE DOPAMINERGIC OR GLUTAMATERGIC AFFERENTS TO THE CAUDATE NUCLEUS PARTICIPATING IN THE INTERACTIONS OF ENKEPHALIN AND GABA-ERGIC NEURONES?

It has been recently demonstrated that striatal dopaminergic terminals contain opiate-receptors (24,25). Moreover, it is generally accepted that opiate receptor agonists increase the turnover rate of dopamine in the caudate nucleus and the striatal levels of homovanillic acid (HVA) and other dopamine metabolites (26,27). Therefore, it seems possible that the above-mentioned modifications of TR_{GABA} might be a consequence of altered dopaminergic function. However, opiate-receptors are not only present on dopaminergic terminals, but also in intrinsic striatal interneurones (25) and in the terminals of afferents to the caudate nucleus which originate in the cerebral cortex and which are probably glutamatergic (Moroni and Guidotti, unpublished observations). Other explanations of the effects of morphine on GABA-ergic neurones are, therefore, possible.

In order to clarify which regulation of opiate-receptors is involved in the interactions of opiates and GABA-ergic neurones, apomorphine (a dopamine agonist) was injected directly into the caudate nucleus using a chronically-implanted cannula. The amount administered was freshly dissolved in 2 μl of saline and delivered over a period of 30 - 50 seconds, and the rats were killed 30 minutes after administration. Table V shows that apomorphine (25 n-moles) did not modify TR_{GABA} in the caudate nucleus. Also, no effects were observed in the globus pallidus or in the substantia nigra (data not shown). In other experiments, activation of dopaminergic-receptors

Table V - Effects of Intra-Striatal Administration of Apomorphine or of Electrical Stimulation of Substantia Nigra on HVA Levels and TR_{GABA} in the Caudate Nucleus

	TR_{GABA} (n-moles/mg protein/hr)	HVA (p-moles/mg protein)
Control animals	330 ± 60	60 ± 4.9
Stimulation of the S.N. (homolateral site)	280 ± 40	110 ± 1.0
Apomorphine (intra-caudate; homolateral site)	380 ± 75	-

Stimulation of the substantia nigra was performed by inserting a bilaterelectrode (Rhodes Medical Instruments Co.; Mod. SNEX 300, Cleveland, Ohio) through chronically-implanted cannulae and by using the following parameters: 15 Hz; 8 volts; 2 m-sec. Apomorphine was freshly dissolved in saline and injected in a volume of 2 μl.

was achieved by stimulating the substantia nigra for various periods of time with bipolar electrodes using the following parameters: 15 Hz, 2 m-sec, 8 volts. A 30-minute stimulation of the substantia nigra caused an increase in striatal HVA content of approximately 100%, but failed to alter TR_{GABA} in the caudate nucleus (Table V). Therefore, it appears that a direct synaptic connection between dopaminergic terminals and GABA-ergic neurones in the striatum does not exist. In order to contribute direct evidence on the cellular location of the action of opioids, and since opiate receptors are also present in the cortico-striatal (probably glutamatergic) afferents, we attempted to reduce the firing rate of glutamatergic afferents to the striatum by lesioning the frontal cortex. In fact, this is the site from which most of the glutamatergic fibers connecting the cortex and the striatum originate (28). Fifteen days before the experiment the gray matter located in the frontal part of the rat cortex

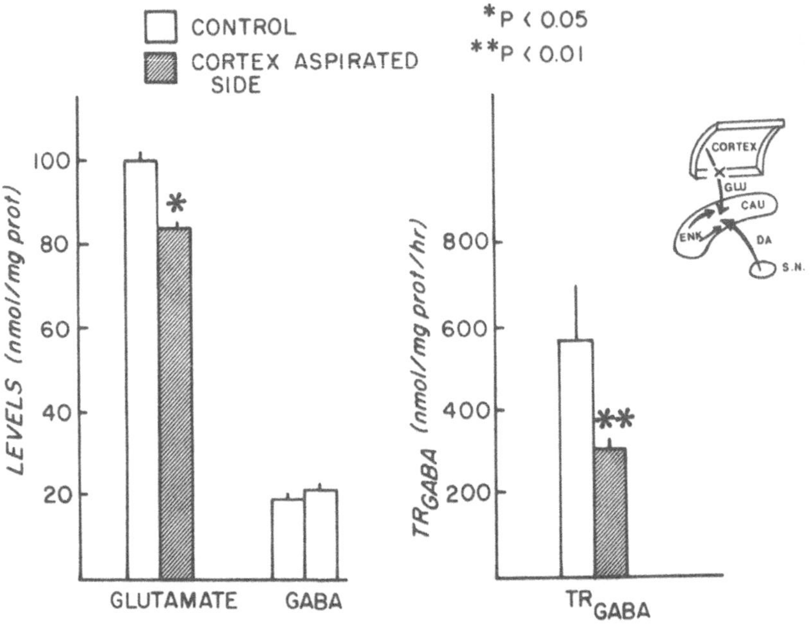

Fig. 3. Effects of frontal cortex aspiration on glutamate and GABA levels and on TR_{GABA} in the caudate nucleus. The frontal cortex was aspirated two to three weeks before the experiment (28) (see legend to Fig. 1).

was aspirated according to the method of Kim and colleagues (28). This surgical procedure reduces striatal glutamate content by nearly 20%. GABA and HVA levels were not modified, but TR_{GABA} in the caudate nucleus ipsilateral to the cortical ablation appeared to be decreased by about 50% (Fig. 3). However, TR_{GABA} in the globus pallidus and in the substantia nigra were not modified by this surgical procedure (Fig. 4). Therefore, it appears that this cortical lesion does not completely mimic the effects of morphine and β-endorphin.

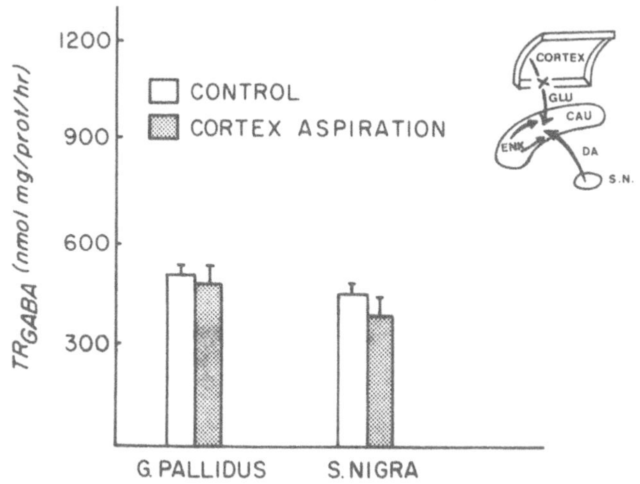

Fig. 4. Effect of cortical aspiration on TR_{GABA} of globus pallidus and substantia nigra (see legends to Figs. 1 and 3).

CONCLUSIONS

Administration of opioids modifies TR_{GABA} in the caudate nucleus, globus pallidus and substantia nigra. Since TR_{GABA} is decreased in the caudate nucleus and increased in the globus pallidus and substantia nigra, one can rule out non-specific metabolic effects of the opioids. Since it is likely that TR_{GABA} reflects the firing rate of GABA-ergic neuronal terminals, our findings suggest that opioids decrease this firing rate in striatum, and probably indirectly increase it in globus pallidus and substantia nigra. The morphine- or β-endorphin-induced decrease of TR_{GABA} in the caudate nucleus is mediated by a stimulation of opiate-receptors located in this nucleus. Opiate-receptors, located in the same brain area, appear responsible for the increase in GABA utilization induced by morphine or β-endorphin in the globus pallidus. These opiate-receptors could be located on the membranes of caudatal GABA-ergic neurones. Opiate-receptors located on caudatal dopaminergic terminals (24) do not seem to be involved in this action since stimulation of these receptors by apomorphine failed to alter TR_{GABA} in the caudate nucleus.

It is also possible that opiate-receptors inhibit the activity of the GABA-ergic neuronal population of the caudate nucleus by a presynaptic mechanism involving the excitatory afferents that link the cortex to the striatum. We have attempted to carry out direct experiments to resolve which of the two possibilities may be operative. Unfortunately, lesioning of the cortico-striatal pathway causes an inhibition of TR_{GABA} in the caudate. However, the effects of ablation of the cortex seem to be different from those caused by morphine and β-endorphin. In fact, ablation of the cortex does not modify TR_{GABA} in the globus pallidus or in the substantia nigra. The present experiments, therefore, favor the idea that enkephalins inhibit neuronal firing rate in the caudatal-pallidal-nigral system. In many areas of the central and peripheral nervous systems (29,30) this has been shown to be the case, but in some brain areas an activation of neuronal firing rate has been observed (31). The increase in GABA metabolism in the globus pallidus after intra-caudatal administration of morphine or β-endorphin may be due to a decrease in GABA-ergic inhibition caused by activation of opiate-receptors in the caudate nucleus. Thus, we propose that caudatal enkephalin-containing neurones synapse with GABA-ergic short-axoned neurones in the striatum and that these interneurones, in turn, regulate the long-axoned GABA-ergic neurones that link the caudate nucleus with the globus pallidus (13). Furthermore, an increase in GABA-ergic activity of the globus pallidus may be operative in opioid-induced catalepsy. Finally, our studies indicate that through these complex interactions with GABA-ergic neurones in the caudate nucleus, globus pallidus and substantia nigra, enkephalins may participate in the regulation of extrapyramidal function.

ACKNOWLEDGEMENTS

The work reported here was done at the Laboratory of Preclinical Pharmacology, NIMH, Saint Elizabeths Hospital, Washington, D.C. 20032.

REFERENCES

(1) Elliott, K. A. C. (1959) Proc. 4th Int. Congr. Biochem. 3, 251.
(2) Toffano, G., Guidotti, A. and Costa, E. (1978) Proc. Nat. Acad. Sci., U.S.A. (in press).
(3) Ticku, M. K. and Olsen, R. W. (1978) Life Sci. 22, 1643-1652.
(4) Bertilsson, L., Mao, C. C. and Costa, E. (1977) J. Pharmacol. Exp. Therapeut. 200, 277-284.
(5) Mao, C. C., Peralta, E., Moroni, E. and Costa, E. (1978) Brain Res. (in press).
(6) Costa, E. and Meek, J. L. (1974) Ann. Rev. Pharmacol. Toxicol. 14, 491-511.
(7) Cheney, D. L. and Costa, E. (1977) Ann. Rev. Pharmacol. Toxicol. 17, 369-386.
(8) Moroni, F., Malthe-Sorenssen, D., Cheney, D. L. and Costa, E. (1978) Brain Res. (in press).
(9) Moroni, F., Cheney, D. L. and Costa, E. (1978) Neuropharmacology (in press).
(10) Bertilsson, L. and Costa, E. (1976) J. Chromatog. 118, 395-399.
(11) Guidotti, A., Cheney, D. L., Trabucchi, M., Doteuchi, M., Wang, C. T. and Hawkins, R. A. (1974) Neuropharmacology, 13, 1115-1122.
(12) Koslow, S. H., Racagni, G. and Costa, E. (1974) Neuropharmacology, 13, 1123-1130.
(13) Fonnum, F., Grofova, I., Rinvik, E., Storm-Mathisen, J. and Walberg, F. (1974) Brain Res. 71, 77-92.
(14) Gale, K., Hong, J. S. and Guidotti, A. (1972) Brain Res. 136, 371-375.
(15) Bunney, B. S. and Aghajanian, G. K. (1976) Brain Res. 117, 423-435.
(16) Precht, W. and Yoshida, W. (1971) Brain Res. 32, 229-233.
(17) McGeer, P. L. and McGeer, E. G. (1975) Brain Res. 91, 331-335.
(18) Enna, S. J., Kuhar, M. J. and Snyder, S. H. (1975) Brain Res. 93, 168-174.
(19) Hong, J. S., Yang, Y. T., Fratta, W. and Costa, E. (1977) Brain Res. 134, 383-386.
(20) Cuello, A. C. (1978) In: Advances in Biochemical Psychopharmacology (E. Costa and M. Trabucchi, eds.) Vol. 18, pp. 111-123, Raven Press, New York.
(21) Meek, J. L., Yang, H.-Y. T. and Costa, E. (1977) Neuropharmacology, 16, 151-154.
(22) Konig, J. F. R. and Klippel, R. A. (1963) The Rat Brain: A Stereotaxic Atlas of the Forebrain and Lower Parts of the Brain-Stem. Kipper, Huntington, N.Y.
(23) Bloom, F., Sega, D., Ling, N. and Guillemin, R. (1976) Science, 194, 530-532.
(24) Schwartz, J. E., Pollard, H., Lloreus, E., Malfroy, B., Gros, C., Prodelles, P. H. and Dray, F. (1978) In: Advances in Biochemical Psychopharmacology (E. Costa and M. Trabucchi, eds.) Vol. 18, pp. 245-264, Raven Press, New York.
(25) Carenzi, A., Frijeni, V. and Della Bella, D. (1978) In: Advances

in Biochemical Psychopharmacology (E. Costa and M. Trabucchi, eds.) Vol. 18, pp. 265-271, Raven Press, New York.

(26) Kuschinsky, K. and Hornykiewicz, O. (1972) Eur. J. Pharmacol. 19, 119-122.

(27) Costa, E., Carenzi, A., Guidotti, A. and Revuelta, A. (1973) In: Frontiers in Catecholamine Research (E. Usdin and S. H. Snyder, eds.) pp. 1003-1010, Pergamon Press, Oxford.

(28) Kim, J. S., Hassler, R., Haug, P. and Paik, K. S. (1977) Brain Res. 132, 370-374.

(29) Zieglgansberger, W., Fry, J. P., Herz, A., Moreover, L. and Wiunsh, E. (1976) Brain Res. 115, 160-164.

(30) North, R. A. and Williams, J. T. (1976) Nature (Lond.) 264, 460-461.

(31) Iwatsubo, K. and Clouet, D. (1977) J. Pharmacol. Exp. Therapeut. 202, 429-436.

(32) Moroni, F., Cheney, D. L. and Costa, E. (1977) Nature (Lond.) 267, 267-268.

(33) Costa, E., Bejovic, M. and Koslow, S. H. (1974) In: Psychoneuroendocrinology (E. Hatotani, ed.) pp. 304-312, Karger, Basel.

(34) Hong, J. S., Costa, E. and Yang, Y. T. (1976) The Pharmacologist, 18, 230.

AUTORADIOGRAPHIC STUDIES ON THE CELLULAR LOCALIZATION OF GABA AND β-ALANINE UPTAKE BY NEURONES AND GLIA IN TISSUE CULTURE

L. Hösli and Elisabeth Hösli

Department of Physiology, University of Basel,
Vesalgasse 1, CH-4051 Basel, Switzerland

INTRODUCTION

A great number of biochemical and electrophysiological studies have provided evidence that GABA is a transmitter substance in various regions of the mammalian nervous system (1,2). More recently, a transmitter role has also been proposed for β-alanine (3). Biochemical studies on the uptake kinetics of GABA and β-alanine in slices or homogenates of nervous tissue have shown that both amino acids are taken up by high-affinity transport mechanisms (3-9). Autoradiographic studies on the uptake of these amino acids in brain slices have revealed that GABA is accumulated by neurones and by glial cells whereas β-alanine is only taken up by glial elements (6, 10,11). The present paper is concerned with studies on the uptake patterns of [^{3}H]GABA and [^{3}H]β-alanine in cultures of cerebellum, brain stem, spinal cord and dorsal root ganglia (DRG) of fetal and newborn rats.

The method of nervous tissue culture provides a useful tool for studying the cellular localization of the uptake of transmitter substances using autoradiography. Diffusion of the transmitter in cultures occurs more rapidly than in slices, allowing shorter incubation times. Furthermore, the various cell types are better preserved in cultures than in slices (12-15).

MATERIALS AND METHODS

Explants were prepared from the cerebellum, brain stem (medulla oblongata - pons) and spinal cord (with attached dorsal root ganglia) of fetal (17 - 18 days *in utero*) and newborn rats. The cultures were placed on collagen-coated coverslips, fed with nutrient medium, and grown in Maximov double coverslip assemblies for 8 - 35 days at 35° C (for details see ref. 16). For the autoradiographic investigations, the cultures were briefly rinsed in Tyrode's solution and incubated for 30 sec - 15 min at 37° C in Hank's solution containing [^{3}H]GABA New England Nuclear; specific activity, 5.6 Ci/m-mole) or [^{3}H]β-alanine (same source; specific activity, 31.5 Ci/m-mole) at concentrations of 5×10^{-7} and 10^{-6} M. In order to determine whether [^{3}H]GABA and [^{3}H]β-alanine are taken up by active transport mechanisms, some cultures were incubated in Na^+-free incubation medium (Na^+ being replaced by choline and Tris) or at 0° C. After incubation, the cultures were rinsed in Tyrode's solution, fixed in 3% glutaraldehyde in 0.1 M phosphate-buffer, dehydrated and air-dried. Ilford L-4 emulsion was placed over the cultures by the loop technique (for details see ref. 16). The autoradiographs were stored at 4° C for 2 weeks and then developed with a Kodak D-19 developer. The histochemical demonstration of GABA-transaminase (GABA-T) was performed using the method of Van Gelder (17). Controls were run in the absence of GABA substrate.

RESULTS AND DISCUSSION

Uptake of [^{3}H]GABA and [^{3}H]β-Alanine by Neurones

Cerebellum. It has been shown that cerebellar neurones possess a specific high-affinity uptake mechanism for GABA, but not for other amino acid transmitters such as glycine (18). After incubation of cerebellar cultures with [^{3}H]GABA, a great number of neurones had accumulated the amino acid. The majority of labelled neurones were multipolar and appeared to be stellate cells or other interneurones, such as basket or Golgi cells (Fig. 1D) (see also references 19,20). [^{3}H]GABA was distributed over the cell bodies and processes of these neurones, the nucleus often remaining unlabelled (Fig. 1D). A considerable number of large cells showing morphological features of Purkinje cells were also intensely labelled by [^{3}H]GABA (Fig. 1A; see also refs. 19,20). The finding that cultured Purkinje cells can take up [^{3}H]GABA (19,21,22) contrasts with results obtained from autoradiographic studies on the cerebellum *in vivo* or in slices (6,12,13) which revealed no labelling of Purkinje cells (Fig. 1B). Since Bergman glia, which tightly surround Purkinje cells *in vivo*, take up [^{3}H]GABA rapidly (12,13), it has been suggested that they form a barrier which prevents uptake of the amino acid by Purkinje cells. This glial barrier is probably disrupted or absent in tissue culture and therefore Purkinje cells are able to accumulate [^{3}H]GABA (19,20).

In contrast to the marked accumulation of [^{3}H]GABA by many cultured cerebellar neurones, [^{3}H]β-alanine was taken up only by glial cells. Similar results have been obtained in slices of cerebellum, thus demonstrating that β-alanine was exclusively accumulated by glial elements (11). After incubation of cerebellar cultures with [^{3}H]glycine, only a few neurones had taken up the amino acid and the labelling was much weaker than that which occurred after incubation with [^{3}H]GABA (19,20). Our findings on the uptake of [^{3}H]GABA by cerebellar Purkinje cells and interneurones are consistent with the proposal that GABA is a transmitter substance in the cerebellum, whereas there is little support that glycine and β-alanine are transmitters in this region (4).

Brain stem and spinal cord. High-affinity uptake systems for GABA have also been found in the brain stem and in the spinal cord (4). Autoradiographic studies on cultures of rat brain stem and spinal cord have shown that a large number of neurones were labelled after incubation with [^{3}H]GABA (Fig. 1E,F; see also refs. 14-16,20). In spinal cord cultures [^{3}H]GABA was accumulated mainly by small neurones (14), whereas in brain stem cultures both large and small neurones were labelled (15,16,20). However, in cultures of both

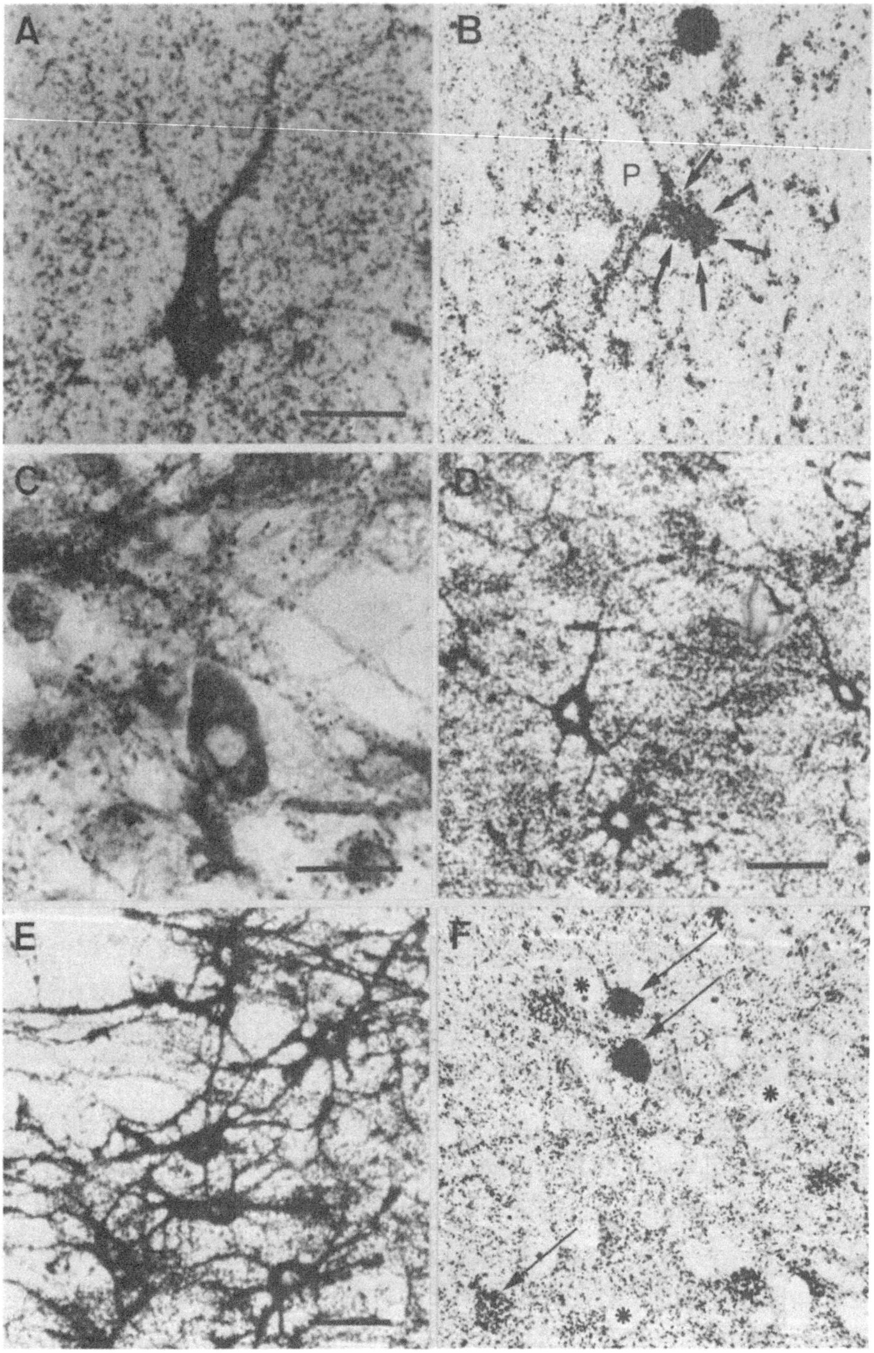

Fig. 1

Fig. 1. A: Light-microscopic autoradiograph of a cerebellar culture after incubation with [^{3}H]GABA (10^{-6} M) for 2 min. The soma and dendrites of the Purkinje cell show a heavy accumulation of the amino acid. Culture: 17 days _in vitro_. Bar: 20 μm (from Hösli and Hösli, 19). B: Light-microscopic autoradiograph showing the grain distribution after incubation of a cerebellar slice with [^{3}H]GABA (AOAA pre-treatment). Note the close correlation between grain distribution basal to the Purkinje cell body (P) (arrows), probably representing the axon terminals of basket cells. Magnification: X 535 (from Hökfelt and Ljungdahl, 13). C: Histochemical localization of GABA-T in a cultured Purkinje cell (rat cerebellar culture, 13 days _in vitro_). Bar: 20 μm. D: Cerebellar culture (17 days _in vitro_) after incubation with [^{3}H]GABA (10^{-6} M) for 5 min. Cell bodies and processes of the neurones, probably stellate cells, are intensely labelled, whereas the nuclei appear to be free of label. Bar: 30 μm (from Hösli and Hösli, 19). E: Neurones of a rat brain stem culture (16 days _in vitro_) exhibiting a strong autoradiographic reaction after incubation with [^{3}H]GABA (10^{-6} M) for 5 min. Bar: 30 μm (from Hösli and Hösli, 15). F: Light-microscopic autoradiograph of a rat spinal cord culture (26 days _in vitro_) incubated with [^{3}H]GABA (10^{-6} M) for 5 min. Several cell bodies are heavily labelled (arrows), whereas the majority of cells (asterisks) are covered only by a few grains. Magnification: X 535 (from Hösli et al., 14).

regions, a large proportion of neurones did not accumulate the amino acid (Fig. 1F, asterisks). Although it has been suggested that GABA is taken up only by neurones utilizing the amino acid as a transmitter substance (12,13,23), it was not possible in our studies on cultures to determine whether [^{3}H]GABA was accumulated by a specific neuronal population (14-16,20).

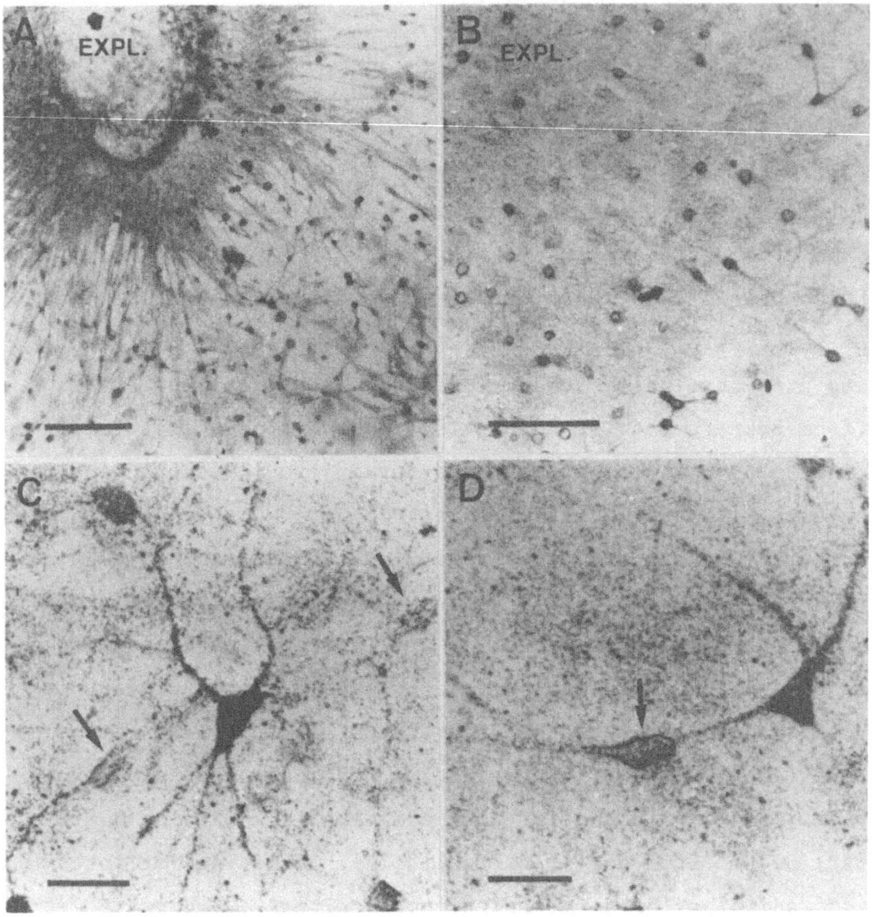

Fig. 2. A, B: Autoradiographs of brain stem and spinal cord cultures (23 days in vitro) after incubation with [^{3}H]β-alanine (10^{-6} M) for 3 min. Both neurones and glial cells have accumulated the amino acid. A greater number of neurones is labelled in the brain stem culture (A) than in the spinal cord culture (B). EXPL. = explant. Bars: A: 150 μm. B: 100 μm. C, D: Cultured brain stem (C) and spinal neurone (D) showing intense label over their cell bodies and processes, whereas other neurones (arrows) have not accumulated [^{3}H]β-alanine (10^{-6} M for 3 min). Bars: 30 μm (from Hösli and Hösli, 24).

After incubating spinal cord and brain stem cultures with [^{3}H]β-alanine, it was found that a much smaller number of neurones had taken up the amino acid than after incubation with [^{3}H]GABA. Furthermore, in brain stem cultures a greater proportion of neurones were labelled by [^{3}H]β-alanine (Fig. 2A) than in cultures of spinal cord (Fig. 2B), the amino acid being distributed over the cell bodies and processes of the neurones (Fig. 2C,D; see also ref. 24). Also, a relatively great number of neurones did not accumulate [^{3}H]β-alanine. Our results which have demonstrated that neurones present in cultures of spinal cord and brain stem can take up [^{3}H]β-alanine provide further evidence that this amino acid might act as a transmitter in these regions (3,24).

Dorsal root ganglia (DRG). In organotypic cultures of DRG it was observed that neurones lying in the dense zones of the culture were surrounded by satellite glial (SG) cells (Fig. 3A), whereas neurones which had migrated into the outgrowth zone had lost their glial envelopments (Fig. 3B; see also ref. 25). Autoradiographic studies have shown that DRG neurones in the dense zone did not accumulate [^{3}H]GABA, whereas the surrounding SG cells were heavily labelled (Fig. 3C; see ref. 25). Similar observations were made on the intact DRG where [^{3}H]GABA was accumulated exclusively by SG cells (26). In contrast, almost all DRG neurones which were deprived of their glial envelopments revealed a strong accumulation of [^{3}H]GABA (Fig. 3D; see ref. 25). Similar observations have been made with L-[^{3}H]glutamate (Fig. 3F). Figures 3D and F illustrate isolated DRG neurones with intense labelling over their cell bodies and their bifurcated axons after incubation with [^{3}H]GABA (Fig. 3D) or L-[^{3}H]glutamate (Fig. 3F). The finding that DRG neurones which were deprived of their glial envelopments were able to accumulate [^{3}H]GABA is consistent with observations made on cultured Purkinje cells, and suggests that glial cells form a barrier which prevents the accumulation of [^{3}H]GABA into these neuronal cell bodies (15,19,20,25).

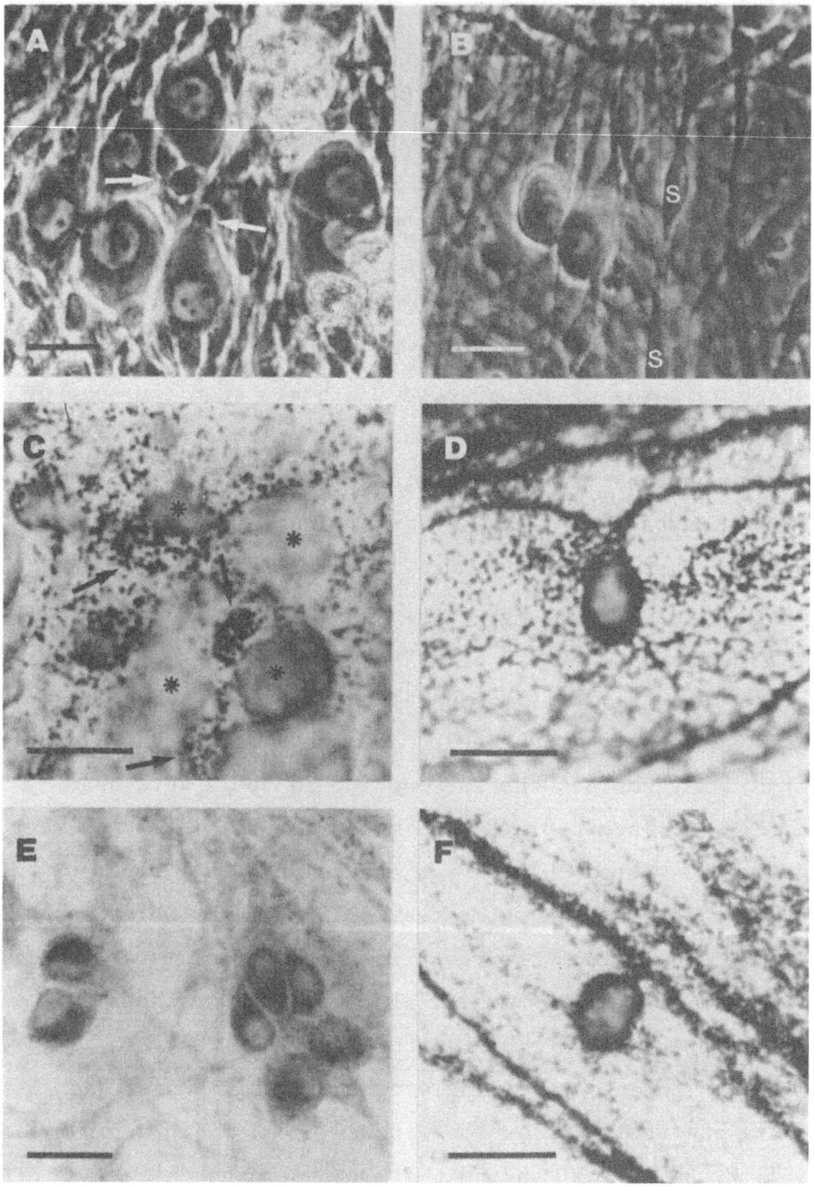

Fig. 3. A, B: Phase-contrast pictures of cultured rat dorsal root ganglia (DRG; 8 days *in vitro*). A: Group of DRG neurones surrounded by satellite glial (SG) cells (arrows) lying in the explant of the culture. B: DRG neurones which are deprived of SG cells in the outgrowth zone of the culture.

S: Schwann cells. Bars: 20 μm. C: Autoradiograph of a cultured DRG after incubation with [^{3}H]GABA. DRG neurones (*) lying in the explant did not take up [^{3}H]GABA, whereas the surrounding SG cells (arrows) are heavily labelled (concentration of GABA, 10^{-6} M; incubation time, 2 min). Bar: 20 μm. D: Accumulation of [^{3}H]GABA by an isolated DRG neurone in tissue culture. Culture: 11 days *in vitro*; incubation with [^{3}H]GABA (10^{-6} M) for 3 min. The neurone has migrated approximately 400 μm away from the explant. Bar: 20 μm. E: Cultured DRG neurones (17 days *in vitro*) which are intensely stained for GABA-T. Bar: 30 μm. F: Isolated neurone in the outgrowth zone which is intensely labelled after incubation with L-[^{3}H]glutamate (10^{-6} M, for 1 min). Bar: 20 μm. A, B, C and F (from Hösli and Hösli, 25); D (from Hösli and Hösli, 20).

After incubating DRG cultures with [^{3}H]β-alanine, only glial cells had taken up the amino acid, in accord with results obtained in the intact DRG where only glial elements were labelled (10).

Uptake of [^{3}H]GABA and [^{3}H]β-Alanine into Glial Cells

Biochemical studies have demonstrated that glial cells also possess a high-affinity transport system for the uptake of certain amino acid transmitters (26-29). After incubation of cultures of cerebellum, brain stem, spinal cord and DRG with [^{3}H]GABA, almost all glial cells were labelled (15,16,19,20,25), the amino acid being distributed over the cell bodies and processes (Fig. 4C,D). The observation that almost all glial cells had accumulated [^{3}H]GABA differed from the uptake pattern observed in neurones, where only some cells were labelled, whereas other neurones did not take up the amino acid (14-16). Furthermore, there was a difference in time course of the uptake between neurones and glial cells. Neurones usually became labelled after a shorter incubation period than glial cells (15).

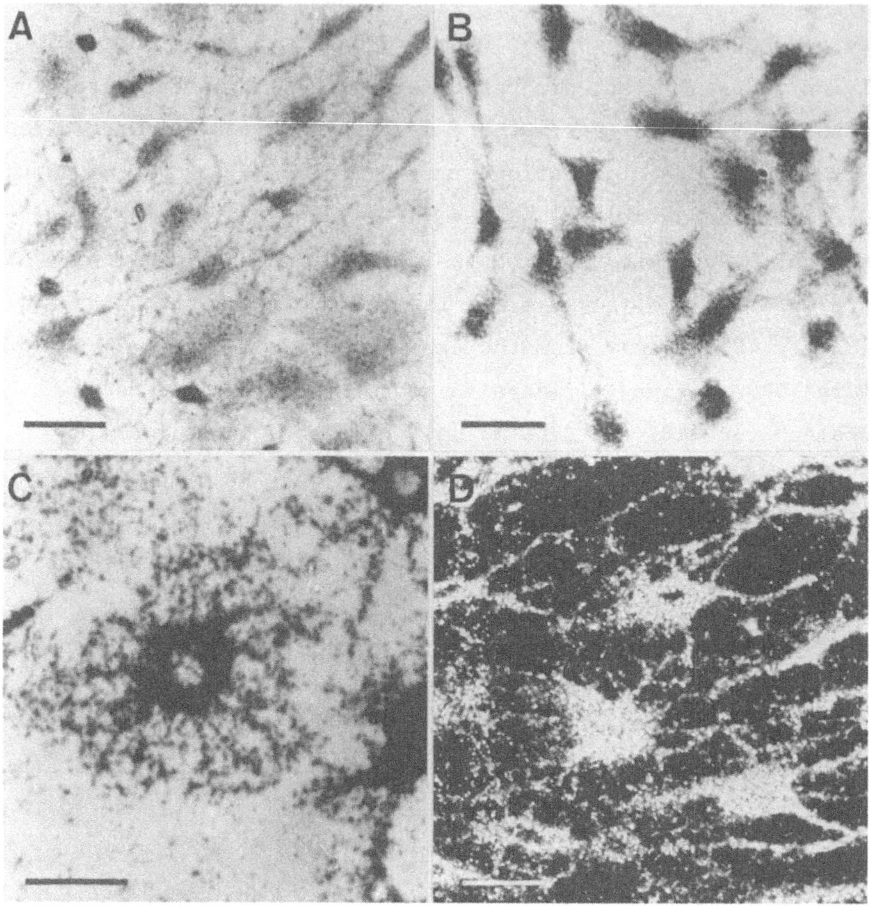

Fig. 4. A, B: Labelled glial cells in the outgrowth zone of brain stem cultures (23 days in vitro) after incubation with [^{3}H]β-alanine (10^{-6} M). Incubation time was 5 min for A and 10 min for B. Note the difference in the intensity of labelling between A and B. Bars: 30 μm (from Hösli and Hösli, 24). C: Astrocyte showing a strong accumulation of [^{3}H]GABA (10^{-6} M) after an incubation time of 5 min (cerebellar culture, 17 days in vitro). Bar: 20 μm. D: Intensely labelled glial cells in the outgrowth zone of an 18-day-old cerebellar culture after incubation with [^{3}H]GABA (10^{-6} M) for 10 min. Dark-field illumination, bar: 30 μm (from Hösli and Hösli, 19).

This observation, together with studies of Schon and Kelly (28) which demonstrated that the rate of GABA uptake into satellite glial cells of DRG is slower than that into cortical neurones, suggests that different transport systems might be involved in the uptake of GABA into neurones and glial cells.

[^{3}H]β-alanine, which was also accumulated by almost all glial cells (Fig. 4A,B; see ref. 24), exhibited a slower time course of uptake than that observed with [^{3}H]GABA (15,24). Figure 4A shows that glial cells had become only weakly labelled with [^{3}H]β-alanine after incubation for 5 minutes but that these cells were intensely labelled after incubation for 10 minutes (Fig. 4B; see ref. 24). In contrast, the protoplasmic astrocyte illustrated in Figure 4C had already become intensely labelled with [^{3}H]GABA after incubation for 5 minutes. Our finding that [^{3}H]β-alanine was taken up by glial cells at a slower rate than other amino acid transmitters is consistent with biochemical studies in the frog spinal cord (9) and in rat brain slices (8) which demonstrated a slower rate of uptake for β-alanine than for other amino acids such as GABA and taurine.

The uptake of [^{3}H]GABA and [^{3}H]β-alanine into neurones and glial cells was dependent on the presence of Na^+ in the incubation medium. After incubation in Na^+-free solution (Na^+ being replaced by choline and Tris), the uptake of [^{3}H]GABA and [^{3}H]β-alanine was markedly reduced or abolished. Incubation at 0°C also caused marked reductions in the uptake of both amino acids. Our studies which demonstrated that glial cells can actively accumulate [^{3}H]GABA and [^{3}H]β-alanine suggest that glial elements are also involved in removing amino acid transmitters after their release from nerve terminals (see also refs. 6,15).

Histochemical Demonstration of GABA-Transaminase (GABA-T) in Cultures of Cerebellum and DRG

Localization of the enzyme involved in the degradation of GABA, GABA-transaminase (GABA-T), was studied in cultures of cerebellum and DRG using the histochemical method of Van Gelder (17). In cultures of DRG, almost all neurones showed a strong activity for GABA-T (Fig. 3E). In cerebellar cultures, only Purkinje cells were intensely stained for GABA-T (Fig. 1C), whereas stellate cells and other interneurones showed only a weak enzymatic reaction or no reaction at all. Some small cells, probably granule cells, also possessed a moderate GABA-T activity (Hösli and Hösli, unpublished observations). Similar studies performed with sections of rat cerebellum revealed pronounced enzymatic activity in Purkinje cells and granule cells, but not in interneurones (30,31). Although glial cells accumulated [^{3}H]GABA to a great extent, almost no GABA-T was observed in cerebellar glial elements in vivo (31) or in tissue culture (Hösli and Hösli, unpublished observations). These results contrast with those obtained with retina (32) and with other CNS structures (30) where glial cells showed relatively high GABA-T activities.

SUMMARY

Nervous tissue cultures have proved to be an excellent tool for investigating the cellular localization of the uptake of transmitter substances using autoradiography. The uptake of [^{3}H]GABA and [^{3}H]β-alanine was studied in cultures of cerebellum, brain stem, spinal cord and dorsal root ganglia (DRG) of the rat. In cultures of brain stem and spinal cord, [^{3}H]GABA was taken up by a great number of neurones and by almost all glial cells, suggesting that glial elements might also be involved in the inactivation of this amino acid. In cerebellar cultures, [^{3}H]GABA was accumulated by many interneurones, by Purkinje cells and by a great number of glial cells. The finding that cultured Purkinje cells accumulated [^{3}H]GABA contrasts with observations made in the cerebellum in vivo or in slices where no labelling of Purkinje cells could be demonstrated. Since GABA is taken up to a great extent by Bergman glia which tightly surround Purkinje cells in vivo, it has been suggested that the barrier formed by these glial cells prevents the uptake of the amino acid into

Purkinje cells. In cultured cerebellum this glial barrier might be disrupted or absent and therefore Purkinje cells are able to accumulate [^{3}H]GABA. A similar glial barrier preventing the uptake of [^{3}H]GABA might also exist in DRG, where neurones that were completely surrounded by satellite glial cells did not take up the amino acid, whereas isolated neurones that were deprived of their glial envelopments became intensely labelled. After incubation with [^{3}H]β-alanine, it was observed that only glial cells were labelled in cerebellar and DRG cultures, whereas in spinal cord and brain stem both neurones and glial cells took up the amino acid providing further evidence that β-alanine might act as a transmitter substance in these two regions.

ACKNOWLEDGMENTS

We are grateful to Miss U. Kyburz for skillful technical assistance, to Mr. M. Wymann for the photographic work and to Miss F. Maeder for typing the manuscript. This study was supported in part by the Sandoz-Stiftung zur Förderung der medizinisch-biologischen Wissenschaften, Basel, Switzerland.

REFERENCES

(1) Curtis, D. R. and Johnston, G. A. R. (1974) Rev. Physiol. Biochem. Pharmacol. 69, 97-188.
(2) Roberts, E., Chase, T. N. and Tower, D. B. (eds.) (1976) GABA in Nervous System Function. Raven Press, New York.
(3) DeFeudis, F. V. and Martin del Rio, R. (1977) Gen. Pharmacol. 8, 177-180.
(4) Iversen, L. L. and Johnston, G. A. R. (1971) J. Neurochem. 18, 1939-1950.
(5) Balcar, V. J. and Johnston, G. A. R. (1973) J. Neurochem. 20, 529-539.
(6) Iversen, L. L. and Kelly, J. S. (1975) Biochem. Pharmacol. 24, 933-938.
(7) Lodge, D., Johnston, G. A. R. and Stephenson, A. L. (1976) J. Neurochem. 27, 1569-1570.
(8) Riddall, D. R., Leach, M. J. and Davison, A. N. (1976) J. Neurochem. 27, 835-839.
(9) Adair, R. and Davidoff, R. A. (1977) J. Neurochem. 29, 213-220.
(10) Schon, F. and Kelly, J. S. (1975) Brain Res. 86, 243-257.

(11) Kelly, J. S. and Dick, F. (1976) Cold Spring Harbor Symposia on Quantitative Biology, 40, 93-106.
(12) Hökfelt, T. and Ljungdahl, Å. (1972) Exp. Brain Res. 14, 354-362.
(13) Hökfelt, T. and Ljungdahl, Å. (1972) Adv. Biochem. Psychopharmacol. 6, 1-36.
(14) Hösli, E., Ljungdahl, Å., Hökfelt, T. and Hösli, L. (1972) Experientia, 28, 1342-1344.
(15) Hösli, L. and Hösli; E. (1978) Rev. Physiol. Biochem. Pharmacol. 81, 136-188.
(16) Hösli, L., Hösli, E., Andrès, P. F. and Wolff, J. R. (1975) In: Perspectives in Neurobiology. Golgi Centennial Symposium (M. Santini, ed.) pp. 473-488, Raven Press, New York.
(17) Van Gelder, N. M. (1965) J. Neurochem. 12, 231-237.
(18) Iversen, L. L. (1972) In: Perspectives in Neuropharmacology (S. H. Snyder, ed.) pp. 75-111, Oxford University Press, New York.
(19) Hösli, E. and Hösli, L. (1976) Exp. Brain Res. 26, 319-324.
(20) Hösli; E. and Hösli, L. (1978) In: Maturation of Neurotransmission (A. Vernadakis, E. Giacobini and G. Filogamo, eds.) pp. 108-115, Karger, Basel.
(21) Sotelo, C., Privat, A. and Drian, M.-J. (1972) Brain Res. 45, 302-308.
(22) Lasher, R. S. (1974) Brain Res. 69, 235-254.
(23) Iversen, L. L. and Bloom, F. E. (1972) Brain Res. 41, 131-143.
(24) Hösli, E. and Hösli, L. (1978) Experientia, 11, 1519-1521.
(25) Hösli, E. and Hösli, L. (1978) Neuroscience Lett. 7, 173-176.
(26) Schon, F. and Kelly, J. S. (1974) Brain Res. 66, 275-288.
(27) Henn, F. A. and Hamberger, A. (1971) Proc. Nat. Acad. Sci., U.S.A., 68, 2686-2690.
(28) Schon, F. and Kelly, J. S. (1974) Brain Res. 66, 289-300.
(29) Schrier, B. K. and Thompson, E. J. (1974) J. Biol. Chem. 249, 1769-1780.
(30) Robinson, N. and Wells, F. (1973) J. Anat. (Lond.) 114, 365-378.
(31) Hyde, J. C. and Robinson, N. (1974) Brain Res. 82, 109-116.
(32) Hyde, J. C. and Robinson, N. (1974) Nature (Lond.) 248, 432-433.

EFFECTS OF GABA-ANALOGUES ON THE HIGH-AFFINITY UPTAKE

OF GABA IN ASTROCYTES IN PRIMARY CULTURES

Arne Schousboe

Department of Biochemistry A, Panum Institute,
University of Copenhagen, 3c Blegdamsvej, DK-2200
Copenhagen N, Denmark

INTRODUCTION

γ-Aminobutyric acid (GABA) is one of the most abundant amino acids in the brain, occurring in concentrations of 1-10 μ-mole/g wet weight, with great topographical variations (2,69). Moreover, except for the pancreatic islets of Langerhans, in which the GABA content is comparable to that of brain (66), this amino acid is unique to the central nervous system (CNS), being present in only trace amounts in other tissues (26). GABA has been shown to be a relatively important constituent of brain metabolism (1,99), and the enzymes which are involved in its metabolism, glutamate decarboxylase (GAD) and GABA-transaminase (GABA-T), have recently been purified and characterized (3,4,7,15,41,57,58,87,88,101,102).

In addition to being an important constituent of brain metabolism, GABA is a major inhibitory transmitter in the mammalian CNS (16, 17,47). Accordingly, there must be an inactivation mechanism, and it is generally agreed that this is brought about by transport processes (cf. ref. 17) which may occur either in neurones or in glial cells (cf. refs. 40,79,80,85).

Judged from electron microscopic studies (8,38,73) of radioautographs of brain sections that had been incubated with [^{3}H]GABA, or from uptake experiments performed after specific lesions of GABAergic pathways (97), there is little doubt that GABA is taken up into presynaptic terminals. However, the extent to which this uptake may be responsible for the inactivation of GABA remains to be determined, since it has been clearly demonstrated that the major part of high-affinity GABA uptake into synaptosomes is due to a 1:1 homoexchange process (56,57) which can be of no physiological importance for removal of GABA from synaptic clefts. Under in vivo conditions there may, however, be a concentrative net uptake since the membrane potential is preserved (94).

Using bulk-prepared glial cells, Henn and Hamberger (27) first demonstrated that this cell type also possesses a high-affinity uptake system for GABA. Such a glial transport system has since been demonstrated in peripheral ganglia (12,70,71,74,75,103), rat retina (45,64), glioma cell lines (35,91,92), spinal cord explant-cultures (37) and primary cultures of glial cells from cerebellum (14,36,54) and cerebrum (30,85), but only the latter preparation exhibited a transport with a V_{max} comparable to those previously reported for GABA uptake into brain slices (cf. refs. 82,85). From the latter studies of GABA uptake into cultured astrocytes it could accordingly be concluded that such cells may be involved in the inactivation of GABA. A crucial point in such a conclusion is the existence of a net inward transport. In this regard, Hertz et al. (30) have shown that GABA uptake into astrocytes indeed represents a net uptake, in contrast to what has been found for glial cells of peripheral ganglia (70) and bulk-prepared glial cells (93). In this context it should be noted also that GABA uptake into C_6 astrocytoma cells (91) and rat retina glial cells (53) seems to represent a net uptake rather than a homoexchange process.

CHARACTERISTICS OF CULTURED ASTROCYTES

Astrocytes were cultured as described by Booher and Sensenbrenner (11) with the modifications introduced by Schousboe et al. (83,84,85,86). Briefly, cerebral hemispheres from newborn NMRI mice were passed through a sterile nylon sieve (80 µm pore size) into a modified Eagle's minimum essential medium (MEM) containing 20% (v/v) foetal calf serum (21,84), inoculated into plastic T-flasks (NUNC A/S, Denmark) and cultured for 2 weeks in this medium. Subsequently, the cultivation was continued for a third week in an analogous serum-free medium containing 0.1 mM dibutyryl-cyclic-AMP (dB-c-AMP). The medium was changed three times a week over the entire culture period. In this manner, well differentiated astrocytes with a morphology (32,84) resembling that of their in vivo counterparts (100) were obtained. In addition to the morphology, the astrocytic character of the cells was indicated by a high content of the glial fibrillary acidic protein (9,10) which is believed to be specific for astrocytes (22). Moreover, they have high activities of enzymes such as carbonic anhydrase (29,46,95) and glutamine synthetase (Table I) which are glial or astrocyte marker enzymes (cf. ref. 81).

Another characteristic of the astrocytes is that the development of some glutamate- and GABA-metabolizing enzymes (Table I) during the course of cultivation mimics that seen in the brain in vivo (cf. refs. 31, 80) except for the glutamine synthetase which seems to develop somewhat more slowly in cultures than in the brain in vivo (cf. ref. 31). In spite of this, the activity of this enzyme in the cultures is higher than in brain homogenates (72). Furthermore, differentiated astrocytes have recently been shown to contain all five isoenzymes of lactate dehydrogenase (65) in contrast to C_6 astrocytomas which exhibit only the isoenzymes containing the M-subunit (19). For further information concerning the validity of using cultured astrocytes as models for their in vivo counterparts, the reader is referred to discussions by Hertz (28) and Schousboe (78).

Table I - Development of GABA-Transaminase (GABA-T), Glutamic-Oxaloacetate-Transaminase (GOT), Glutamate Dehydrogenase (GLDH) and Glutamine Synthetase (Glu-S) Activities in Astrocytes from Dissociated Mouse Brain Hemispheres Cultured for 1, 2 or 3 Weeks in a Modified Eagle's MEM plus 20% (v/v) Foetal Calf Serum

Weeks in culture	Enzyme activity (n-mole x min^{-1} x mg^{-1} protein)			
	GABA-T	GOT	GLDH	Glu-S
0[a]	1.10 ± 0.05 (10)	143 ± 7.2 (8)	3.0 ± 0.5 (8)	13.0 ± 0.7 (8)
1	0.21 ± 0.03 (10)	117 ± 14 (7)	9.7 ± 1.8 (7)	9.0 ± 1.1 (9)
2	0.94 ± 0.08 (8)	184 ± 17 (8)	10.3 ± 1.6 (16)	13.8 ± 0.8 (10)
3	0.93 ± 0.06 (9)	194 ± 9 (10)	3.0 ± 0.3 (9)	20.2 ± 1.2 (7)
3[b]	1.64 ± 0.18 (7)	206 ± 22 (20)	12.3 ± 2.1 (15)	25.9 ± 4.1 (16)

Values were taken from Schousboe et al. (85,86) and Hertz et al. (31) and represent means ± S.E.M.; numbers of experiments in parentheses. GLDH was assayed at 27°C and the other enzymes at 37° as described by Schousboe et al. (85,86).

[a] Neonatal brain homogenate.

[b] During the third week of cultivation the cultures were grown in a serum-free medium to which 0.1 mM dB-c-AMP was added.

GABA UPTAKE IN ASTROCYTES AND IN THIN BRAIN SLICES

Astrocytes

The kinetics of GABA uptake into astrocytes was studied as detailed by Schousboe *et al.* (85). Figure 1 shows that the uptake consisted of a non-saturable component ($K = 7.3 \times 10^{-4}$ ml x min^{-1} x mg^{-1} cell protein) plus a saturable component which obeyed Michaelis-Menten kinetics (K_m = 40 μM; V_{max} = 0.35 n-mole x min^{-1} x mg^{-1} cell protein). To study the effects of analogues on high-affinity GABA uptake, small portions of cultures (30 - 50 μg protein) or brain cortex "prisms" (cf. below) were incubated for 5 minutes at 37°C in phosphate-buffered saline (86) containing 6 mM glucose and 1 uM [^{3}H]GABA (specific radioactivity, 37 Ci/m-mole, New England Nuclear) plus the analogue at a concentration of 10, 50, 100, 500, 1000 or 5000 μM (89). The incubation was terminated by centrifugation (10 sec) and rapid washing of the pellet (10 sec) with non-radioactive medium. From the calculated inhibitions, IC_{50} values were obtained as described by Beart *et al.* (5). The IC_{50} value is defined as that concentration of analogue which inhibits the uptake of 1 μM GABA by 50%.

Brain Cortex Slices

As a model for the neuronal transport system for GABA, mini-slices ("prisms") of brain cortex were used. Experiments by Riddall *et al.* (68) have indicated that the neuronal transport system is preserved whereas the glial transport system is progressively destroyed as the dimensions of a brain slice decrease. "Prisms" were prepared from 0.5 mm mouse cerebral cortex slices obtained from the outer convexity of each hemisphere using a longitudinally-vibrating nylon string (25,77). Subsequently, the slices were chopped into 0.1 mm "prisms" by 2 consecutive choppings (angle 60°) on a McIlwain tissue chopper (61), and the "prisms" were suspended in phosphate-buffered saline containing 6 mM glucose. The incubation procedure was similar to that described above for cultured astrocytes.

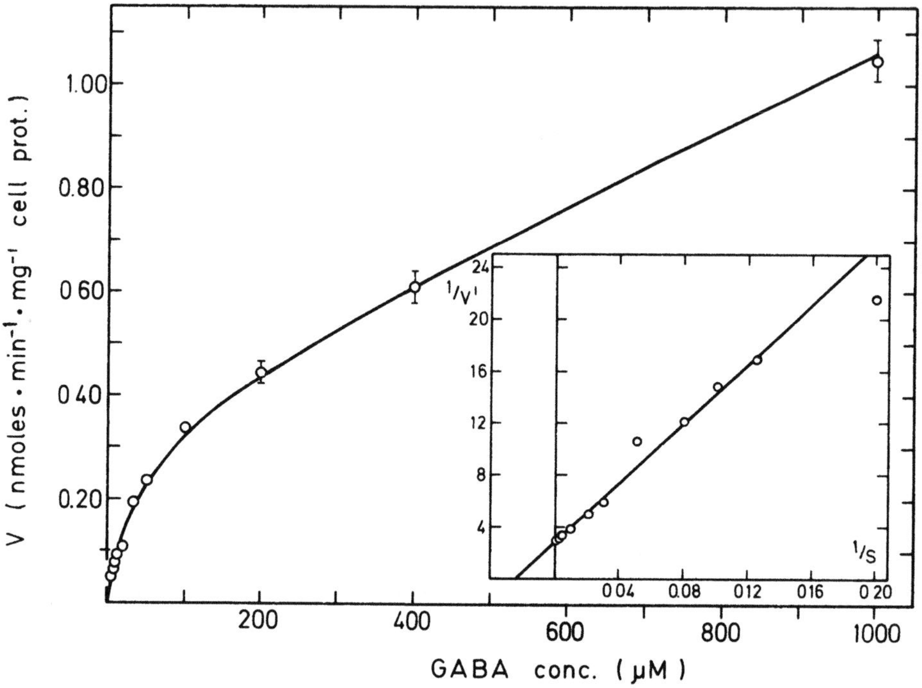

Fig. 1. GABA uptake into cultured astrocytes as a function of external GABA concentration. Means of 9 - 10 individual experiments with S.E.M. indicated by vertical bars if they extend beyond the symbols. The curve was fitted to the experimental points by computer analysis and is described by the equation:

$$V = V_{max}\ [1/(1 + K_m/S)] + k \times S \quad ,$$

in which V = velocity (n-mole x min^{-1} x mg^{-1} cell protein), K_m = Michaelis constant (μM), k = rate constant of an unsaturable component (ml x min^{-1} x mg^{-1} cell protein) and S = substrate concentration (μM). The insert shows 1/V' versus 1/S where V' is the observed V corrected for the unsaturable transport component (k x S) calculated from S and the k value obtained by computer analysis. From Schousboe et al. (85).

Table II - Effects of Aliphatic Compounds on GABA Uptake in Astrocytes and Cortical Mini-Slices

Compound	IC_{50} (μM)	
	Astrocytes	Mini-slices
R-4-Methyl-GABA	120	200
S-4-Methyl-GABA	1,000	750
R-Trans-4-methyl-4-aminocrotonic acid	500	160
S-Trans-4-methyl-4-aminocrotonic acid	> 5,000	> 5,000
4-R-3-Hydroxy-4-methyl-GABA	> 5,000	2,500
4-S-3-Hydroxy-4-methyl-GABA	> 5,000	> 5,000
(±)-3-Hydroxy-5-aminovaleric acid	1,400	430
Trans-5-amino-pentenoic-2-acid	> 5,000	4,000
β-Alanine	1,000	21,000
S-Diaminobutyric acid	> 5,000	50
Aminooxyacetic acid	> 5,000	—

Values are from unpublished results of A. Schousboe and P. Krogsgaard-Larsen and from Iversen and Johnston (39), Schon and Kelly (75) and Schousboe *et al*. (89).

EFFECT OF ANALOGUES ON GABA TRANSPORT

Aliphatic Compounds

The effects of a series of aliphatic GABA-analogues, representing both folded and extended conformations, on high-affinity GABA uptake by astrocytes and "prisms" are shown in Table II. The 4-methyl derivatives of GABA and trans-4-aminocrotonic acid, representing GABA in its extended conformation, show strict stereospecificity for the transport carriers, the R-isomers being much more potent inhibitors than the S-isomers. This means that the transport carriers are stereoselective for the hydrogens on C-4 although the GABA molecule is symmetrical. This is analogous to what is known for several enzymes that exhibit stereoselectivity although their substrates are symmetrical molecules (20). It should be noted that the postsynaptic GABA-receptor has been shown to have an opposite stereoselectivity for trans-4-methyl-4-aminocrotonic acid, the S-isomer being a much more potent agonist than the 4-isomer (44,51,90). The observed IC_{50} value for R-trans-4-methyl-4-aminocrotonic acid on GABA uptake in "prisms" is somewhat higher than a corresponding value obtained by Johnston et al. (51) in a study on "prisms" in which a much lower GABA concentration was used. This is analogous to the higher IC_{50} for R-4-methyl-GABA found in this study, compared to that reported by Beart and Johnston (6) from a similar study in which, again, a much lower GABA concentration was used. Such a difference is to be expected if the inhibition by these compounds is of the competitive type, which is the case at least for 4-methyl-GABA (6). The somewhat higher IC_{50} value found for R-trans-4-methyl-4-aminocrotonic acid in astrocytes compared to "prisms" may indicate that this compound selectively inhibits the neuronal uptake system, but a more detailed kinetic study is needed before any definite conclusion can be reached. This would represent an analogy to cis-1,3-aminocyclohexane carboxylic acid which has been shown to interfere selectively with neuronal GABA transport (13,63) and which has a structure resembling that of trans-4-methyl-4-aminocrotonic acid. A similar selectivity was observed for (±)-3-hydroxy-5-aminovaleric acid which inhibited the neuronal transport system but had essentially no effect on glial transport. It should be noted that although the only difference between 3-hydroxy-5-aminovaleric acid and the 4-R- and 4-S-3-hydroxy-4-methyl GABA is that the amino group has been moved from C-4 to C-5, the former compound selectively inhibited neuronal transport whereas the latter compounds were without effect on the transport systems (Table II).

In addition to these methylated analogues, aminooxyacetic acid, an extremely powerful inhibitor of GABA-T (88), was found to be a weak inhibitor of GABA transport in brain slices, in agreement with previous findings by Snodgrass and Iversen (96) and Johnston and Balcar (42), and it had no effect on GABA transport in astrocytes (89). Also, diaminobutyric acid, which has been previously classified as a selective inhibitor of the neuronal GABA carrier (40, 55, 75, 98), had no effect on GABA uptake in astrocytes (89) The observed IC_{50} for β-alanine, which is believed to be more inhibitory for glial GABA transport than for neuronal transport (13, 40,76), was higher than the corresponding value reported for GABA uptake into glial cells of sympathetic ganglia but relatively close to the value reported for rat retinal glial cells (13). A detailed kinetic analysis of the inhibition by β-alanine of GABA transport in astrocytes may shed more light on this apparent discrepancy.

Five-Membered Rings

The effects of different heterocyclic compounds with 5-membered rings on high-affinity GABA transport in astrocytes and "prisms" are shown in Table III. The most interesting finding is that β-proline seems to selectively inhibit the glial transport system (89). From the IC_{50} values obtained for (±)β-proline and S-(+)β-proline a theoretical IC_{50} value of 100 - 150 μM for R-(-)β-proline may be obtained. This means that the active optical isomer of β-proline is 10 times as potent as β-alanine at inhibiting GABA uptake into glial cells of central origin. This is interesting since the structure of β-proline combines the structures of GABA and β-alanine (Fig. 2), and, as shown in Figure 3, the inhibition produced by β-proline is of the competitive type. Whether this is also the case for β-alanine is presently unknown (cf. above).

The inhibitory effect of β-proline was found to be restricted to the parent compound since both the introduction of a double bond (3,4-position) and a hydroxyl group (4-position) led to compounds that had no effect on the transport systems. This differs from what has been observed for analogous substitutions in the 6-membered ring

Table III - Effects of Heterocyclic Compounds (5-Membered Rings) on GABA Uptake in Astrocytes and Cortical Mini-Slices

Compound	IC_{50} (μM)	
	Astrocytes	Mini-slices
(±)β-Proline	320	> 1,000
S-(+)β-Proline	1,900	1,700
3,4-Dehydro-β-proline	1,200	1,800
(±)Cis-4-hydroxy-β-proline	1,800	2,500
(±)Isoxazolidine-5-carboxylic acid	> 5,000	1,000
Imidazoleacetic acid	> 5,000	> 5,000
Muscimol	2,000	> 1,000
(±)4,5-Dihydromuscimol	4,000	1,500

Values are from unpublished results of A. Schousboe and P. Krogsgaard-Larsen and from Krogsgaard-Larsen and Johnston (49), Krogsgaard-Larsen (48) and Schousboe *et al*. (89).

of nipecotic acid; such substitutions on nipecotic acid did not interfere with the inhibitory action on the GABA transport systems (43,48,89). It should be noted that β-proline, in addition to being a selective inhibitor of GABA transport in astrocytes, interferes with GABA binding at the postsynaptic receptor (90).

Isoxazolidine-5-carboxylic acid, which represents aminooxyacetic acid in an extended conformation, had no effect on GABA transport in astrocytes but was a weak inhibitor of neuronal transport, findings which agreed with the corresponding effects of aminooxyacetic acid (42,89,96). Table III shows also that imidazoleacetic acid, muscimol and 4,5-dihydromuscimol, which are GABA-agonists interacting with the bicuculline-sensitive postsynaptic receptor or the slowly adapting stretch receptor (18,23,24,33,48,51,90) had little or no effect on GABA transport systems. It appears especially noteworthy that 4,5-dihydromuscimol, an even more powerful GABA-agonist

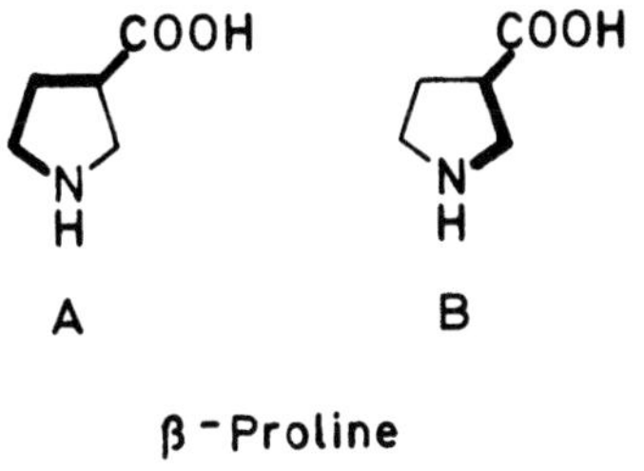

Fig. 2. Formulas of β-proline illustrating its structural relation to GABA (A) and β-alanine (B).

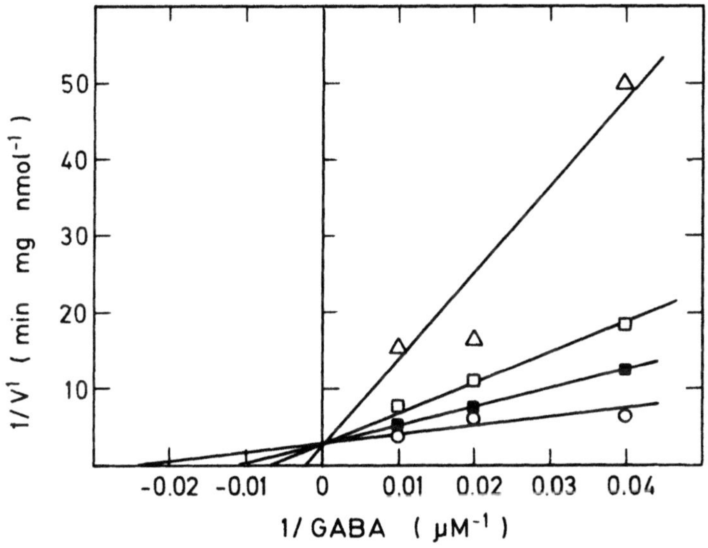

Fig. 3. Double reciprocal plot of the velocity (V') of high-affinity GABA uptake in astrocytes versus GABA concentration at different concentrations of (±)β-proline. The control (no β-proline) in indicated by (o); (■), (□) and (Δ) indicate the presence of 0.5 mM, 2.0 mM and 5.0 mM β-proline, respectively. From Schousboe (82).

than muscimol itself (51), had no effect on the transport systems in contrast to muscimol which was found to be a weak inhibitor of both transport systems. Thus, the former compound seems to be the ideal tool with which GABA receptor interaction may be specifically manipulated.

Six-Membered Rings

The effects of different compounds with 6-membered heterocyclic rings on GABA uptake in astrocytes and "prisms" are shown in Table IV. These 6-membered rings may be classified in two groups, one having the carboxyl group in the 3-position (e.g., nipecotic acid; guvacine) and the other having this substituent in the 4-position (e.g., isonipecotic acid; isoguvacine). The structures of some of these compounds are shown in Figure 4. Compounds belonging to the former group generally interact specifically with GABA transport, whereas those of the latter group specifically interact with GABA binding to bicuculline-sensitive postsynaptic sites (cf. ref. 48). As can be seen from Table IV, this was also the case in the present study in which both glial and neuronal transport systems were examined. The most interesting finding was that the optical isomers of nipecotic acid had different effects on the two transport systems. S-Nipecotic acid was found to be a more potent inhibitor of GABA uptake in brain slices than in astrocytes, whereas only a slight difference in potency was observed for R-nipecotic acid. This probably means that the former compound may be regarded as a selective inhibitor of neuronal GABA transport, in analogy with cis-1,3-aminocyclohexane carboxylic acid (13). This result also indicates that the glial transport system shows a higher degree of stereoselectivity than does the neuronal transport system. As mentioned above, neutral substitutions on the ring were found not to affect the ability of the molecule to interact with the transport systems, in contrast to what was observed for the β-proline deriva-

Table IV - Effects of Heterocyclic Compounds (6-Membered Rings) on GABA Uptake in Astrocytes and Cortical Mini-Slices

Compound	IC_{50} (μM)	
	Astrocytes	Mini-slices
S-(±)Nipecotic acid	1,000	30
R-(−)Nipecotic acid	30	5
(±)Cis-4-hydroxynipecotic acid	10	12
(±)Cis-4-aminonipecotic acid	> 5,000	> 5,000
(±)Trans-4-aminonipecotic acid	> 5,000	> 5,000
(±)3-Aminonipecotic acid	> 5,000	> 5,000
(±)N-Methylnipecotic acid	70	46
Guvacine	25	8
Isoguvacine	> 5,000	> 1,000
Isonipecotic acid	> 5,000	> 5,000
(±)Cis-3-hydroxyisonipecotic acid	> 5,000	> 5,000
4-Hydroxyisonipecotic acid	> 5,000	> 5,000
(±)Cis-3-aminoisonipecotic acid	> 5,000	> 5,000
4-Aminoisonipecotic acid	> 5,000	> 5,000

Values are from unpublished results of A. Schousboe, P. Krogsgaard-Larsen and G. A. R. Johnston and from Johnston et al. (43,44,45), Krogsgaard-Larsen et al. (52), Krogsgaard-Larsen (48) and Schousboe et al. (89).

tive (cf. Table III). On the other hand, the introduction of a positively-charged amino group on the ring completely abolished the inhibitory effect on both glial and neuronal uptakes. This result was somewhat surprising since such compounds combine the structures of nipecotic acid and diaminobutyric acid, the latter of which is a selective inhibitor of neuronal GABA uptake (40,55,75,89,98).

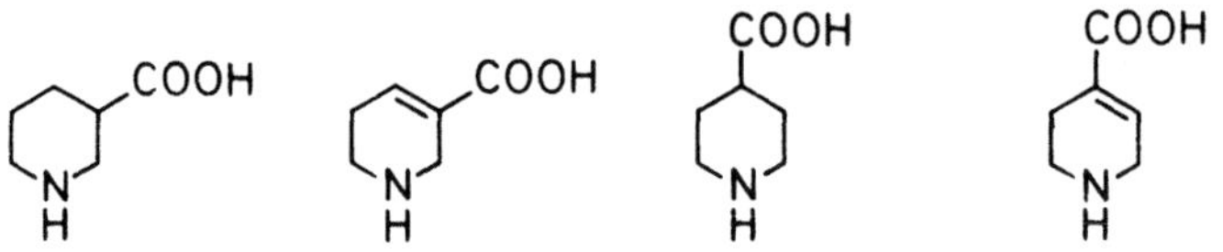

RS-Nipecotic acid   Guvacine   Isonipecotic acid   Isoguvacine

Fig. 4. Structural formulas of nipecotic acid, isonipecotic acid, guvacine and isoguvacine. The numbering in the ring starts at the nitrogen atom and proceeds counter-clockwise.

Isoguvacine, isonipecotic acid and derivatives of isonipecotic acid, which have been shown to interfere with GABA binding to the postsynaptic receptor (50,52), did not affect GABA transport into astrocytes or brain slices. Again, it was surprising that 4-amino-isonipecotic acid, whose structure resembles very closely that of diaminobutyric acid, had no effect on GABA uptake into brain slices (cf. above).

CONCLUDING REMARKS

Employing primary cultures of astrocytes which seem to constitute a valid model of their *in vivo* counterparts, it has been demonstrated that this cell type is likely to be of importance in terminating the transmitter activity of GABA.

It has been shown that the transport carrier in astrocytes is stereospecific for the C-4 hydrogens of the GABA molecule and that its structural requirements are distinct from those exhibited by the neuronal GABA carrier. β-Proline was a selective inhibitor of GABA transport in astrocytes, whereas *R*-trans-4-methyl-4-aminocrotonic acid and *S*-nipecotic acid seemed to be selective inhibitors of neuronal GABA transport, as studied using very thin slices ("prisms") of brain cortex.

These findings may be important for studies on the relative significance of the two transport systems in GABA-mediated neurotransmission, and thus for future pharmacological manipulations of the GABA system. This may eventually be beneficial for the treatment of neurological disorders such as epilepsy, Huntington's chorea and Parkinson's disease in which the GABA system seems to be disturbed (34,60,62).

ACKNOWLEDGEMENTS

Dr. P. Krogsgaard-Larsen, Department of Chemistry, The Royal Danish School of Pharmacy, is cordially thanked for performing the syntheses of a variety of GABA-analogues. The expert technical assistance of Miss Hanne Fosmark, Mrs. Inge Damgaard and Mrs. Grete Rössing is gratefully acknowledged. The Danish State Research Council is cordially thanked for a travel grant (511-10310) and for financial support (511-10567).

REFERENCES

(1) Balázs, R., Machiyama, Y., Hammond, B. J., Julian, T. and Richter, D. (1970) Biochem. J. (Lond.) 116, 445-467.
(2) Baxter, C. F. (1970) In: Handbook of Neurochemistry (A. Lajtha, ed.) Vol. 3, pp. 289-353, Plenum Press, New York.
(3) Bayón, A., Possani, L. D. and Tapia, R. (1977) J. Neurochem. 29, 513-517.
(4) Bayón, A., Possani, L. D., Tapia, M. and Tapia, R. (1977) J. Neurochem. 29, 519-525.
(5) Beart, P. M., Johnston, G. A. R. and Uhr, M. L. (1972) J. Neurochem. 19, 1855-1861.
(6) Beart, P. M. and Johnston, G. A. R. (1973) J. Neurochem. 20, 319-324.
(7) Bloch-Tardy, M., Rolland, B. and Gonnard, P. (1974) Biochimie, 56, 823-832.
(8) Bloom, F. E. and Iversen, L. L. (1971) Nature (Lond.) 229, 628-630.
(9) Bock, E., Jørgensen, O. S., Dittmann, L. and Eng, L. F. (1975) J. Neurochem. 25, 867-870.
(10) Bock, E., Møller, M., Nissen, C. and Sensenbrenner, M. (1977) FEBS Lett. 83, 207-211.
(11) Booher, J. and Sensenbrenner, M. (1972) Neurobiology 2, 97-105.
(12) Bowery, N. G. and Brown, D. A. (1972) Nature New Biol. 238, 89-91.

(13) Bowery, N. G., Jones, G. P. and Neal, M. J. (1976) Nature (Lond.) 264, 281-284.
(14) Burry, R. W. and Lasher, R. S. (1975) Brain Res. 88, 502-507.
(15) Cash, C., Maitre, M., Ciesielski, L. and Mandel, P. (1974) FEBS Lett. 47, 199-203.
(16) Curtis, D. R. and Johnston, G. A. R. (1970) In: Handbook of Neurochemistry (A. Lajtha, ed.) Vol. 4, pp. 115-134, Plenum Press, New York.
(17) Curtis, D. R. and Johnston, G. A. R. (1974) Ergb. Physiol. 69, 97-188.
(18) Curtis, D. R., Duggan, A. W., Felix, D. and Johnston, G. A. R. (1971) Brain Res. 32, 69-96.
(19) De Vellis, J. and Brooker, G. (1973) In: Tissue Culture of the Nervous System (G. Sato, ed.) pp. 231-245, Plenum Press, New York.
(20) Dixon, M. and Webb, E. C. (1964) Enzymes. 2nd Ed., pp. 199-258, Longmans, London.
(21) Eagle, H. (1959) Science, 130, 432-437.
(22) Eng, L. F., Van der Haeghen, J. J., Bignami, A. and Gerstl, B. (1971) Brain Res. 28, 351-354.
(23) Enna, S. J. and Snyder, S. H. (1975) Brain Res. 100, 81-97.
(24) Enna, S. J., Collins, J. F. and Snyder, S. H. (1977) Brain Res. 124, 185-190.
(25) Franck, G., Cornette, M. and Schoffeniels, E. (1968) J. Neurochem. 15, 843-857.
(26) Haber, B., Kuriyama, K. and Roberts, E. (1970) Biochem. Pharmacol. 19, 1119-1136.
(27) Henn, F. A. and Hamberger, A. (1971) Proc. Nat. Acad. Sci., U.S.A. 68, 2686-2690.
(28) Hertz, L. (1977) In: Cell Tissue and Organ Cultures in Neurobiology (S. Fedoroff and L. Hertz, eds.) pp. 39-71, Academic Press, New York.
(29) Hertz, L. and Sapirstein, V. (1978) In Vitro, 14, 376.
(30) Hertz, L., Wu, P. H. and Schousboe, A. (1978) Neurochem. Res. 3, 313-323.
(31) Hertz, L., Bock, E. and Schousboe, A. (1979) Devel. Neurosci. (in press).
(32) Hertz, L., Schousboe, A., Boechler, N., Mukerji, S. and Fedoroff, S. (1978) Neurochem. Res. 3, 1-14.
(33) Hori, N., Ikeda, K. and Roberts, E. (1978) Brain Res. 14, 364-370.
(34) Hornykiewicz, O., Lloyd, K. G. and Davidson, L. (1976) In: GABA in Nervous System Function (E. Roberts, T. N. Chase and D. B. Tower, eds.) pp. 479-485, Raven Press, New York.
(35) Hutchison, H. T., Werrbach, K., Vance, C. and Haber, B. (1974) Brain Res. 66, 265-274.
(36) Hösli, E. and Hösli, L. (1976) Exp. Brain Res. 26, 319-324.
(37) Hösli, E., Ljungdahl, Å., Hökfelt, T. and Hösli, L. (1972) Experientia, 28, 1342-1344.

(38) Iversen, L. L. and Bloom, F. E. (1972) Brain Res. 41, 131-143.
(39) Iversen, L. L. and Johnston, G. A. R. (1971) J. Neurochem. 18, 1939-1950.
(40) Iversen, L. L. and Kelly, J. S. (1975) Biochem. Pharmacol. 24, 933-938.
(41) John, R. A. and Fowler, L. J. (1976) Biochem. J. (Lond.) 155, 645-651.
(42) Johnston, G. A. R. and Balcar, V. J. (1974) J. Neurochem. 22, 609-610.
(43) Johnston, G. A. R., Krogsgaard-Larsen, P. and Stephanson, A. (1975) Nature (Lond.) 258, 627-628.
(44) Johnston, G. A. R., Allan, R. D., Kennedy, S. M. E. and Twitchin, B. (1979) In: GABA Neurotransmitters — Pharmacochemical, Biochemical and Pharmacological Aspects (P. Krogsgaard-Larsen, J. Scheel-Krüger and H. Kofod, eds.). pp. 149-164. Munksgaard, Copenhagen.
(45) Johnston, G. A. R., Krogsgaard-Larsen, P., Stephanson, A. L. and Twitchin, B. (1976) J. Neurochem. 26, 1029-1032.
(46) Kimelberg, H. K., Narumi, S., Biddlecome, S. and Bourke, R. S. (1978) Trans. Am. Soc. Neurochem. 9, 63.
(47) Krnjević, K. (1974) Physiol. Rev. 54, 418-540.
(48) Krogsgaard-Larsen, P. (1978) In: Amino Acids as Neurotransmitters (F. Fonnum, ed.) pp. 305-321. Plenum Press, New York.
(49) Krogsgaard-Larsen, P. and Johnston, G. A. R. (1975) J. Neurochem. 25, 797-802.
(50) Krogsgaard-Larsen, P. and Johnston, G. A. R. (1978) J. Neurochem. 30, 1377-1382.
(51) Krogsgaard-Larsen, P., Honoré, T. and Thyssen, K. (1979) In: GABA Neurotransmitters — Pharmacochemical, Biochemical and Pharmacological Aspects (P. Krogsgaard-Larsen, J. Scheel-Krüger and H. Kofod, eds.) pp. 201-216. Munksgaard, Copenhagen.
(52) Krogsgaard-Larsen, P., Johnston, G. A. R., Lodge, D. and Curtis, D. R. (1977) Nature (Lond.) 268, 53-55.
(53) Lake, N. and Voaden, M. J. (1976) J. Neurochem. 27, 1571-1573.
(54) Lasher, R. S. (1975) J. Neurobiol. 6, 597-608.
(55) Leach, M. J., Riddall, D. R. and Winkley, C. M. (1976) J. Neurochem. 27, 1281-1282.
(56) Levi, G. and Raiteri, M. (1974) Nature (Lond.) 250, 735-737.
(57) Lippert, B., Metcalf, B. W., Jung, M. J. and Casara, P. (1977) Eur. J. Biochem. 74, 441-445.
(58) Maitre, M., Ciesielski, L., Cash, C. and Mandel, P. (1975) Eur. J. Biochem. 52, 157-169.
(59) Marshall, J. and Voaden, M. (1974) Exp. Eye Res. 18, 367-370.
(60) McGeer, P. L. and McGeer, E. G. (1976) In: GABA in Nervous System Function (E. Roberts, T. N. Chase and D. B. Tower, eds.) pp. 487-495, Raven Press, New York.
(61) McIlwain, H. and Rodnight, R. (1962) Practical Neurochemistry, pp. 109-133, Churchill, London.
(62) Meldrum, B. S. (1975) Int. Rev. Neurobiol. 17, 1-36.

(63) Neal, M. J. and Bowery, N. G. (1977). Brain Res. 138, 169-174.
(64) Neal, M. J. and Iversen, L. L. (1972) Nature New Biol. 235, 217-218.
(65) Nissen, C. and Schousboe, A. (1979) J. Neurochem. (in press).
(66) Okada, Y., Taniguchi, H. and Shimada, C. (1976) Science, 194, 620-622.
(67) Raiteri, M., Federico, R., Coletti, A. and Levi, G. (1975) J. Neurochem. 24, 1243-1250.
(68) Riddall, D. R., Leach, M. J. and Davison, A. N. (1976) J. Neurochem. 27, 835-839.
(69) Roberts, E. (1962) In: Neurochemistry — The Chemistry of Brain and Nerve (K. A. C. Elliott, I. H. Page and J. H. Quastel, eds.) 2nd Ed., pp. 636-656. Charles C. Thomas, Springfield, Ill., U.S.A.
(70) Roberts, P. J. (1976) Brain Res. 113, 206-209.
(71) Roberts, P. J. (1976) In: Transport Phenomena in the Nervous System (G. Levi, T. L. Battistin and A. Lajtha, eds.) pp. 165-178, Plenum Press, New York.
(72) Sadasivudu, B., Rao, T. I. and Murthy, C. R. (1977) Neurochem. Res. 2, 639-655.
(73) Schon, F. and Iversen, L. L. (1972) Brain Res. 42, 503-507.
(74) Schon, F. and Kelly, J. S. (1974) Brain Res. 66, 275-288.
(75) Schon, F. and Kelly, J. S. (1974) Brain Res. 66, 289-300.
(76) Schon, F. and Kelly, J. S. (1975) Brain Res. 86, 243-257.
(77) Schousboe, A. (1972) Exp. Brain Res. 15, 521-531.
(78) Schousboe, A. (1977) In: Cell, Tissue and Organ Cultures in Neurobiology (S. Fedoroff and L. Hertz, eds.) pp. 441-446, Academic Press, New York.
(79) Schousboe, A. (1978) In: Dynamic Properties of Glia Cells (E. Schoffeniels, G. Franck, L. Hertz and D. B. Tower, eds.) pp. 173-182. Pergamon Press, London.
(80) Schousboe, A. (1978) Thesis, University of Copenhagen. pp. 1-83.
(81) Schousboe, A. (1979) In: Nervous System Specific Proteins; Scand. J. Immunol., Suppl. (E. Bock, ed.) (in press).
(82) Schousboe, A. (1979) In: GABA Neurotransmitters — Pharmacochemical, Biochemical and Pharmacological Aspects (P. Krogsgaard-Larsen, J. Scheel-Krüger and H. Kofod, eds.) pp. 263-280. Munksgaard, Copenhagen.
(83) Schousboe, A., Fosmark, H. and Hertz, L. (1975) J. Neurochem. 25, 909-911.
(84) Schousboe, A., Fosmark, H. and Svenneby, G. (1976) Brain Res. 116, 158-164.
(85) Schousboe, A., Hertz, L. and Svenneby, G. (1977) Neurochem. Res. 2, 217-229.
(86) Schousboe, A., Svenneby, G. and Hertz, L. (1977) J. Neurochem. 29, 999-1005.
(87) Schousboe, A., Wu, J.-Y. and Roberts, E. (1973) Biochemistry, 12, 2868-2873.

(88) Schousboe, A., Wu, J.-Y. and Roberts, E. (1974) J. Neurochem. 23, 1189-1195.
(89) Schousboe, A., Krogsgaard-Larsen, P., Svenneby, G. and Hertz, L. (1978) Brain Res. 153, 623-626.
(90) Schousboe, A., Thorbeck, P., Hertz, L. and Krogsgaard-Larsen, P. (1979) Submitted to J. Neurochem.
(91) Schrier, B. K. and Thompson, E. J. (1974) J. Biol. Chem. 249, 1769-1780.
(92) Schubert, D. (1975) Brain Res. 84, 87-98.
(93) Sellström, Å and Hamberger, A. (1976) Acta Physiol. Scand. 98, 94-102.
(94) Sellström, Å., Venema, R. and Henn, F. (1976) Nature (Lond.) 264, 652-653.
(95) Sensenbrenner, M. (1977) In: Cell, Tissue and Organ Cultures in Neurobiology (S. Fedoroff and L. Hertz, eds.) pp. 191-213, Academic Press, New York.
(96) Snodgrass, S. R. and Iversen, L. L. (1973) J. Neurochem. 20, 431-439.
(97) Storm-Mathisen, J. (1975) Brain Res. 84, 409-427.
(98) Sutton, I. and Simmonds, M. A. (1974) J. Neurochem. 23, 273-274.
(99) Van den Berg, C. J., Matheson, D. F., Ronda, G., Reijnierse, G. L. A., Blokhuis, G. G. D., Kroon, M. C., Clarke, D. D. and Garfinkel, D. (1975) In: Metabolic Compartmentation and Neurotransmission. Relation to Brain Structure and Function (S. Berl, D. D. Clarke and D. Schneider, eds.) pp. 515-543, Plenum Press, New York.
(100) Wolff, J. R. (1970) Triangle, 9, 153-164.
(101) Wu, J.-Y. and Roberts, E. (1974) J. Neurochem. 23, 759-767.
(102) Wu, J.-Y., Matsuda, T. and Roberts, E. (1973) J. Biol. Chem. 248, 3029-3034.
(103) Young, J. A. C., Brown, D. A., Kelly, J. S. and Schon, F. (1973) Brain Res. 63, 479-486.

CELLULAR MATURATION AND GABA UPTAKE BY NEURONAL AND GLIAL PRIMARY CULTURES

J. Mark, J. Borg, N. Ramaharobandro and P. Mandel

Centre de Neurochimie et Institut de Chimie Biologique, Faculté de Médecine de Strasbourg, Strasbourg, France

INTRODUCTION

Uptake of γ-aminobutyric acid (GABA), a putative inhibitory neurotransmitter, has been extensively studied in nervous cell cultures (1-5). These cultures do not, however, display all the characteristics of fully mature cells. The purpose of this chapter is to examine the influence of cellular maturation on GABA transport.

Previous experiments have revealed some age-related modifications of GABA uptake in brain slices from rat (6), mouse (7,8) or chick (9). However, in these complex tissues the respective involvement of neuronal and glial cells in the development of GABA transport could not be separated. In the present study we used primary cultures of neuronal or glial cells originating from rat brain at different ages (embryos or newborn animals). These cells were allowed to grow during various culture periods either in a standard culture medium or in a medium containing cyclic nucleotide derivatives. Cyclic-AMP (c-AMP) and cyclic-GMP (c-GMP) have been shown to induce morphological and biochemical differentiation in cultured cells (10, 11); we therefore used dibutyryl-cyclic-AMP (dB-c-AMP) and dibutyryl-cyclic-GMP (dB-c-GMP) in order to investigate changes which occur during such differentiation.

MATERIALS AND METHODS

Cerebral hemispheres from newborn rats or 13-15-day-old embryos were dissected free of meninges under sterile conditions and placed in Tyrode solution. Cultures from rat embryos were prepared as described by Yavin and Yavin (12). Briefly, this involved gentle dissociation with a syringe and centrifugation of the cell suspension at 200 g for 2 minutes in 10 ml Eagle Dulbecco medium containing 20% fetal calf serum (Gibco) and 10% egg ultrafiltrate (Gibco). The supernatant was discarded, and the pellet was resuspended and centrifuged as described above. The resulting pellet was redispersed in fresh medium. Aliquots (3 ml) of the cell suspension (5 hemispheres/60 ml of medium) were transferred into Petri dishes (35 mm diameter) that had been pre-coated with poly-L-lysine (Sigma). Primary cultures of dissociated newborn rat hemispheres were prepared as follows: after dissociation, by pressing the tissue through a nylon sieve (87 mesh) with a glass rod, the cells were collected into Eagle Dulbecco medium containing 10% fetal calf serum (2 hemispheres/80 ml of medium). Aliquots (3 ml) of the cell suspension were seeded into Petri dishes. Cells were allowed to attach to the floor of the dishes for at least 48 hours before changing the medium. Thereafter, the medium was changed every two days.

The dishes and flasks were maintained at 37°C in a humidified atmosphere consisting of 95% air/5% CO_2. The differentiation process was initiated 48 hours before the uptake experiments. The growth medium was removed and the fresh medium containing 10% fetal calf serum and both dB-c-AMP and dB-c-GMP, each at 0.5 mM, was added.

Uptake was studied by a previously described method (13). Briefly, the medium of the Petri dishes was removed and Krebs-Ringer-phosphate incubation medium (3 ml) was substituted. The dishes were then pre-incubated in a shaking water bath (New Brunswick Girotory Water Bath Shaker G 76) for 5 minutes at 37°C. Radioactive GABA was then added and the incubation was continued for 4 minutes. The incubation was terminated by removing the medium, and the cells were washed and then frozen at -20°C overnight. The radioactivity was later extracted with distilled water, and protein contents were determined by the method of Lowry et al. (14). After alkaline hydrolysis with 1 N NaOH for 2 hours at 37°C, ribonucleotides were separated from DNA by perchloric acid precipitation. DNA was determined by the diphenylamine reaction (15) Kinetic parameters were calculated using the Cleland program (16) and a Univac 100 computer. V_{max} is expressed as n-moles/mg DNA/minute.

γ-Aminobutyric acid was purchased from Sigma Chemical Company, and N^6-2'-O dibutyryl cyclic-adenosine-3'5' monophosphate (dB-c-AMP) and N^6-2'-O-dibutyryl cyclic-guanosine-3'5' monophosphate (dB-c-GMP) were purchased from Boehringer, Germany. [2,3-^{3}H]γ-Aminobutyric acid (35.4 Ci/m-mole) was purchased from New England Nuclear Chemicals GmbH, Dreieichenhain, Germany.

RESULTS

Cellular Morphology

Untreated neuronal cultures were characterized by round cell aggregates connected with neurites which were arranged in bundles of variable lengths and thicknesses (Fig. 1a). Isolated neurones had scant cytoplasm and thin neurites, usually bipolar. When the cultures grew older, neuronal cells were more numerous and the network of processes appeared more dense (Fig. 1b). The glial cells in primary cultures appeared undifferentiated and grew in epithelial-like sheets (Fig. 2a). After 2 — 3 weeks in culture, glial cells were confluent and few processes could be seen (Fig. 2b).

When the cells were grown in the presence of both dB-c-AMP and dB-c-GMP (0.5 mM of each) for two days, morphological changes occurred in these primary cultures. The processes which connected the neuronal aggregates increased in number and thickness and formed a dense network that extended throughout the whole culture (Fig. 1c and d). These neuronal aggregates were rarely seen in control cultures where

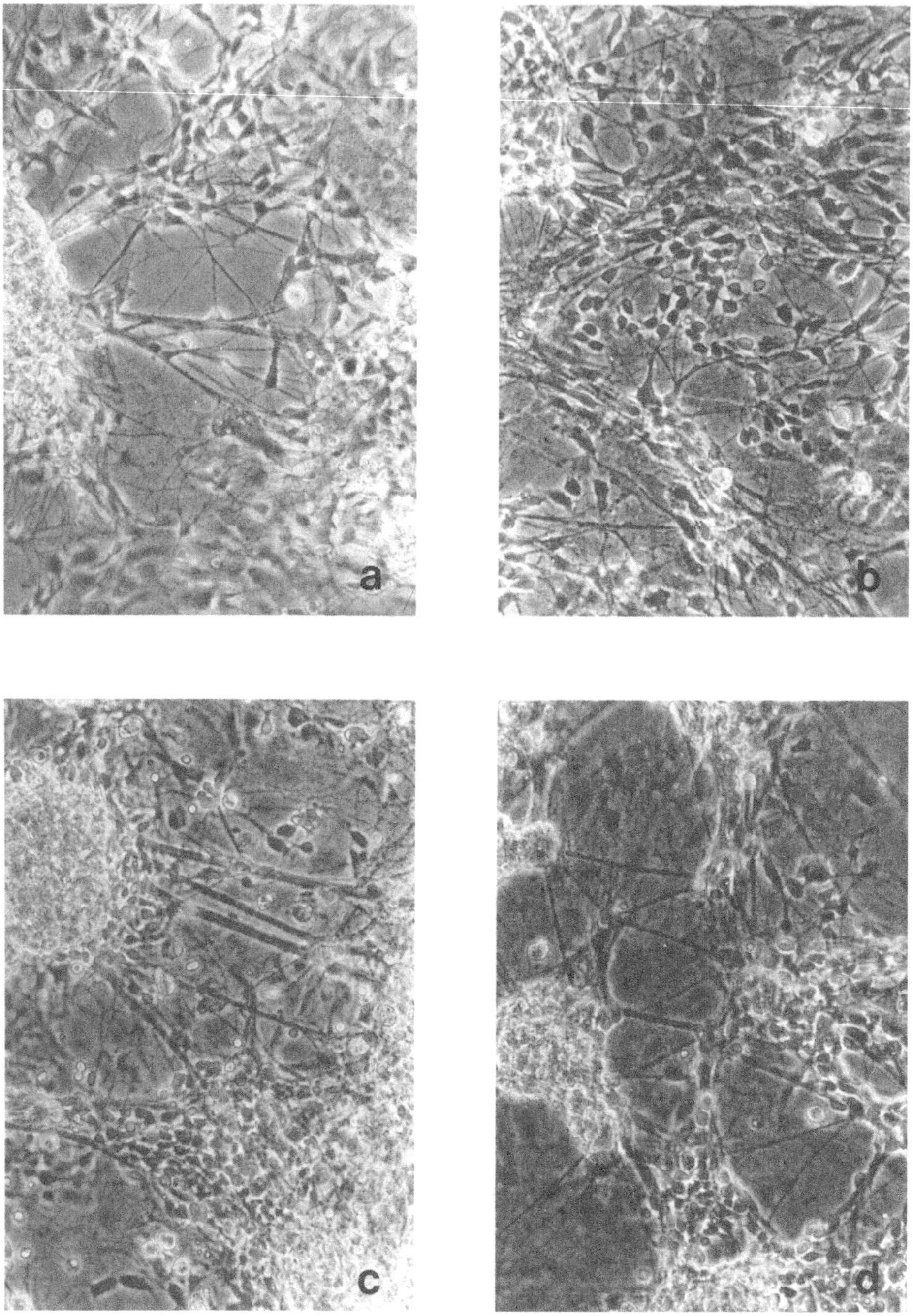

Figure 1

Fig. 1. Phase-contrast microscopy of neuronal cultures originating from dissociated cerebral hemispheres of 14-day-old rat embryos; magnification, X 30. Photographs were taken after 6 days (a and c) or 8 days (b and d) in culture in either control conditions (a and b) or in medium containing cyclic nucleotide derivatives (c and d).

---

the cell layer was more evenly distributed. The neurite bundles were also somewhat thinner in the control cultures. Some of these processes were axo-dendritic and formed synapses (see Fig. 3). The neurones appeared morphologically well differentiated with large, phase-bright soma and an extensive network of neurites. The differentiated glial cells were characterized by abundant processes and the appearance of large astrocyte-like cells (Fig. 2c and d). Such astrocyte-like cells were rarely seen in control cultures.

GABA Uptake

Uptake of [^{3}H]GABA in primary cultures was previously shown to occur linearly with time during at least 4 minutes (5). Kinetic studies were performed by measuring the rate of [^{3}H]GABA uptake at 9 different concentrations ranging between 1 μM and 2 mM. At least two components of uptake were found; one was saturable, characterized by a low K_m (high affinity) and observable at low substrate concentrations, and the other was non-saturable (attributed to passive diffusion) and became predominant at high substrate concentrations. Another component, characterized by a high K_m (low affinity) was sometimes observed.

The kinetic parameters for GABA transport calculated at the earlier stages of neuronal (six days) and glial (eight days) control cultures were respectively 2.06 and 13.63 n-moles/mg DNA/minute for V_{max} and 4.95 and 12.97 μM for K_m (Tables I and II). Further kinetic analyses of GABA influx were performed at other selected developmental

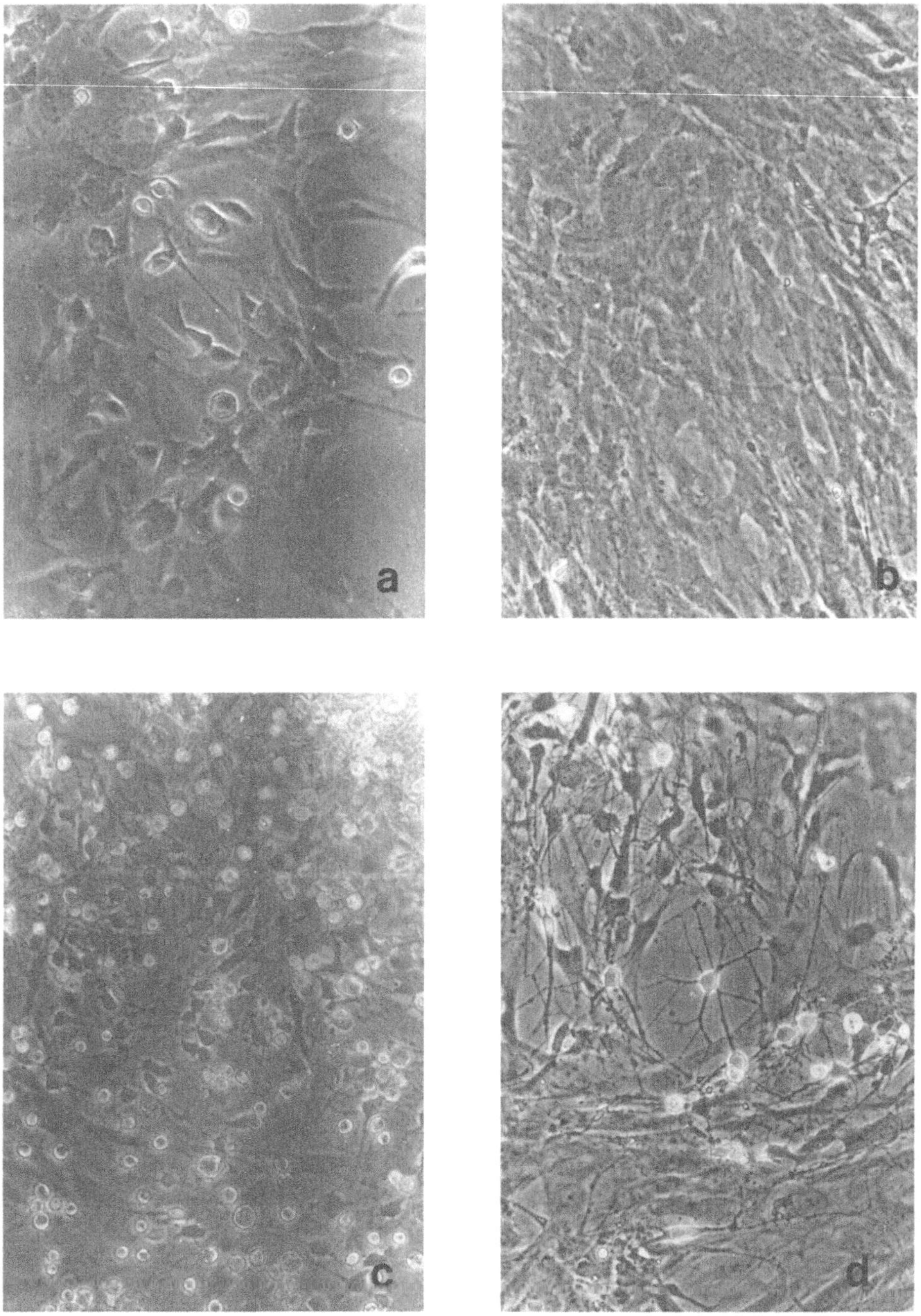

Figure 2

Fig. 2. Morphological aspects of glial cultures originating from dissociated cerebral hemispheres of newborn rat; phase-contrast microscopy (X 30). Cells are shown after 8 days (a and c) or 14 days (b and d) in culture in either control conditions (a and b) or in medium containing cyclic nucleotide derivatives (c and d).

---

stages in neurones (8 and 14 days) and in glial cells (14, 19, and 21 days). In neurones, the transport efficiency increased with time in culture, the greatest changes in V_{max} being observed from the 8th to the 14th day of culture (Table I). The highest uptake rate was observed in 14-day-old cultures which showed a seven-fold increase when compared with the V_{max} in 6-day-old cultures; the K_m increased to 8.61 μM. It should be noted that a low affinity uptake was present in these cells, which disappeared after 14 days in culture. The GABA transport in non-treated glial cells decreased as the cells grew older and was entirely lost at the 21st day of culture (Table II). We could detect first a diminution in the V_{max} of the high-affinity transport system at 14 days, then at the 19th day a rise in V_{max} related to the appearance of a low-affinity mechanism, and thereafter a complete disappearance of the transport systems (i.e., if present, these were below the level of experimental measurement).

In treated neurones, K_m values calculated at each stage of culture did not differ significantly and these were always slightly below the control values (Table I). On the other hand, significant changes were evident in the V_{max} values: it increased 20 times from the 6th to the 14th day. However, the rate of uptake was still below the levels reached at these developmental stages in control cultures, and no low-affinity mechanism could be detected. In treated glial cells, the V_{max} value followed a course which resembled that of controls; i.e., it first decreased, then increased and disappeared after 21 days. After 19 days in culture, the V_{max} was more than twice as high as in controls.

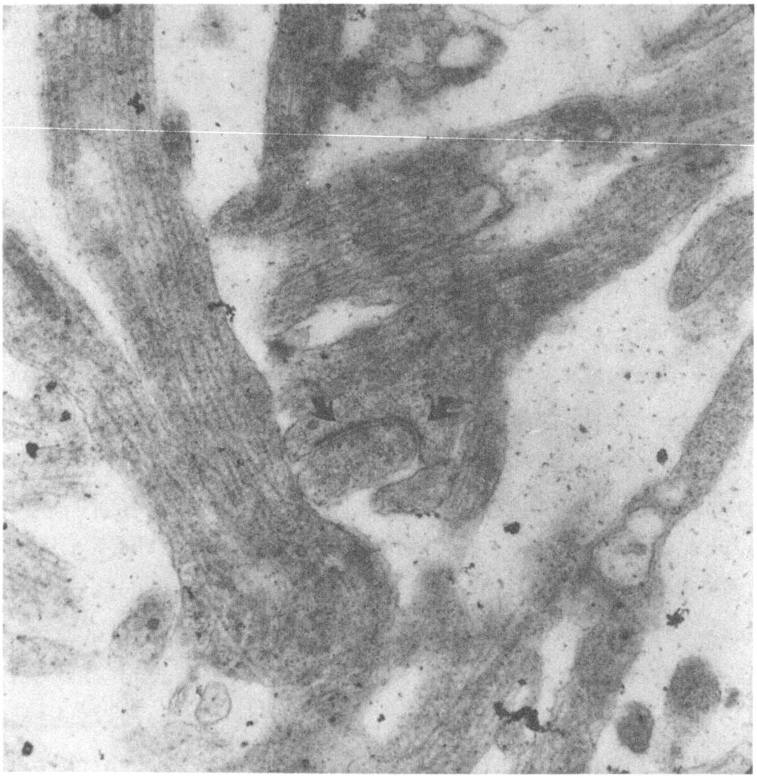

Fig. 3. Electron microscopy of neuronal cells grown for 8 days in the presence of cyclic nucleotide derivatives (magnification, X 27,000). An axo-dendritic synapse can be seen (arrow) with typical asymmetric densification and vesicles in the presynaptic ending. Note the numerous microtubules in the neuritic processes throughout the photograph.

DISCUSSION

The studies reported here were aimed at investigating a possible correlation between cellular maturation and the efficiency of GABA transport. An increase in the V_{max} of high-affinity GABA uptake in neuronal cultures was shown to be related to the culture period. Similarly, Sershen and Lajtha (7) observed an increase in net GABA uptake by mouse brain slices from embryos or new-born animals, and

Table I - GABA Uptake by Neuronal Cultures

Culture conditions	Culture period	K_{diff}	K_m (μM)		V_{max} (n-mole/mg DNA/min	
			High-affinity	Low-affinity	High-affinity	Low-affinity
Control	6 days	0.94	4.95 ± 0.99	254.2 ± 75.4	2.06 ± 0.15	15.57 ± 1.33
Treated	"	0.95	4.49 ± 3.73		0.48 ± 0.18	
Control	8 days	0.63	5.45 ± 0.80	316.0 ± 161.5	4.18 ± 0.15	13.57 ± 2.42
Treated	"	1.22	2.22 ± 0.67		2.42 ± 0.15	
Control	14 days	1.22	8.61 ± 1.57		15.07 ± 0.79	
Treated	"	1.23	3.76 ± 2.23		9.76 ± 0.14	

K_m and V_{max} (± S.D.) were determined using a Fortran program (16) in control cultures and in cultures treated with cyclic nucleotide derivatives (0.5 mM). K_{diff} = diffusion coefficient in p-moles.mg $protein^{-1}.min^{-1}.\mu M^{-1}$.

Table II - GABA Uptake by Glial Cultures

Culture conditions	Culture period	K_{diff}	K_m (μM) High-affinity	K_m (μM) Low-affinity	V_{max} (n-mole/mg DNA/min) High-affinity	V_{max} (n-mole/mg DNA/min) Low-affinity
Control	8 days	0.615	12.97 ± 5.54		13.63 ± 2.43	
Treated	"	2.25	3.59 ± 2.15		6.30 ± 1.12	
Control	14 days	1.63	4.49 ± 3.73		1.13 ± 0.43	
Treated	"	1.38	10.73 ± 7.16		1.51 ± 0.64	
Control	19 days	0.375		269.77 ± 34.27		18.48 ± 0.85
Treated	"	0.053		168.33 ± 66.93		47.33 ± 10.05

See legend to Table I. No uptake could be measured after 21 days in culture in either control or treated cultures.

Johnston and Davies (6) showed a 20-fold increase of V_{max} in rat brain slices from 1-day-old to adult animals.

On the other hand Levi (9) found a low-affinity uptake system in brain slices originating from grown chicks, while the high-affinity mechanism, present at the embryonic stage, had disappeared. Similarly, we observed a low-affinity GABA uptake in 19-day-old glial cultures, while the high-affinity mechanism present at the earlier stages had disappeared. Thus, the observation by Levi (9) could be due predominantly to uptake by the glial cells, if the brains of both rat and chicken develop similarly.

The disappearance of the high-affinity GABA uptake in mature rat astrocytes is not in accord with other studies on 21-day-old dispersed non-neuronal cultures of post-natal rat cerebellum (2) or on mouse astrocytes cultivated for 3 weeks (4). In fact, some results on mouse astrocytes in culture showed that under our experimental conditions a high-affinity GABA uptake (K_m = 2.28 ± 1.74 μM; V_{max} = 47.66 ± 9.07 p-mole/mg protein/min) was still present after 21 days in culture. These discrepancies may thus be attributed to different maturational patterns, depending on the species and on the brain area studied. However, the V_{max} values for high-affinity GABA uptake found with 21-day-old dispersed neuronal cell cultures of post-natal rat cerebellum (2) agree with those obtained with our 14-day-old neuronal control cultures. Thus, one might suggest that in mature cells GABA is predominantly taken up by neurones, whereas GABA transport by glial cells decreases as the cells grow older (1).

The present study did not show a clear correlation between morphological and biochemical mechanisms of differentiation. Similarly, Lasher (17) reported that the development of GABA uptake by stellate neurones in cultures of dissociated post-natal rat cerebellum was not related to the morphological differentiation of these cells. Our cultures that were treated with cyclic nucleotide derivatives showed a more pronounced morphological differentiation than the control cultures, with regard to both neuronal and glial cells.

On the other hand, V_{max} of high-affinity GABA uptake was always lower in the treated cultures than in control cultures, except in the case of 14-day-old glial cells. Thus, with our experimental conditions it seems that the cyclic nucleotides do not produce, by themselves, the expected maturation of the transport systems.

REFERENCES

(1) Schrier, B. K. and Thompson, E. J. (1974) J. Biol. Chem. 249, 1769-1780.
(2) Lasher, R. S. (1975) J. Neurobiol. 6, 597-608.
(3) Schubert, D. (1975) Brain Res. 84, 87-98.
(4) Schousboe, A., Hertz, L. and Svenneby, G. (1977) Neurochem. Res. 2, 217-229.
(5) Balcar, V. J., Mark, J., Borg, J. and Mandel, P. (1979) Neurochem. Res. (in press).
(6) Johnston, G. A. R. and Davies, L. P. (1974) J. Neurochem. 22, 101-105.
(7) Sershen, H. and Lajtha, A. (1976) Neurochem. Res. 1, 417-428.
(8) Kelly, P., Luttges, M., Johnson, T. and Grove, W. (1974) Brain Res. 68, 267-280.
(9) Levi, G. (1972) Archs. Biochem. Biophys. 151, 8-21.
(10) Prasad, K. N., Sahu, S. K. and Sinha, P. K. (1976) J. Nat. Cancer Inst. 57, 619-631.
(11) Zwiller, J., Goridis, C., Ciesielski-Treska, J. and Mandel, P. (1977) J. Neurochem. 29, 273-278.
(12) Yavin, E. and Yavin, Z. (1974) J. Cell Biol. 62, 540-546.
(13) Balcar, V. J., Borg, J. and Mandel, P. (1977) J. Neurochem. 28, 87-93.
(14) Lowry, O. H., Rosebrough, N. J., Farr, A. L. and Randall, R. J. (1951) J. Biol. Chem. 193, 265-275.
(15) Burton, K. (1956) Biochem. J. (Lond.) 62, 315-323.
(16) Cleland, W. W. (1967) Adv. Enzymol. 29, 1-32.
(17) Lasher, R. S. (1974) Brain Res. 69, 235-254.

PERIPHERAL SENSORY STIMULATION AND THE RELEASE OF TRANSMITTER AMINO ACIDS *IN VIVO* FROM SPECIFIC REGIONS OF CEREBRAL CORTEX

A.-S. Abdul-Ghani, H. F. Bradford, D. W. G. Cox*
and P. R. Dodd

Department of Biochemistry, Imperial College of Science
and Technology, London SW7 2AZ, Great Britain

INTRODUCTION

The compelling case for considering certain amino acids as major chemical transmitters in various regions of the CNS has been set out in detail several times in the past few years (10,22), and a strong and essential part of this case concerns the stimulus-evoked release of these compounds from neural tissues. These studies have mainly been *in vitro* using the application of electrical or chemical depolarizing stimuli to cortical slices (4,26,33) or synaptosomes (8,11) though photic stimuli have been employed to evoke release from isolated retina (28). In contrast, there have been few convincing studies of amino acid release *in vivo* (7,19,20) following synaptic activation through neural pathways, because of the technical difficulties of the animal preparation, the problems associated with measuring the small quantities of amino acids released, and the unattractive constraints of being unable to accurately control experimental conditions such as local Ca^{2+} concentrations. However, the importance of demonstrating *in vivo* release of amino acid

* Present address: Department of Biochemistry, St Thomas's Hospital Medical School, London SE1, Great Britain.

neurotransmitter candidates during specific neural stimulation remains of cardinal importance to their case.

Through the development of a special chronic superfusion cannula (15,16), and a fast and very sensitive amino acid autoanalyzer (5), we have been able to overcome some of these problems.

This present study concerns the effect of electrical stimulation of the brachial plexus of the forelimb, and of light flash stimuli applied to the retina on the release of free amino acids to fluid superfusing the appropriate regions of cerebral cortex; i.e., sensorimotor and visual cortex, respectively.

MATERIALS AND METHODS

The experiments were carried out on adult female hooded Rowett rats (220 - 250 g body weight). Anesthesia was induced with Avertin (250 mg/kg) and atropine (1.7 mg/kg) given intraperitoneally and repeated at smaller doses to maintain light anesthesia as tested by a positive corneal reflex. The rats were kept in a 0° – 4°C cold room until the body temperature decreased to 22°C. This greatly reduced the risk of brain edema during surgery.

After implantation of the superfusion cannula (ref. 16 for full details), the animals were either left at least 1 hour for recovery before all _acute_ experiments, or left for 24 hours before _chronic_ experiments were begun.

In some experiments normal saline was employed as the superfusion fluid, and in others artificial cerebrospinal fluid (CSF) solution was employed, containing the following comⁿonents (ref. 14) (mM): NaCl 123.5; KCl 2.4; KH_2PO_4, 0.5; Na_2SO_4, 0.16; $MgCl_2$, 0.9; $CaCl_2$, 1.47; $NaHCO_3$, 22.0; glucose, 5.0; fructose, 0.22; urea, 4.2; adenosine, 0.1; guanosine, 0.1; creatine, 0.1; _m_-inositol, 0.15: Na-citrate, 0.32; Na-lactate, 1.34; Na-pyruvate, 0.08; Na-ascorbate, 0.06. Neutral pH was maintained by gassing the solution with a 5% CO_2 : 95% O_2 gas mixture, and the colloidal osmotic pressure was maintained by adding 20 mg% inulin instead of protein.

Cannulation

For these experiments a special superfusion method was used which is described in detail by Dodd et al. (16). This method used a specially constructed "Swivel" cannula which allows study of the release of cerebral amino acid transmitters in sub-acute and chronic experiments from the cortex of unrestrained, behaviorally normal rats.

The cannula (4-mm diameter) is made from non-toxic materials (Teflon) and has a very small dead space (15 μl). It withstands normal intracranial pressure without leaking, and embodies facilities which prevent the tubing leads from twisting whilst allowing the animal to move freely. The techniques used to fix the cannula firmly to the animal's skull, thereby allowing continuous superfusion with saline or artificial CSF at neutral pH for periods of at least 2 weeks, are fully described elsewhere (16). All the surgery and subsequent experiments were performed in a temperature-stabilized room (24°C) which was also on a 12 hr "light-on", 12 hr "light-off" cycle (16).

The small size of the cannula allows superfusion of small discrete areas of the brain surface and therefore permits study of the effects of specific forms of sensory stimulation on the release of amino acids from these small localized regions of cerebral cortex. The sensori-motor strip and the visual cortex were chosen for study. The cortical area superfused was the S_1 forelimb sensori-motor area (14) which was located by drilling a 4 mm diameter hole in the skull with a center 3 mm from the Central (Sagittal) suture, and centered on the Bregma suture. The primary visual cortex was located by positioning the cannula with its center 3 mm from the Sagittal suture and at 3 mm from the Lambda suture. The accurate location of the cannula over the forelimb sensori-motor area was checked by stimulating the cortex either chemically (50 mM KCl) or electrically with pulses applied to an incorporated stimulating wire (16). This evoked clear contra-lateral forelimb movements. Also, cobalt powder or discs implanted on this area caused clonic limb-jerks of principally the forelimb (13).

Sensory Cortical Stimulation

Brachial Plexus Stimulation. After exposing the brachial plexus of both forelimbs and separating the nerve trunks from adjacent blood vessels a small electrode of the kind developed by Dobkin (12) was placed around all the nerves of the plexus on each side. Stimulation applied to the forelimb contra-lateral to the cannulated hemisphere activated the cortex being superfused through the cannula. Ipsi-lateral stimulation should affect mainly the cortex in the uncannulated hemisphere, though a later interhemispherically conducted response could have occurred.

By comparing the effects of these two forms of stimulation, it was possible to test whether any changes in transmitter release was the direct result of stimulation of the cortical area, or due to other effects such as changed cerebral blood flow or blood pressure, or due to other bodily stresses caused by the stimulation. Electrical stimulation of the brachial plexus was by square wave pulses of the following specification: amplitude 5 - 12 volts; current 2 - 8 mA; pulse duration 1 m-sec; frequency 2 - 3 Hz or 50 Hz. Five-minute periods of stimulation were applied and always evoked clearly visible muscular jerking in the ipsi-lateral forelimb. Clonic seizures were never seen to occur.

Light-Flash Stimuli. For stimulation of the visual cortex, flash stimuli were applied to both eyes of anesthetized Rowett rats kept in darkness for at least 60 minutes prior to the experiment (7). The flash stimuli (2/sec) were given from a dry cell battery torch (1.2 W) held at a distance of 15 cm from the nose on the long axis of the head. The front of the torch was covered with thin black-white checkered paper to give projection of a structured image. This is likely to be a more effective stimulus for the retina. Unanesthetized hooded rats were observed to move their heads in the direction of this torch stimulus when it was moved laterally at a distance of 15 - 30 cm around the circumference of a circular glass cage, indicating that it was clearly visible to them.

In these experiments γ-acetylenic GABA was administered (100 mg/kg body weight; intraperitoneal injection) as an aqueous solution (0.6 ml/100 g body weight). The drug was synthesized and supplied by Centre de Recherche Merrell International, Strasbourg, France. Its purpose was to raise brain GABA levels and possibly enhance the efflux of GABA in parallel.

Anesthetics. The effects of brachial plexus stimulation were examined under anesthesia and in awake animals. Retinal stimulation of the visual cortex was only performed on anesthetized animals. The anesthetic used was Avertin mixed with atropine (250 mg/kg and 1.7 mg/kg, respectively) repeated at smaller doses to maintain light anesthesia as monitored by a positive corneal reflex.

Determination of Amino Acids. The amino acids were measured by autoanalysis using the analyzer described by Bradford and Thomas (5). This technique allows rapid (35 min) measurement of 1 - 5 p-mole of of amino acids in 200 μl sample volumes. Calculation of the results was aided by a small on-line computer (Digico 16S) by relating the values to standard norleucine signals added to the samples. The high sensitivity of the technique allowed direct measurement of amino acids in the superfusion fluid after acidification, without concentration or purification. The latter was collected in 6 - 8-min fractions (about 1 ml) in an automatic fraction collector from the output tube of the superfusion line.

The concentration of the amino acids present in the superfusate was expressed as p-mole/min/cm^2 of the cortical surface exposed. The statistical significance of the difference between stimulated and control in the same animals was evaluated using "Paired t-test" (36).

In some of the experiments reported, analysis conditions (i.e., the elution program) were such that the very small aspartate peak was fused with the methionine sulphoxide peak. Aspartate data for these experiments are therefore omitted. However, aspartate was clearly separable in many analyses, and the data are therefore included where appropriate (e.g., Table III). Glutamine and serine or threonine were also sometimes eluted in a single peak. Threonine was relatively constant at about one-third of the total peak size. In some experiments the glutamine content was measured by difference following its acid hydrolysis to pyrrolidone carboxylic acid which was not detectable in the analysis system (14). This is indicated in the text. GABA was not completely resolved from γ-acetylenic GABA in our analysis system and the data for GABA therefore include an uncertain proportion of γ-acetylenic GABA.

RESULTS

Brachial Plexus Stimulation and Amino Acid Release from Sensori-motor Cortex

Response to Stimulation. When square-wave pulses of 2 - 8 mA and 2 - 3 Hz were applied to the brachial plexus of anesthetized animals, the principal observable response was a clonic jerking of the stimulated limb without visible movement of other body musculature.

Contra-lateral Stimulation. Electrical stimulation of the contra-lateral forelimb of anesthetized animals increased significantly the release of glutamate and glutamine to the superfusion fluid (Table I). These responses consistently followed the first stimulus and were readily detectable when the pulses were around 7.5 mA and at a frequency of 2 - 3 Hz. No significant increase in the rate of release of other amino acids was observed. The increased rates of glutamate and glutamine release were reversible, reducing to control levels immediately after cessation of stimulation. Many amino acids were already being effluxed to superfusion fluid washing control cortex before brachial plexus stimulation was applied. How-

Table I - Amino Acid Release from Sensori-motor Cortex During Contra-lateral Stimulation

Amino acid	No. Exp.	Release rate (p-mole/cm^2/min)			Change (%)
		Control	Stimulated	Difference between control and stimulated release	
Glutamine	(18)	204 ± 27	240 ± 30	36 ± 6**	18
Glutamate	(27)	48 ± 9	61 ± 11	12 ± 6*	27
Glycine	(19)	156 ± 22	166 ± 21	10 ± 7	7
Alanine	(25)	119 ± 14	130 ± 15	8 ± 5	7
Valine	(18)	46 ± 8	48 ± 8	2 ± 3	5
Threonine	(17)	190 ± 24	203 ± 27	12 ± 11	7
Isoleucine	(22)	64 ± 12	69 ± 12	4 ± 2	7
Leucine	(20)	100 ± 20	103 ± 19	3 ± 2	3
Phenylalanine	(19)	32 ± 4	36 ± 4	3 ± 3	12
Tyrosine	(17)	55 ± 10	58 ± 10	2 ± 6	5

Release of amino acids from sensori-motor cortex was determined in 8-minute fractions (0.8 ml) of superfusate collected immediately prior to (control) or during contra-lateral stimulation of the brachial plexus. Values are means ± S.E.M. for the numbers of experiments indicated. Differences are expressed as means ± S.E.M. for the number of pairs shown, each stimulus being paired with its own control.

* $p < 0.05$ Paired comparison *t*-test; ** $p < 0.001$ Paired comparison *t*-test.

ever, GABA was never detected in the superfusion fluid, either before or after stimulation.

Ipsi-lateral Stimulation. When electrical stimuli of similar strength were applied to the brachial plexus of the ipsi-lateral forelimb, no significant change was observed in the rate of release of glutamate, glutamine or any other amino acid (Table II).

Table II - Amino Acid Release from Sensori-motor Cortex During Ipsi-lateral Stimulation

Amino acid	No. Exp.	Release rate (p-mole/cm^2/min)		
		Control	Stimulated	Difference between control and stimulated release
Glutamine	(14)	176 ± 26	177 ± 27	0.98
Glutamate	(13)	34 ± 8	30 ± 6	-3.9
Glycine	(11)	117 ± 18	121 ± 20	4.2
Alanine	(11)	94 ± 12	94 ± 13	0.0
Valine	(12)	54 ± 10	53 ± 12	-0.5
Isoleucine	(8)	37 ± 7	35 ± 6	-2.6
Leucine	(8)	50 ± 9	52 ± 12	1.6

Experimental details are given in the legend to Table I. Values represent means ± S.E.M. for the numbers of experiments indicated.

When stimulation was applied to brachial plexus of unanesthetized animals, clonic and vibrational movements of the stimulated limbs occurred as for the anesthetized animals, but these were accompanied by several other whole body responses including movement of trunk musculature, running movements within the cage, and self-grooming movements employing the unstimulated limb. The putative transmitter amino acids displayed increased rates of release to the superfusion fluid. However, these increases were not specific, and were accompanied by parallel increases in efflux of most of the other amino acids measured (Table III).

Photic Stimulation and Amino Acid Release from Visual Cortex. With a cannula positioned over the primary visual area of the occipital cortex animals were left in complete darkness to dark-adapt for 18 hours. They were then anesthetized and given γ-acetylenic GABA before experimentation began. Weak light-flash stimuli were

Table III - Effect of Afferent Electrical Stimulation on Release of Amino Acids from Sensori-motor Cortex of Unanesthetized Animals

Amino acid	No. Exp.	Release rate (p-mole/cm^2/min) Control	Difference between control and stimulated release	Change (%)
Aspartate	(18)	37 ± 6	17 ± 5	46
Threonine	(14)	144 ± 12	130 ± 25	90
Glutamine	(15)	430 ± 57	419 ± 83	97
Glutamate	(17)	101 ± 13	69 ± 17	68
Glycine	(18)	341 ± 35	153 ± 21	45
Alanine	(7)	194 ± 30	123 ± 32	63
Valine	(18)	106 ± 11	69 ± 19	65
Isoleucine	(15)	47 ± 4	34 ± 11	72
Leucine	(15)	75 ± 6	62 ± 19	83
Tyrosine	(11)	23 ± 3	26 ± 6	113
Phenylalanine	(11)	17 ± 2	18 ± 5	106

Release of amino acids from sensori-motor cortex of 3 unanesthetized animals was determined in 8-minute fractions of superfusate collected immediately prior to, or during contra-lateral afferent stimulation. All the changes in amino acids were significant ($p < 0.05$).

then applied to both eyes for 6-minute periods. Equivalent periods (6 min) of photic stimulation and dark adaptation then followed for a total of about an hour. Light-flashing applied for 6-minute periods caused no measurable change in glutamine, serine, alanine, glycine or leucine efflux from the visual cortex. There was a visible and significant increase (30%; $p < 0.001$) in the level of glutamate measured in the superfusion fluid (Figs. 1 and 2). The same results were obtained using either artificial CSF or normal saline as perfusate. In 20 light-flashing stimulations with associated controls,

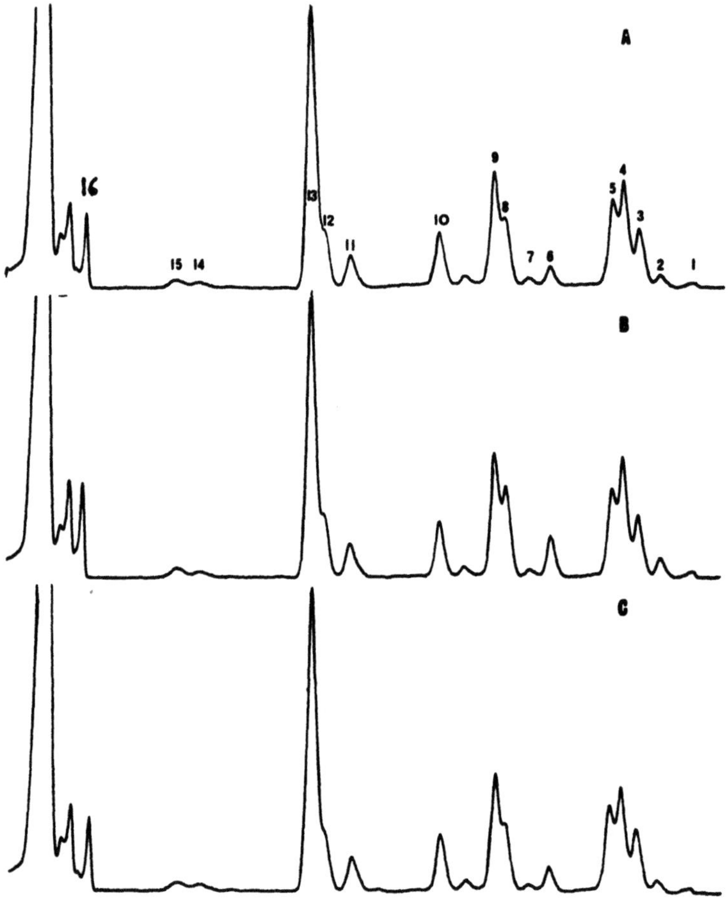

Fig. 1. Typical chromatograms of superfusion fluid from visual cortex showing effects of photic stimulation. Fractions of superfusate (8 min, 1 ml) were collected from the visual cortex of an animal treated with γ-acetylenic-GABA (100 mg/kg, i.p.) 49 minutes before photic stimulation. A, Before stimulation; B, during stimulation; C, after stimulation. Amino acids are as follows: 1, methionine sulphoxide; 2, aspartate; 3, threonine; 4, serine; 5, glutamine; 6, glutamate; 7, citrulline; 8, glycine; 9, α-alanine, 10, valine; 11, isoleucine; 12, norleucine added standard of 100 p-mole; 14, tyrosine; 15, phenylalanine; 16, GABA plus γ-acetylenic GABA.

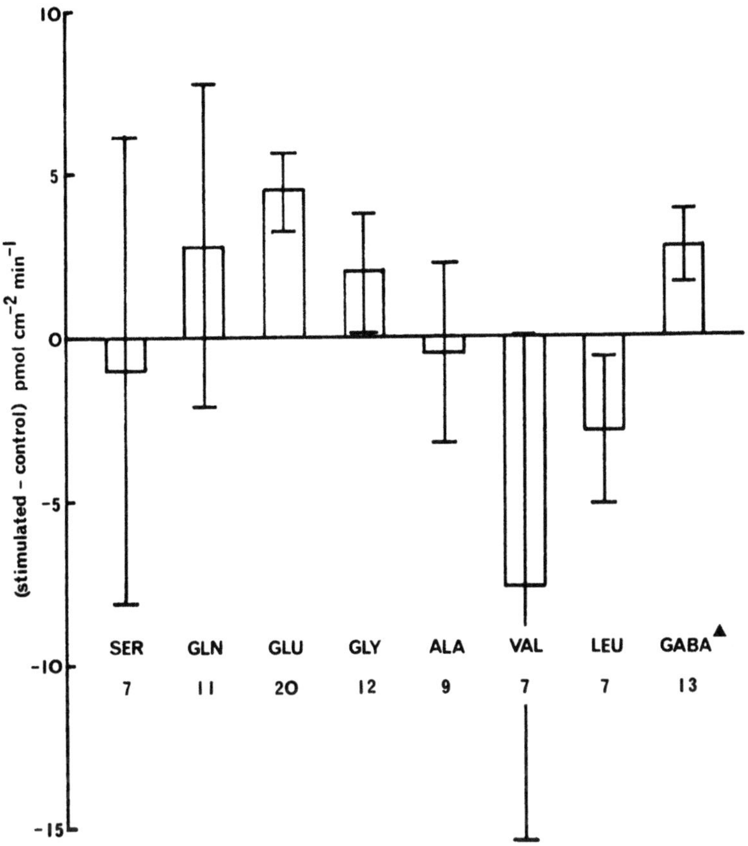

Fig. 2. Effect of photic stimulation on release of amino acids from visual cortex following administration of γ-acetylenic-GABA. Six-minute periods of light stimulation (2 Hz, 1.2 W) were given after injection of γ-acetylenic GABA (100 mg/kg, i.p.). Results shown are the means of the differences between release of amino acids during stimulation and release during an equivalent period of dark preceding stimulation expressed as mean ± S.E.M. of the numbers of pairs shown below abbreviations. Only glutamate release was significantly increased by photic stimulation (30%, $p < 0.001$). The average level of glutamate was 14.9 ± 1.2 (20) p-mole/min/cm^2 during the pre-stimulus period. SER = serine; GLN = glutamine; GLY = glycine; ALA = α alanine; VAL = valine; LEU = leucine; GABA▲ = GABA plus γ-acetylenic GABA.

performed in 5 animals, 7 stimulations evoked a glutamate increase of more than 40%, 7 evoked a change between 15 and 40%, and 6 evoked small changes between ± 5%. The only other peak showing any changes due to light stimulation contained a mixture of GABA and γ-acetylenic GABA in uncertain proportions. During 13 stimulations, 3 responses in this peak were more than 30%, 6 were between 8 and 30% and 4 produced either no change or a small decrease (< 30%).

DISCUSSION

We have previously reported (15) the finding that superfusion under these conditions does not result in tissue damage or a change in the neuronal population in the tissue below the cannula which receives the stream of superfusion fluid. It can be assumed, therefore, that the superfused tissue is normal and retains its responsiveness. Also, the slight positive pressure of the superfusion fluid allows a small net fluid entry across the brain surface and ensures collection of transmitters only from the local regions being superfused (16).

Brachial Plexus Stimulation

As stimulation of sensori-motor cortex by this route in anesthetized animals caused consistent (first stimulus) increases in glutamine and glutamate release when applied to the contra-lateral but not the ipsi-lateral forelimb, it must be concluded that interhemispheric conduction of the activation is minimal under these conditions. Alternatively, the intercallosal fibers may not be using these substances as transmitters, in which case activation could occur without detection in these experiments. At the same time, the absence of any glutamine or glutamate release from this cortical area or from superfused visual cortex following ipsi-lateral brachial plexus stimulation shows that the increased efflux of these compounds is not due to systemic changes caused by responses to this stimulation.

Awake versus Anesthetized Animals

The non-specific release of all amino acids from the cortex of awake rats during contra-lateral brachial plexus stimulation suggests that anesthesia is essential for a successful study, presumably because the amino acid transmitter signals are measured against a much reduced noise level. The latter may be due to the neural activity, or to unidentified systemic effects associated with the increased and wide-ranging muscular or bodily movements observed in this experimental situation. However, the significance of this wide spectrum of release remains to be established.

Comparable Studies

Glutamate and glutamine release to cerebral venous blood during brachial plexus stimulation has been reported by others (1). Jasper and Koyama (20) used an equivalent system in anesthetized cats but employed open Mitchell (25) cups on the cortical surface, a *cerveau isolé* preparation, and cortical stimulation via the mesencephalic reticular formation, and glutamate, aspartate and taurine release occurred during its stimulation. Several studies have been made *in vivo* on amino acid transmitter release to cerebral ventricles. Thus, Obata and Takeda observed GABA release to the fourth ventricle (27) of cats, following cerebellar stimulation. Glutamate release from the surface of hippocampus of cats was observed by Crawford and Connor (9) to follow stimulation of its entorhinal input. Electrical stimulation *in vivo* of the dorsal columns of rats was shown to cause activation of the gracile and cuneate nuclei, and, in parallel, a Ca^{2+}-dependent release of pre-loaded glutamate and GABA (30).

Glutamine Release

It is an attractive possibility, consistent with previous suggestions on the relationship between glutamate and glutamine (2,6), that the glutamine release could result from the conversion of released glutamate to glutamine in glial cells, subsequent to its uptake, followed by the passive release of this glutamine to the extracellular space where it would be collected by superfusion fluid. We are now extensively exploring this possibility. Previous workers (1), using brachial plexus stimulation detected an increased efflux of both glutamine and glutamate (GABA not detected) to cerebral venous blood which was correlated with a reduction in the concentrations of glutamine in the cerebral tissue (12).

Release from Visual Cortex

The finding that glutamate specifically showed an increased rate of efflux from visual cortex to the superfusion fluid in response to retinal stimulation, but not to brachial plexus stimulation, suggests a transmitter function for this amino acid in this cortical region, linked to the input from the visual system. Whether this would be at the synapses of the primary input of the visual radiation, i.e. layer IV (18,34) or at subsequent or other synaptic interconnections in the vertically organized columns of neurones remains uncertain.

Unequivocal conclusions about GABA release from visual cortex cannot be made due to the presence of γ-acetylenic GABA, but the significant increase in release of the mixture suggests either a separate GABA release or a combined release, with γ-acetylenic GABA acting as a false transmitter. Jung *et al.* (21) calculated that at the dose used, γ-acetylenic GABA would be present at less than 10% of total brain GABA, though any GABA compartmentation would restrict the usefulness of this calculation.

Iversen *et al.* (19) observed increased efflux of pre-loaded [^{3}H]GABA from rabbit visual cortex in response both to stimuli applied to the visual cortex itself or to the lateral geniculate nucleus. No glutamate release was reported. Reubi and Cuenod (29) observed *in vivo* release of pre-loaded radio-labelled glycine and

GABA from pigeon optic tectum during stimulation of a mid-brain nucleus shown to project to the tectum. There was also a late (6 min) but correlated release of glutamate. There have been several studies of the pharmacology of visual cortical neurones and these have concluded that there is unlikely to be a cholinergic (32) link in the transmission of specific impulses to the neurones of the primary visual cortex but rather an involvement of GABA and glutamate as key neurotransmitters (35). This conclusion is supported by the data reported here and by that of Iversen *et al.* (19).

Origin of Released Compounds

Whether the increased rates of efflux of amino acids evoked by neural activation originate in neurones or glial cells is not easily answered. Thus, neuronal activity and concomitant K^+ release can lead to glial depolarization, and the latter in turn to transmitter release from pre-loaded (24) glial cells. Dependence on Ca^{2+} has been employed *in vitro* to distinguish neuronal and glial amino acid release (3,17,31) in response to high-K^+. Iversen *et al.* (19) and Clark and Collins (7) were able to demonstrate Ca^{2+}-dependent *in vivo* GABA release from visual cortex stimulated directly at its surface by electrodes, or by interhemispheric stimulation. Calcium-dependence is clearly an important criterion for the likely physiological relevance of release signals and we are examining this in our preparation. However, it is not unequivocal in this respect, since damped neural activity due to low Ca^{2+} concentrations would anyway lead to reduced glial stimulation.

SUMMARY

The effects of sensory stimulation on the release of amino acids from sensori-motor and visual cortices have been studied using a superfusion technique. Electrical stimulation of the brachial plexus contra-lateral to the superfusion cannula increased significantly the release of glutamate and glutamine from the sensori-motor cortex of anesthetized rats. No clear effect was observed with the other amino acids. Stimulation of the ipsi-lateral plexus had no effect on glutamate and glutamine release. In unanesthetized animals, stimulation of the contra-lateral brachial plexus raised the levels of all the amino acids in sensori-motor cortex superfusate. Weak photic stimulation of the eyes of dark-adapted rats increased glutamate release from the visual cortex but caused no significant change in the release of other amino acids. All evoked increases in amino acid release were reversible at the cessation of the stimuli.

ACKNOWLEDGEMENTS

A.-S. Abdul-Ghani was supported by an EMBO Fellowship and D. W. G. Cox by a Wellcome Grant. This work was also supported by an MRC Programme Grant. We are extremely indebted to Dr. J. Wilkins and his colleagues of Centre de Recherche Merrell International, Strasbourg (France), for supplying us with γ-acetylenic GABA.

REFERENCES

(1) Abdul-Ghani, A.-S., Marton, M. and Dobkin, J. (1978) J. Neurochem. 31, 541-546.
(2) Benjamin, A. M. and Quastel, J. H. (1974) J. Neurochem. 23, 457-464.
(3) Bowery, N. G. and Brown, D. A. (1972) Brit. J. Pharmacol. 45, 160-161.
(4) Bradford, H. F. and Richards, C. D. (1976) Brain Res. 105, 168-172.
(5) Bradford, H. F. and Thomas, A. J. (1969) J. Neurochem. 16, 1459-1504.
(6) Bradford, H. F., Ward, H. K. and Thomas, A. J. (1978) J. Neurochem. 30, 1453-1459.
(7) Clark, R. M. and Collins, G. G. S. (1976) J. Physiol. (Lond.) 262, 383-400.
(8) Cotman, C. W. and Hamberger, A. (1978) In: Amino Acids as Chemical Transmitters (F. Fonnum, ed.) pp. 379-412, Plenum Press, New York.
(9) Crawford, I. L. and Connor, J. D. (1973) Nature (Lond.) 244, 442-443.

(10) Curtis, D. R. and Johnston, G. A. R. (1974) Ergebn. Physiol. 69, 98-188.
(11) De Belleroche, J. S. and Bradford, H. F. (1972) J. Neurochem. 19, 585-602.
(12) Dobkin, J. (1970) J. Neurochem. 17, 237-246.
(13) Dodd, P. R. and Bradford, H. F. (1974) J. Neurochem. 23, 289-292.
(14) Dodd, P. R. and Bradford, H. F. (1976) Brain Res. 111, 377-388.
(15) Dodd, P. R. and Bradford, H. F. (1976) Biochem. Soc. Trans. 4, 13-19.
(16) Dodd, P. R., Pritchard, M. J., Adams, R. C. F., Bradford, H. F., Hicks, G. and Blanshard, K. C. (1974) J. Scient. Instr. E7, 897-901.
(17) Hamberger, A., Hanson, H. A. and Sellström, Å. (1975) Exp. Cell Res. 92, 1-10.
(18) Hubel, D. H. and Wiesel, T. N. (1972) J. Comp. Neurol. 146, 421-450.
(19) Iversen, L. L., Mitchell, J. F. and Srinivasan, V. (1971) J. Physiol. (Lond.) 212, 519-534.
(20) Jasper, H. H. and Koyama, I. (1969) Can. J. Physiol. Pharmacol. 47, 889-905.
(21) Jung, M. J., Lippert, B., Metcalf, B. W., Schechter, P. J., Böhlen, P. and Sjoerdsma, A. (1977) J. Neurochem. 28, 717-723.
(22) Krnjević, K. (1974) Physiol. Rev. 54, 418-540.
(23) Metcalf, B. W. and Casara, P. (1975) Tetrahedron Lett. 38, 3337-3340.
(24) Minchin, M. C. W. and Beart, P. M. (1975) Brain Res. 83, 437-449.
(25) Mitchell, J. F. (1963) J. Physiol. (Lond.) 165, 98-116.
(26) Neal, M. J. (1971) J. Physiol. (Lond.) 215, 103-117.
(27) Obata, K. and Takeda, K. (1969) J. Neurochem. 16, 1043-1047.
(28) Pasantes-Morales, H., Klethi, J., Urban, P. F. and Mandel, P. (1974) Exp. Brain Res. 19, 131-141.
(29) Reubi, J. C. and Cuenod, M. (1976) Brain Res. 112, 347-361.
(30) Roberts, P. J. (1974) Brain Res. 67, 419-428.
(31) Sellström, Å. and Hamberger, A. (1977) Brain Res. 119, 189-198.
(32) Spehlman, K., Daniels, J. C. and Smathers, C. C. (1971) Brain Res. 94, 125-138.
(33) Srinivasan, V., Neal, M. J. and Mitchell, J. F. (1969) J. Neurochem. 16, 1235-1244.
(34) Toyama, K., Matsunami, K., Ohno, T. and Tokashiki, Z. (1974) Exp. Brain Res. 21, 45-66.
(35) Wallingford, E., Ostdahl, R., Zarzecki, P., Kaufman, P. and Somjen, G. (1973) Nature New Biology (Lond.) 242, 210-212.
(36) Winer, B. J. (1962) Statistical Principles in Experimental Design. McGraw-Hill Book Co., Inc., London, p. 377.

SECTION III

GABA-RECEPTOR

PRE- AND POST-SYNAPTIC INHIBITION

K. Krnjević

Departments of Physiology and Anesthesia Research,
McGill University, Montreal, Quebec, Canada

INTRODUCTION

Inhibition is now a well-accepted mechanism of control in the central nervous system (CNS), and, as the present symposium attests, in one way or another is keeping busy a very large number of scientists. But this has not always been the case.

Inhibition was recognized as an important aspect of brain function over a hundred years ago (1). Its significance for cortical activity was much emphasized, especially by Russian authors (2,3); and of course, Sherrington (4) demonstrated its critical role in spinal reflex action. However, it was practically ignored by some very influential scientists (such as Ramón y Cajal, ref. 5), and many western neurophysiologists, who by and large were ready to believe that inhibition might be significant for reflex action in the spinal cord, nevertheless felt that brain function could be accounted for without invoking any inhibitory processes at all (6,7). However, the rapid development of intracellular recording after the Second World War forced investigators to recognize that inhibitory phenomena (especially IPSPs) are probably the most common and widespread manifesta-

---

Abbreviations: GABA, γ-aminobutyric acid; GAD, glutamate decarboxylase; IPSP and EPSP, inhibitory and excitatory post-synaptic potential; PAD, primary afferent depolarization; NA, noradrenaline; 5-HT, 5-hydroxytryptamine.

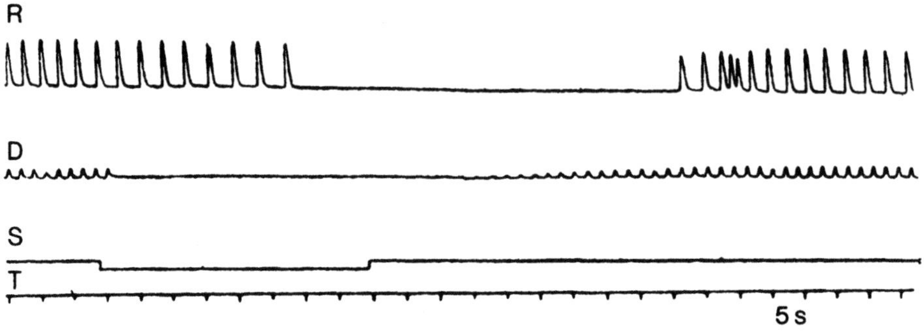

Fig. 1. Chemical transmission of inhibition in the heart demonstrated by modified Loewi method. Two frog hearts are perfused in series. Beating of first "donor" heart is recorded on middle trace (D), and of second ("recipient") heart on upper trace (R). Stimulation of vagal supply to donor heart for 40 sec (signal on third trace, S) causes stoppage of donor heart, followed after about 15 sec by first slowing, and then stoppage of recipient heart (from Bain, 58).

tion of neural activity (8-12), and that inhibition plays not just a significant, but perhaps the dominant role in the elaboration and control of central neural activity and its realization in behavior (9,13,14).

How does inhibition operate? It is now widely agreed that, like other forms of synaptic transmission, inhibition is predominantly if not exclusively mediated by chemical transmitters. This idea, though supported by the demonstration that vagal inhibition of the heart was mediated by a humoral agent (Loewi, ref. 15; cf. Fig. 1), was long resisted by most neurophysiologists (16,17).

The best-known mechanisms of inhibition in the vertebrate CNS are those mediated by the liberation of GABA. They may act both pre- and post-synaptically. Since we know more about post-synaptic inhibition, this will be discussed first.

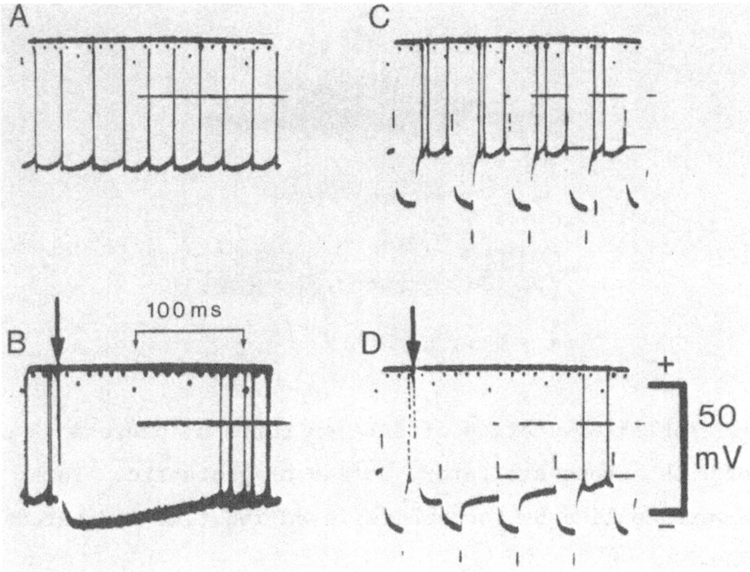

Fig. 2. Inhibition in cerebral cortex of cat. Intracellular records show A, spontaneous firing of cortical neurone; B, prolonged hyperpolarizing IPSP associated with cessation of firing (several traces are superimposed), evoked by brief stimulus applied at time indicated by arrow; C, measurement of cell resistance with 5 equal pulses of hyperpolarizing current; and D, diminished hyperpolarizing responses, evoked by identical current pulses during IPSP, indicating reduced membrane resistance. Current pulses are monitored in C and D on intermediate trace displaced to right (from Dreifuss et al., 59).

POST-SYNAPTIC INHIBITION

Essentially, this can be defined as a process which reduces excitability, independently of previous activity. (It has to be distinguished from various forms of post-activity depression.) There may be only rather minimal changes in membrane potential: in contrast to excitation, where the EPSP is usually a well-defined

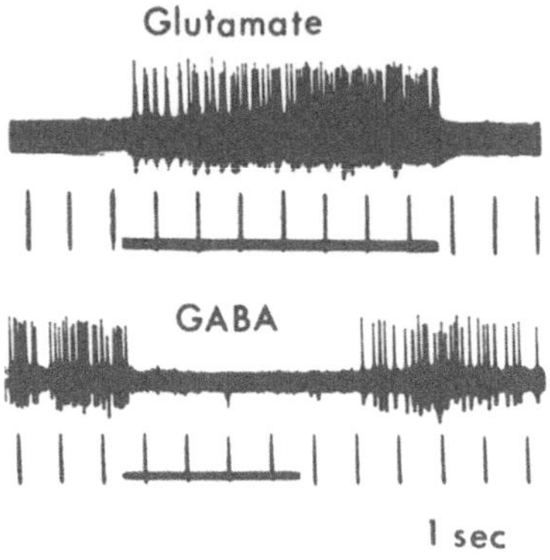

Fig. 3. Inhibitory action of GABA on cortical neurone is contrasted with strong excitatory action of glutamate. Both agents were applied by iontophoresis during time indicated by signal.

depolarizing potential, the IPSP may be quite inconspicuous. As a rule, however, some hyperpolarization is visible (Figs. 2 and 4). Thus, in Figure 2 the IPSP recorded from a cortical neurone is seen as a prolonged hyperpolarization lasting >100 m-sec (typically much longer than in the spinal cord); it is accompanied by a reduction in excitability, manifested by a cessation of ongoing firing. The most striking other feature of inhibition is that it is usually associated with a large fall in cell membrane resistance — note the smaller voltage-deflections evoked by constant current pulses in Figure 2D — which is caused by a large increase in Cl^- permeability.

The strongest reason for believing that GABA is an important inhibitory agent in regions such as the cerebral cortex and the cerebellum is that it is the only substance known to be present there (18) with a strong inhibitory action which appears to be identical to synaptic inhibition (19). Not only does GABA strongly inhibit cell firing (Fig. 3), but, like synaptic inhibition, it produces a fall in resistance and hyperpolarization (Fig. 4), with a reversal potential that is practically identical to that of the IPSP.

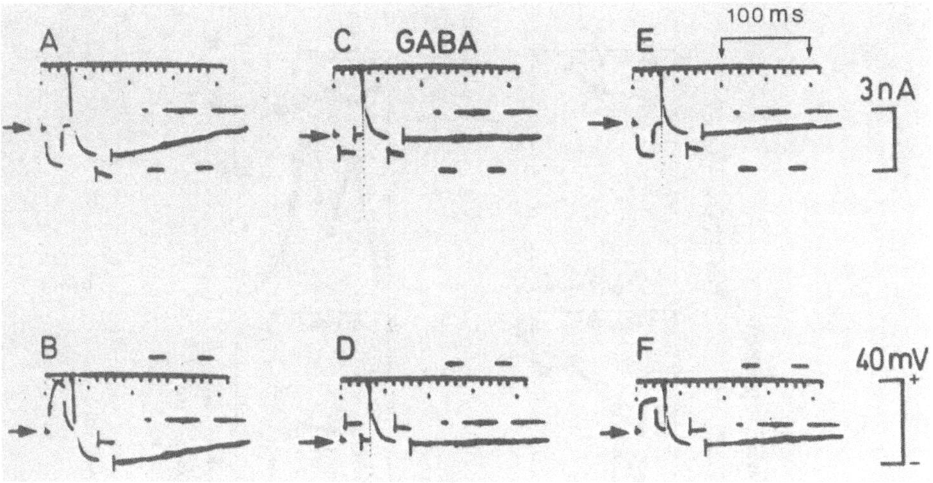

Fig. 4. GABA causes a hyperpolarization and fall in membrane resistance just like synaptic inhibition. Each intracellular record from neurone in cerebral cortex of cat displays resting potential (arrows) and IPSP, as well as tests of membrane resistance with 20 m-sec pulses of current, one at rest and one near negative peak of IPSP (in A, C and E, current pulses are in negative direction, and in B, D and F, in positive direction; they are monitored on second set of traces displaced to right). A,B: initial controls; C,D: during application of GABA; E,F: after end of GABA release (from Dreifuss et al., 59). Note downward (negative) shift of resting potential and lowered resistance during release of GABA.

Of course, there is now a great deal of further supporting evidence; for example, that GABA is released by nerve stimulation (20, 21), that the enzyme needed for its production from glutamate (GAD) is present in inhibitory nerve endings (22) and that GABA is actively removed from synaptic regions, by Na^+-dependent uptake into nerve endings and neuroglia (23,24).

In other regions, especially in the spinal cord, glycine is also a powerful inhibitor (25). Its action appears to be identical to

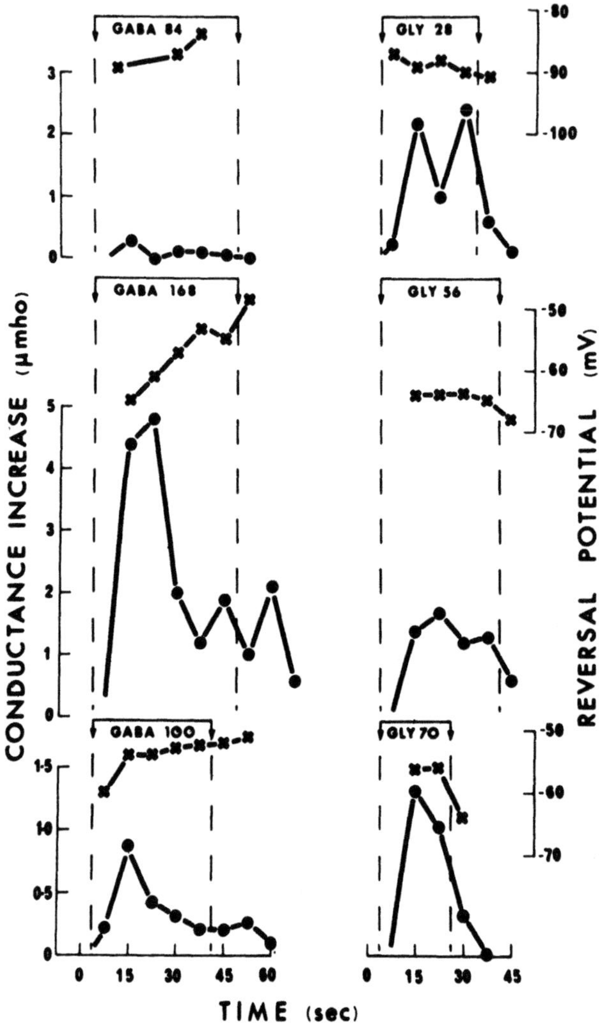

Fig. 5. Conductance increases evoked by GABA or glycine (full circles) in cat motoneurones and reversal level of the associated potential change (crosses) observed at various times during 30-45 sec applications of inhibitory amino acid (iontophoretic currents are indicated in nA). Data show consistent tendency for conductance increase evoked by GABA to fall-off after initial peak, while associated reversal potential becomes more positive. Neither effect is evident during applications of glycine (from Krnjević et al., 29).

that of GABA, since it also activates Cl^- permeability. But it differs in three respects: its time course of action is briefer (26,27); it activates different receptors, which are blocked by strychnine, whereas GABA-receptors are more likely to be blocked by picrotoxin or bicuculline (28), its action shows little or no evidence of any "desensitization," whereas that of GABA does (29; see Fig. 5).

In spinal motoneurones there is some indication that this "desensitization" to GABA is accompanied by a positive shift in reversal potential for the effect of GABA. It has been postulated that both changes could be explained by rapid uptake, which would lower the extracellular concentration of GABA, and, if the uptake can also occur into the post-synaptic cells (cf. ref. 30), it could be linked to an appreciable Na^+ influx (as in the Schultz and Curran (31) model for the uptake of sugars and neutral amino acids). Such an electrogenic (depolarizing) influence of GABA uptake could thus account not only for the apparent "desensitization" but also for some delayed depolarizing effects of GABA (including the well-known post-inhibitory rebound of excitability). Since the uptake mechanism is in principle reversible, a high intra-neuronal level of GABA could have a hyperpolarizing influence by causing an extrusion of Na^+ driven by the steep GABA outward gradient. Some evidence of this has been seen in very recent experiments (32) in which intra-motoneuronal injections of GABA were found to lower excitability and cause hyperpolarization without increasing the membrane conductance.

The reduction of excitability and the depression of EPSPs by post-synaptic inhibition are directly caused by the increased permeability to Cl^-, because any depolarizing tendency due to increased inward current is automatically compensated by an accompanying influx of Cl^-; moreover, the space constant (or electrotonic distance) is greatly reduced by the fall in membrane resistance, and so EPSPs generated at some distance from the site of spike initiation have a lesser chance of evoking a supra-threshold depolarization at this critical site. Furthermore, since Cl^- is not distributed passively across the neuronal membrane, being in fact actively transported out of the cells, its equilibrium potential is at least 10 mV more negative than the resting potential; hence, inhibition is accompanied by an actual hyperpolarization. Although this is not the most important aspect of inhibition, it does contribute significantly to its effectiveness, as shown by the convulsant actions of agents which block Cl^- transport (such as Cu^{2+} (ref. 33), and salts of NH_4^+ (ref. 34)). This phenomenon at least partly accounts for the convulsive aspects of ammonia intoxication (35).

PRESYNAPTIC INHIBITION

This is a more controversial topic. Frank and Fuortes (36) first described a prolonged depression of spinal synaptic transmission by some afferent inputs, which differed from typical post-synaptic inhibition by the absence of an IPSP or change in excitability (Fig. 6). Two possible kinds of mechanism suggested themselves: an electrotonically distant post-synaptic inhibition — at some remote, possibly dendritic site — or a presynaptic inhibition, leading to a diminished release of transmitter from afferent nerve endings.

Eccles and his colleagues subsequently investigated this kind of inhibition in great detail (37,38). The main feature emphasized by Eccles was that the inhibition is correlated with a slow depolarization (the dorsal root potential or primary afferent depolarization (PAD)) and an increase in excitability of afferent terminals. He postulated that afferent terminals are depolarized by presynaptic axo-axonal synapses (cf. refs. 39,40); this causes a reduction in the amplitude of the action potentials in afferent terminals and thus reduces their ability to release the synaptic transmitter. An interesting point is that this kind of inhibition is seen most clearly at the terminals of primary afferent fibers and is much less evident at more central synapses. Presynaptic inhibition of afferent input is elicited both by afferent activity and by activity in descending pathways of supraspinal origin (including the cerebral cortex). The

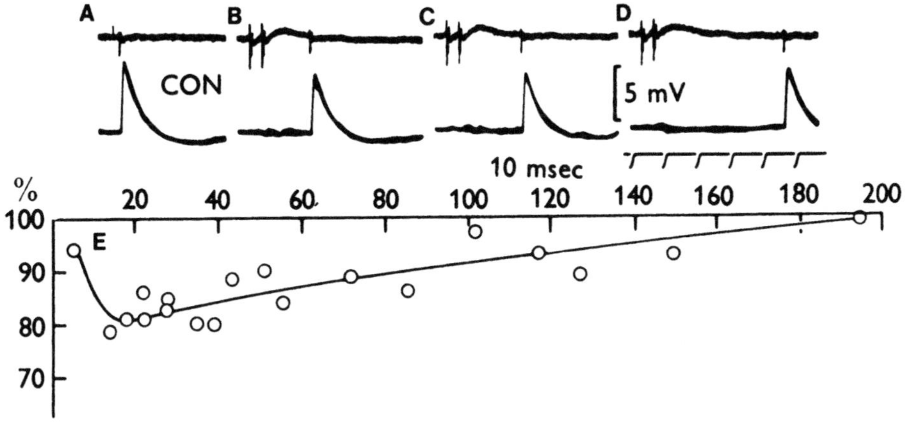

Fig. 6. Presynaptic inhibition of monosynaptic EPSP in an extensor (plantaris) motoneurone of cat. A) control record: upper trace monitors afferent volley in dorsal root near point of entry into spinal cord; lower trace shows EPSP evoked by maximum Ia stimulus to plantaris nerve. B-D) when plantaris nerve stimulus is preceded by two shocks (of maximum group I intensity) applied to afferent fibers from a flexor muscle (posterior biceps-semitendinosis) the EPSP is significantly reduced. Time course of inhibition is plotted in graph E. Note absence of IPSP evoked by conditioning stimulation (from Eccles et al., 60).

whole system, therefore, could well be a very significant part of the sensory gate control proposed by Melzack and Wall (41) and Wall (42).

Some controversial aspects of presynaptic inhibition need to be mentioned. Thus it is by no means certain that all "remote" inhibitions are presynaptic. According to Kellerth (43), many of the "presynaptic" inhibitions can be shown to be associated with a clear reduction in post-synaptic excitability. Eccles had earlier found that presynaptic inhibition (unlike the better known, brief IPSPs) are removed by picrotoxin and he had assumed that all picrotoxin-sensitive, prolonged inhibitory effects must be of a presynaptic nature. It is now clear that, as suggested by Kellerth (43), a number of such inhibitions are post-synaptic and probably mediated by GABA (44).

Another source of disagreement has been the significance of depolarization by extracellular accumulation of K^+. Barron and Matthews (45) had pointed out that the close proximity of large numbers of unmyelinated terminals would necessarily lead to significant accumulation of K^+ (or other depolarizing agents) released from active fibers and thus could well explain the slow dorsal root potentials. That significant accumulations of K^+ take place near active afferent terminals has now been conclusively demonstrated with K^+-selective microelectrodes (46,47); however, it seems that the increased extracellular K^+ contributes mainly to the later portion of the PAD (48), and that a GABA-mediated action is primarily responsible for the early component that lasts approximately 100 m-sec (49).

Neurochemical studies have revealed that a large amount of GABA (18) and a high GAD activity (50) exist in the dorsal horns, especially in the region of the substantia gelatinosa where the presynaptic inhibitory interneurones seem to be concentrated. GABA has consistently proved to have a depolarizing action on dorsal root fibers and their cell bodies (Fig. 7; see also refs. 51 and 52). This might seem paradoxical in view of the hyperpolarizing action of GABA on central neurones and to indicate a quite different mechanism of action. In fact, in both cases there is a marked increase in Cl^- permeability, but since afferent neurones have an excess of Cl^- in their cytoplasm (instead of a deficiency), the result is an efflux of Cl^- and therefore depolarization.

Other evidence that GABA is a major transmitter for presynaptic inhibition (49) is its suppression by GABA-antagonists and by drugs that deplete GABA (such as semicarbazide) and its enhancement by barbiturates and benzodiazepine tranquilizers, which appear to increase the responsiveness of GABA-receptors (e.g., 53). However,

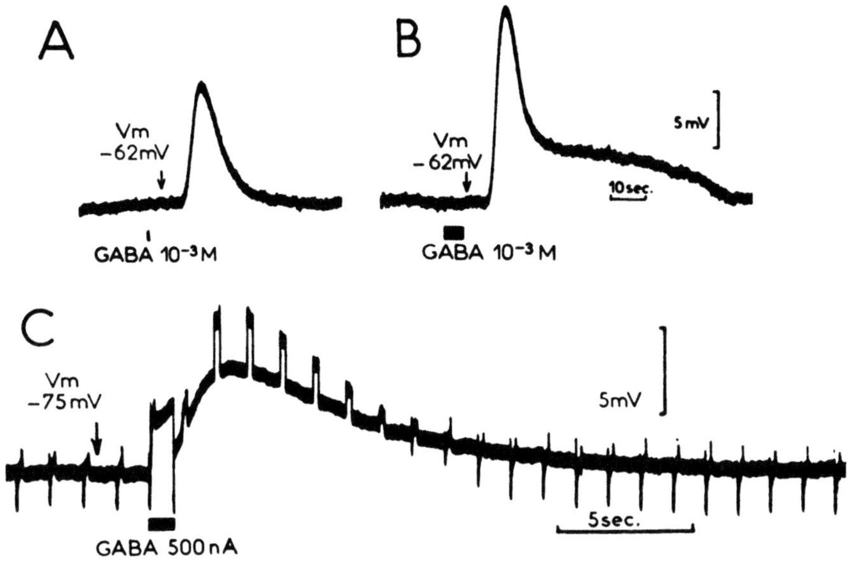

Fig. 7. GABA causes a depolarization of rat dorsal root ganglion cells. A,B: depolarizations induced by superfusion with GABA solution, for 1 and 5 sec, respectively. C: Another cell; depolarizing effect of GABA is associated with fall in membrane resistance. This was measured with current pulses applied at regular intervals; initially the resulting voltage deflections were balanced out by a bridge circuit, but during GABA-induced depolarization pulses appear, thus demonstrating a fall in resistance (in C, GABA was applied by iontophoresis) (from Deschênes et al., 52).

some presynaptic inhibitory actions, especially those evoked from supraspinal centers, may be mediated by other agents, notably 5-HT, NA, and perhaps opioids (49).

A very recent discovery has thrown quite a new light on the presynaptic mechanism of action of GABA (and some other agents). According to Eccles' interpretation (37) the amount of transmitter released by afferent terminals would be a very steep function of the amplitude of the action potential; therefore even a relatively

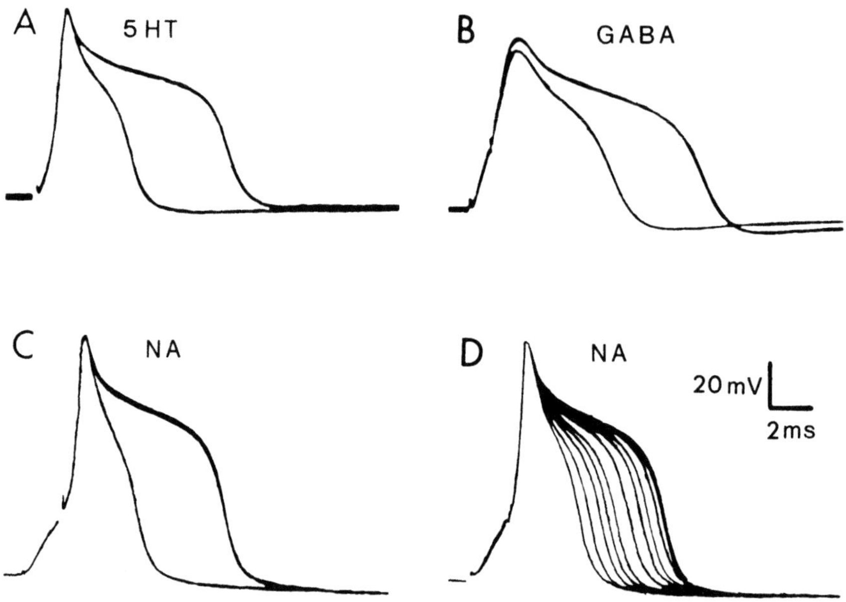

Fig. 8. Putative chemical transmitters of presynaptic inhibition selectively block Ca^{2+} inward current in primary afferent neurones. Action potentials of dorsal root ganglion cells in dissociated cultures normally show a plateau which is due to Ca^{2+} inward current. In each record one can see (superimposed) a control trace of spike (with plateau) and a briefer spike produced by application of: A) 10^{-5}M 5-HT; B) 10^{-4}M GABA; and C,D) 10^{-4}M NA; progressive recovery of Ca^{2+} component is shown in D, where successive traces were recorded at 10-sec intervals. Calibrations indicate 20 mV and 2 m-sec (from Dunlap and Fischbach, 56).

small PAD could have a pronounced inhibitory effect on transmission. In terms of the Ca^{2+} hypothesis of excitation-secretion coupling (54), this would require a corresponding steep relation between the amount of Ca^{2+} that enters the terminal and the spike amplitude. This is not what one would expect, unless the Ca^{2+}-permeability of the terminal membrane is readily inactivated by membrane depolarization. While there is some evidence that axonal depolarization can lead to Ca^{2+}-inactivation (55), according to Dunlap and Fischbach's (56) very recent observations on dorsal root ganglion cells in dissociated cultures – which show a very clear Ca^{2+} component to their action potential – this Ca^{2+}-inward current is rather insensitive to depolarization (Fig. 8). On the other hand, they found that quite low concentrations of GABA (as well as NA, 5-HT and enkephalin) suppress the Ca^{2+}-inward current, without causing any significant change in membrane potential or conductance. So they postulate a more direct mechanism of presynaptic inhibition, by a selective block of Ca^{2+} influx. The only possible reservation is that this action of GABA has so far proved to be insensitive to GABA-antagonists.

CONCLUSIONS

We now know a great deal about central inhibitory mechanisms: how they are organized in various neuronal circuits ("feed-forward" and "feed-back" inhibitions, inhibition of inhibitory cells giving "disinhibition" which releases neuronal activity in a finely graded and particularly safe manner); how they exercise a preponderant control over much of central neural activity (9,13,14,57); and of course, how inhibition operates at the cell membrane, by increasing Cl^- permeability; this has a stabilizing action post-synaptically, but presynaptically, in afferent fibers, it depresses transmitter release from afferent terminals. In addition, there is some evidence that GABA transport may be electrogenic and therefore may significantly modulate membrane excitability, and that GABA may selectively depress Ca^{2+} influx in afferent terminals and thus inhibit transmitter release particularly effectively. However, it is by no means certain that we are fully aware of all the possible ways in which GABA affects neuronal excitability and synaptic transmission, and there may well be further exciting surprises in store.

REFERENCES

(1) Sechenov, I. M. (1863) Physiologische Studien über die Hemmungsmechanismen für die Reflexthatigkeit des Rückenmarks im Gehirne des Frosches. Hirschwalk, Berlin.
(2) Bubnoff, W. and Heidenhain, R; (1881) Pflügers Arch. ges. Physiol. 26, 137-200.
(3) Pavlov, I. P. (1927) Conditioned Reflexes. University Press, Oxford.
(4) Sherrington, C. (1906) The Integrative Action of the Nervous System. Yale University Press, New Haven.
(5) Ramón y Cajal, S. R. (1909) Histologie du Système Nerveux de l'Homme et des Vertébrés. Maloine, Paris.
(6) Dusser de Barenne, J. G. and McCulloch, W. S. (1941) J. Neurophysiol. 4, 311-323.
(7) Bremer, F. and Stoupel, N. (1962) Arch. Ital. Biol. 100, 399-414.
(8) Eccles, J. C. (1957) The Physiology of Nerve Cells. The Johns Hopkins Press, Baltimore.
(9) Eccles, J. C. (1969) The Inhibitory Pathways of the Central Nervous System. Charles C. Thomas, Springfield.
(10) Phillips, C. G. (1961) In: The Nature of Sleep (G. E. W. Wolstenholme and M. O'Connor, eds.) pp. 4-24, Churchill, London.
(11) Krnjević, K., Randić, M. and Straughan, D. W. (1966) J. Physiol. (Lond.) 184, 16-48.
(12) von Euler, C., Skoglund, S., and Söderberg, U. (Eds.) (1968) Structure and Function of Inhibitory Neuronal Mechanisms (Proceedings of the 4th Int. Meeting of Neurobiologists, Stockholm, 1966) Pergamon Press, Oxford.
(13) Eccles, J. C., Ito, M. and Szentagothai, J. (1967) The Cerebellum as a Neuronal Machine. Springer Verlag, New York.
(14) Roberts, E. (1976) In: GABA in Nervous System Function (E. Roberts, T. N. Chase and D. B. Tower, eds.) pp. 515-539, Raven Press, New York.
(15) Loewi, O. (1921) Pflügers Arch. ges. Physiol. 189, 239-242.
(16) Eccles, J. C. (1948) Ann. Rev. Physiol. 10, 93-116.
(17) Fulton, J. F. (1949) Physiology of the Nervous System. 3rd Ed. Oxford University Press, New York.
(18) Fahn, S. (1976) In: GABA in Nervous System Function (E. Roberts, T. N. Chase and D. B. Tower, eds.) pp. 169-186, Raven Press, New York.
(19) Krnjević, K. (1976) In: GABA in Nervous System Function (E. Roberts, T. N. Chase and D. B. Tower, eds.) pp. 269-281, Raven Press, New York.
(20) Obata, K. and Takeda, K. (1969) J. Neurochem. 16, 1043-1047.
(21) Potashner, S. J. (1978) J. Neurochem. 31, 177-186.
(22) McLaughlin, B. J., Wood, J. G., Saito, K., Barber, R., Vaughan, J. E., Roberts, E. and Wu, J. Y. (1974) Brain Res. 76, 377-391.
(23) Elliott, K. A. C. and van Gelder, N. M. (1958) J. Neurochem. 3, 28-40.
(24) Martin, D. L. (1976) In: GABA in Nervous System Function (E.

Roberts, T. N. Chase and D. B. Tower, eds.) pp. 347-386, Raven Press, New York.
(25) Werman, R., Davidoff, R. A. and Aprison, M. H. (1968) J. Neurophysiol. 31, 81-95.
(26) Diamond, J. and Roper, S. (1973) J. Physiol. (Lond.) 232, 113-128.
(27) Barker, J. L. and McBurney, R. N. (1978) J. Physiol. (Lond.) 284, 127-128P.
(28) Johnston, G. A. R. (1976) In: GABA in Nervous System Function (E. Roberts, T. N. Chase and D. B. Tower, eds.) pp. 395-411, Raven Press, New York.
(29) Krnjević, K., Puil, E. and Werman, R. (1977) Can. J. Physiol. Pharmacol. 55, 658-669.
(30) Okada, Y. and Shimada, C. (1976) Brain Res. 107, 658-662.
(31) Schultz, S. G. and Curran, P. F. (1970) Physiol. Rev. 50, 637-718.
(32) Constanti, A., Krnjević, K. and Nistri, A. (1979) J. Physiol. (Lond.) (in press).
(33) Dreifuss, J. J., Kelly, J. S. and Krnjević, K. (1969) Brain Res. 13, 607-611.
(34) Lux, H. D., Loracher, C. and Neher, E. (1970) Exp. Brain Res. 11, 431-447.
(35) Waelsch, H. (1951) Adv. Prot. Chem. 6, 299-341.
(36) Frank, K. and Fuortes, M. G. F. (1957) Fed. Proc. 16, 39-40.
(37) Eccles, J. C. (1964) Progr. Brain Res. 12, 65-89.
(38) Schmidt, R. F. (1971) Ergebn. Physiol. 63, 20-101.
(39) Gray, E. G. (1963) J. Anat. (Lond.) 97, 101-106.
(40) Walberg, F. (1966) Exp. Brain Res. 2, 107-128.
(41) Melzack, R. and Wall, P. D. (1965) Science, 150, 971-979.
(42) Wall, P. D. (1978) Brain, 101, 1-18.
(43) Kellerth, J. O. (1968) In: Structure and Function of Inhibitory Neuronal Mechanisms (C. von Euler, S. Skoglund and U. Söderberg, eds.) pp. 197-212, Pergamon Press, Oxford.
(44) Curtis, D. R. , Duggan, A. W., Felix, D. and Johnston, G. A. R. (1971) Brain Res. 32, 69-96.
(45) Barron, D. H. and Matthews, B. H. C. (1938) J. Physiol. (Lond.) 92, 276-321.
(46) Krnjević, K. and Morris, M. E. (1974) Can. J. Physiol. Pharmacol. 52, 852-871.
(47) Kříž, N., Syková, E., Ujec, E., and Vyklicky, L. (1974) J. Physiol. (Lond.) 238, 1-15.
(48) Krnjević, K. and Morris, M. E. (1975) Can. J. Physiol. Pharmacol. 53, 912-922.
(49) Levy, R. A. (1977) Progr. Neurobiol. 9, 211-267.
(50) Miyata, Y. and Otsuka, M.(1975) J. Neurochem. 25, 239-244.
(51) Nishi, S., Minota, S. and Karczmar, A. G. (1974) Neuropharmacology, 13, 215-219.
(52) Deschênes, M., Feltz, P. and Lamour, Y. (1976) Brain Res. 118, 486-493.

(53) Choi, D. W., Farb, D. H. and Fischbach, G. D. (1977) Nature (Lond.) 269, 342-344.
(54) Rubin, P. (1974) Calcium and the Secretory Process. Plenum Press, New York.
(55) Baker, P. F. (1972) Progr. Biophys. Mol. Biol. 25, 177-223.
(56) Dunlap, K. and Fischbach, G. D. (1978) Nature (Lond.) 276, 837-839.
(57) Ito, M. (1976) In: GABA in Nervous System Function (E. Roberts, T. N. Chase and D. B. Tower, eds.) pp. 427-448, Raven Press, New York.
(58) Bain, W. A. (1933) Quart. J. Exp. Physiol. 22, 269-274.
(59) Dreifuss, J. J., Kelly, J. S. and Krnjević, K. (1969) Exp. Brain Res. 9, 137-154.
(60) Eccles, J. C., Eccles, R. M. and Magni, F. (1961) J. Physiol. (Lond.) 159, 147-166.

STOICHIOMETRY OF GABA-RECEPTOR INTERACTIONS: GABA MODULATES THE GLYCINE-RECEPTOR INTERACTION ALLOSTERICALLY IN A VERTEBRATE NEURON

Robert Werman

Neurobiology Unit, The Hebrew University of Jerusalem, Israel

INTRODUCTION

For synaptic interactions of short and intermediate duration, the immediate electrical consequence of activation of a synaptic receptor by a neurotransmitter is a change - usually an increase - in ionic conductance of the post-synaptic membrane. In general, amino acid transmitters, and particularly γ-aminobutyric acid (GABA), appear to work in this manner and GABA, like glycine (1), appears to operate usually by increasing the membrane conductance to small monovalent anions, which in the physiological situation means a conductance increase for Cl^-. If one assumes (2) that the activation of single receptors results in a mean conductance increase that is normally distributed and that - more certainly with low concentration of transmitter - there are negligible receptor-receptor or ionophore-ionophore interactions, then Ohm's law tells us that the activation of N receptors should produce N times the unitary conductance change.

The conductance increases produced can be measured as a function of transmitter concentration and thereby treated by the formalisms of enzyme-substrate reactions, where conductance change is treated as if it was reaction product, the receptor as enzyme and the transmitter as substrate (2-4). Thus, in general,

$$\Delta g = Ng_a = F(A), \qquad (1)$$

where A is the concentration of transmitter; Δg, the measured change in conductance; N, the number of receptors activated; and g_a, the unitary conductance change produced by activation of one receptor. More specifically, assuming that n molecules of transmitter are needed to activate a receptor, that K_j is the association constant of the jth step and that g' is the change in conductance at steady-state:

$$g' = g'_{max}A^n/(A^n + K_nA^{n-1} + K_nK_{n-1}A^{n-2} + \ldots + K_nK_{n-1}\ldots K_2A + K_nK_{n-1}\ldots K_2K_1) \qquad (2)$$

It should be noted that this function (equation 2) is different from the binding function and reflects only those receptors activated (5).

An important consequence of equation (2) is the fact that:

$$\lim_{A\to 0} d(\log g')/d(\log A) = n \qquad (3)$$

Thus, log-log plotting of steady-state conductance change as a function of concentration of transmitter will give a curve whose slope increases with decreasing concentration and approaches n, the molecularity of the reaction, as a limiting slope (2). In general, n represents the minimal number of molecules of transmitter that will successfully activate the receptor (6).

GABA-RECEPTOR STOICHIOMETRY IN LOCUST MUSCLE

Using the approach outlined above, Brookes and I studied the stoichiometry of the GABA interaction with its receptors in the metathoracic flexor tibialis muscle of the locust (5,7). This invertebrate preparation was particularly favorable, being a flat band of few muscle fibers which showed no apparent desensitization, and the conductance increases produced by exposure to bath-applied GABA grew monotonically to a maximum with time. Voltage changes produced by GABA were minimal, ruling out serious membrane field effects.

The pooled conductance data from 19 experiments are shown in a log-log plot in Figure 1A where GABA concentration was varied from $4 - 10 \times 10^{-5}$ M. The regression line fitting the data obtained with ascending concentrations of GABA had a slope of 3.15 ± 0.06 (SE of estimate). There is some hysteresis in the system and the curve obtained with descending concentrations is shifted to the left and the slope reduced to 2.45 ± 0.11. These data indicate that at least three molecules of GABA are necessary to activate its receptor. The absence of a slope reduction at lower concentrations, even at levels as low as 1% of the extrapolated maximal concentration, is particularly significant: it rules out the possibility that any significant receptor activation can be produced by one or two molecules of GABA and is also incompatible with the allosteric-cooperativity formulation of Monod, Wyman and Changeux (8).

In order to examine the effect of higher doses of GABA on the system and at the same time avoid drastic effects which might lead to irreversible changes in the membrane, we replaced 75% of the Cl^- in the medium with propionate ions. This reduced the average GABA-induced conductance increase to 18% and the effects of GABA concentrations as great as 50×10^{-5} M and beyond could be investigated (5). When a reproducible maximum response is obtainable, a Hill plot can be constructed which provides further stoichiometric information (Fig. 1B). It can be seen that the experimental points are well described by a straight line with a slope of 2.78 ± 0.05 until more than 60% of the receptors are activated. This indicates that the reaction exhibits strong positive cooperativity and involves no more than three molecules of GABA per receptor. The data were well described by the equation:

$$g' = g'_{max} G^3/(G^3 + K_3 G^2 + K_3 K_2 G + K_3 K_2 K_1), \qquad (4)$$

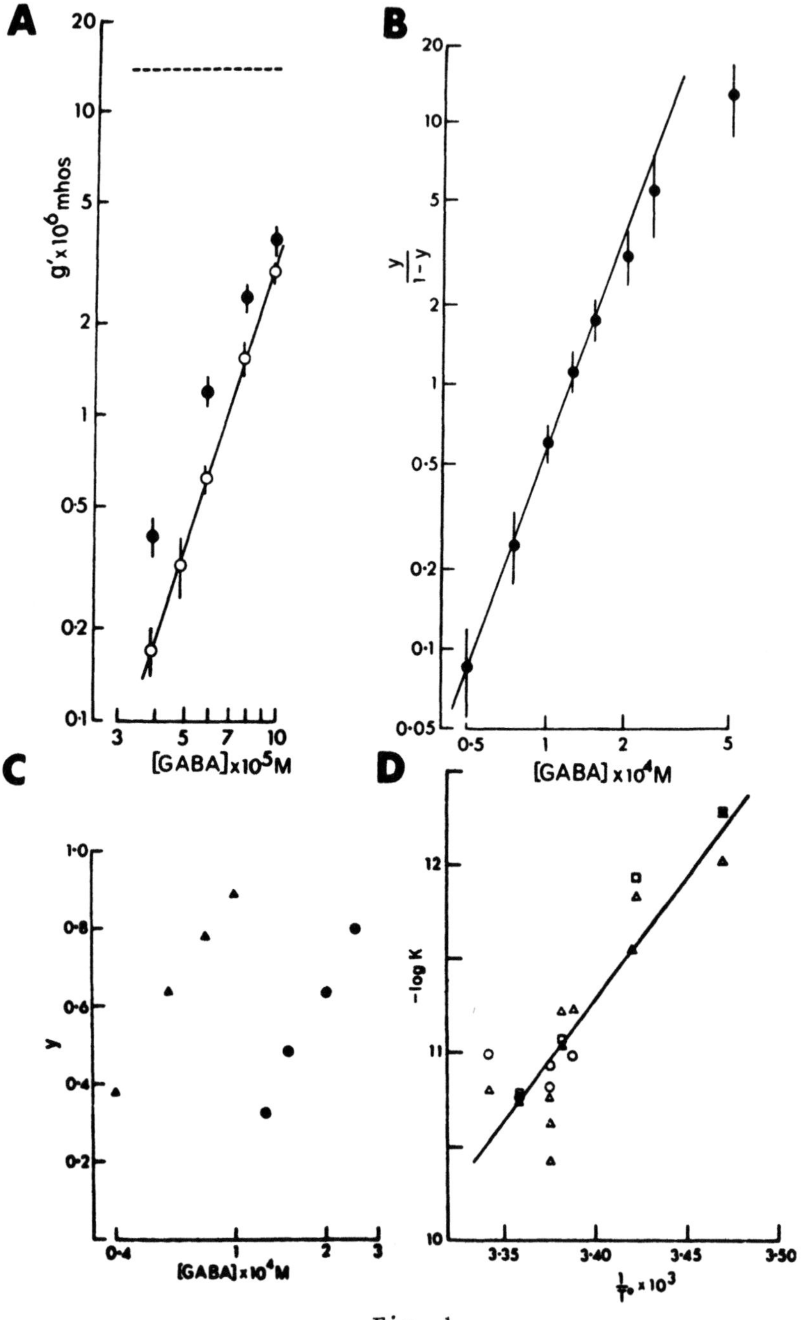
A
g'x10^6 mhos
[GABA]x10^5M
B
y/1-y
[GABA] x10^4M
C
y
[GABA]x10^4M
D
-log K
1/T₀ x10^3

Fig. 1

Fig. 1. Cooperativity and temperature sensitivity of GABA action on locust muscle fibers. A, B, Three molecules of GABA interact with its receptor with strong positive cooperativity to produce a conductance increase to Cl^- ions (5). A, Limiting log-log slope of conductance increase at steady state, g', as a function of GABA concentration. Open circles, means (± S E.M., n = 9) with ascending concentrations; closed circles, means (n = 10) on descent of concentrations. Slope of regression line is 3.15 ± 0.06. Dashed line, estimated mean maximum response. B, Hill plot (log [$g'/(g'_{max} - g')$] vs GABA concentration), proprionate saline (n = 9). Regression line from five lower points = 2.78 ± 0.05. C, D, Affinity for GABA increases with lower temperatures (9). C, Log-log dose response at 15°C (triangles) and at 24.5°C (circles) expressed as fraction of maximum conductance increase. D, van't Hoff plot of interaction of GABA and receptor. The negative log of the concentration necessary to produce a half-maximal response ($A_{\frac{1}{2}}$ = K) is plotted against the reciprocal of absolute temperature. Triangles from experiments with ascending concentrations; circles, descending concentrations; squares, intermediate washout between concentrations (from Brookes and Werman, 7; Brookes et al., 9).

where G is the concentration of GABA and $K_1^{-1} = 7.0 \times 10^{-4}$ M, K_2^{-1} = 1.4 and $K_3^{-1} = 2.8 \times 10^{-5}$ M. It can be seen that the affinity of each successive step increases five-fold.

The kinetics of the GABA-receptor interaction are obtained with ease as serial conductance measurements are made. Indeed, the response was generally biphasic and could be described by two time constants (9). From our modelling of the process of desensitization (4,10-12), it appears that the two time constants reflect the presence of occult desensitization. Our modelling indicates, however, that even in the presence of desensitization, steady-state measurements give an accurate description of the number of molecules of transmitter needed to activate a receptor; the concentration needed to produce a half-maximal response, however, is underestimated (4, 10). The desensitization produced in the continued presence of GABA is quite common, if not universal (13,15).

Since the equilibrium constants of the transmitter-receptor interaction can be obtained by conductance measurement, it is also possible to determine the temperature dependence of the interaction. It can be seen (Fig. 1C) that cooling increases the affinity of GABA to its receptor without significantly changing the cooperativity of the interaction. Assuming a reaction of infinite positive cooperativity with a cooperativity number of 2.78 (see Fig. 1B) gives a good approximation of most of the data (9). Measuring the temperature dependence of the concentration producing a half-maximal response, $A_{\frac{1}{2}}$, a van't Hoff plot shows more or less linear behavior (Fig. 1D). The thermodynamic parameters derived indicate a standard enthalpy change for the overall reaction, H^{o} = -59.3 kcal and are given in Table I (9). The high values seen are compatible with conformational changes in a multi-subunit protein. Indeed, the kinetics of the conductance changes (9) appear to be much too slow to reflect binding of GABA to its receptor and would also appear to reflect the slower conformational changes associated with synaptic conductance channel opening and/or closing.

A further question that can be asked of the stoichiometric analysis is the number of molecules of antagonist that are necessary to inhibit the GABA-receptor interaction. An example of such an investigation is illustrated in Figure 2, where the effect of picrotoxin on the GABA-receptor interaction is examined. A constant dose of GABA was used to produce a steady-state increase in conductance. The conductance decrease produced was measured as a function of picrotoxin concentration and the log-log plot shows that only one molecule of picrotoxin (treating the 1:1 mixture of the more potent picrotoxinin and the less potent picrotin as a single compound) is needed to

Table I - Thermodynamic Parameters for Interaction between GABA and its Receptor in Locust Muscle (H^{o} = -59.3 kcal/mole)

Temperature (°C)	$A_{\frac{1}{2}}$ (M x 10^4)	ΔF^{o} (kcal/mole)	ΔS^{o} (e.u./mole)
23	1.52	-14.5	-151
19	0.63	-15.7	-149
15	0.45	-16.1	-150

ΔF^{o} = Gibbs free energy change; ΔS^{o} = entropy change (from Brookes et al., 9).

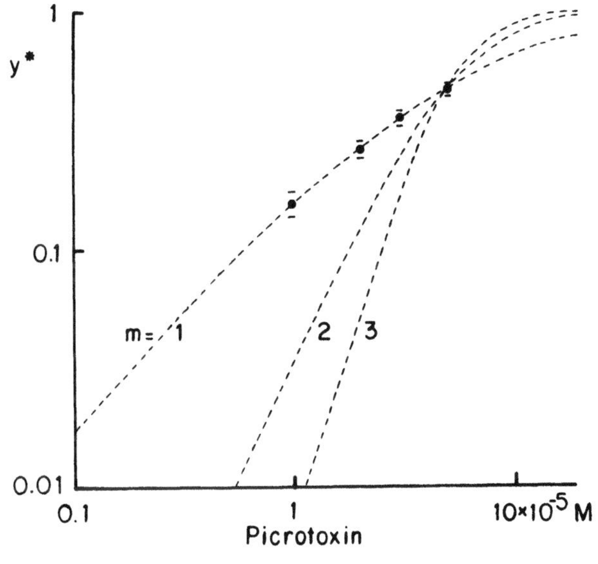

Fig. 2. One molecule of picrotoxin is sufficient to block GABA-receptor interaction in locust muscle. Log-log plot of Y, the conductance decrease (= $[(G/A_{\frac{1}{2}})^3/\{1 + (G/A_{\frac{1}{2}})^3\}] - [(G/A_{\frac{1}{2}})^3/\{1 + (G/A_{\frac{1}{2}})^3 + (P/K_p)^m\}]$) against picrotoxin concentration (P) in the presence of a fixed GABA concentration. Data points show means (± S.E.M.) and dashed lines indicate theoretical curves for m = 1, 2 and 3 (Brookes and Werman, unpublished results).

block the interaction of picrotoxin and GABA with the latter's receptor (Brookes and Werman, 1979; unpublished observations). These data may be interpreted as indicating that there are three and not more than three GABA-reactive subunits in the receptor and that occupation of one of these is sufficient to block a reaction requiring the activation of all three subunits. However, the large size of the picrotoxin moieties is also compatible with a reaction at a site near the receptor but different from GABA identification sites and the new complex blocking the GABA-receptor interaction through steric hindrance. Furthermore, picrotoxin may act by blocking the ionophore (13). General discussions of problems that arise in interpretation of the actions of synaptic blockaders can be found elsewhere (16,17).

STOICHIOMETRY OF GABA AND GLYCINE RECEPTOR INTERACTIONS IN CNS NEURONS

We were greatly interested in a quantitative description of the stoichiometry of amino acid responses in the central nervous system (CNS). Qualitative results indicated that both glycine (1,2,14) and GABA (14) required more than one molecule to activate their receptors in CNS neurons. Mazliah and I (18,19) used the lateral dendrite of the goldfish Mauthner cell in order to make the necessary measurements. This large cell is found in the medulla of the fish and sends axons to the opposite side of the spinal cord. The lateral dendrite is electrically inexcitable and the height of the antidromically-evoked action potential which is electrotonically transmitted from the axon is a convenient and reliable constant current source for the measurement of conductance change (20), thereby eliminating the need for bridge measurements or the use of multiple intracellular electrodes. At the site chosen for recording, the dendrite approximates a cylinder, 10 - 15 μ in diameter. A rigid electrode assembly (1) was used with the tip of the potassium-citrate

Table II - Quantitative Aspects of Glycine and GABA Interactions with Goldfish Mauthner Cell Lateral Dendrite

	Maximum conductance increase	Current (nA) producing half-maximal response ($A_{\frac{1}{2}}$)	
	Glycine/GABA	Glycine	GABA
Observed (n = 10)	4.10 ± 0.72*	60.0 ± 4.7***	101.3 ± 11.1
Calculated (n = 4)	3.86 ± 2.03**	57.8 ± 11.5****	117.6 ± 44.8

Both mean ratio (± S.E.M.) of maximal conductance changes (glycine/GABA) and mean currents (± S.E.M.) necessary to produce half-maximal response in the same cell for each drug are shown. Calculated responses are based on a reaction of infinite positive cooperativity of four amino acid molecules with their receptor (from Werman and Mazliah, 19).
* $p < 0.0005$; ** $p < 0.01$; *** $p < 0.005$; **** $p < 0.05$.

electrode for intracellular recording protruding 20 - 30 μ from the tip of a parallel triple-barrelled electrode containing concentrated solutions of glycine, GABA and NaCl. Iontophoretic currents through the amino acid barrels were controlled by constant current sources and were always delivered push-pull with respect to the NaCl barrel.

Since the formulation used for analysis (equation 2) required steady-state measurements, the proper duration of iontophoretic application had to be determined. Adequate steady-state values were obtained for both drugs in one minute, and this duration was used throughout. Even from extracellular records it was obvious that glycine was appreciably more effective than GABA in shunting the antidromic response. These results were confirmed by intracellular recording (Table II); we can see that the maximum conductance change produced by glycine was 4.10 times that produced by GABA, while the iontophoretic current producing a half-maximal glycine response was

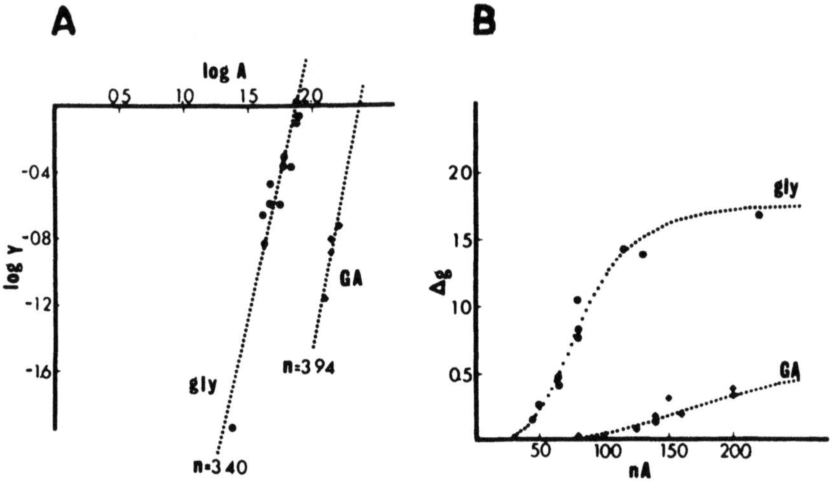

Fig. 3. Four molecules of GABA (GA) or glycine (gly) are needed to activate their receptors in the goldfish Mauthner cell. A, Log-log plot of conductance increase as a function of GABA and glycine concentrations in the same cell. Regression lines with slopes are also shown. B, Dose-response curves in the same cell. Theoretical curves were drawn assuming interactions of infinite positive cooperativity of four molecules of transmitter with its receptor using these derived constants: glycine, $A_{\frac{1}{2}}$ = 83 nA, g'_{max} = 1.76; GABA, $A_{\frac{1}{2}}$ = 185 nA, g'_{max} = 0.56 (from Werman and Mazliah, 19).

only 59.2% of that needed for GABA. Thus, on a dose (current) basis, glycine was 7.3 times more potent than GABA.

Log-log plotting of the increase in conductance as a function of amino acid dose (current) is illustrated in Figure 3A. It can be seen that the regression slopes obtained are greater than three for both amino acids; this finding was consistent and indicates that at least four molecules of each amino acid are needed to activate their receptors. Plotting the reciprocal of the conductance increase as a function of the reciprocal of the fourth power of the amino acid concentration allowed us to determine theoretical maximum responses as well as doses (currents) producing half-maximal responses, $A_{\frac{1}{2}}$ (Table II). Despite the simplifying assumption of a reaction of infinite positive cooperativity:

$$g' = g'_{max}/\{1 + (A_{\frac{1}{2}}/A)^4\}, \qquad (5)$$

the theoretical curves obtained fit the data reasonably well (Fig. 2B, Table II). The dose response curves are clearly sigmoidal, with the GABA curve shifted to the right and depressed with respect to the glycine curve (Fig. 2B).

Both GABA and glycine appear to utilize four binding sites in the activation of a unitary conductance change in the Mauthner cell. Moreover, the strong positive cooperativity indicates that the energy barrier for binding the first molecule of amino acid is appreciably higher than that of subsequent reactions for both amino acids. The action of glycine is appreciably greater than that of GABA, showing a lower affinity and a greater maximal conductance change. The difference in maximal conductances might be the result of different open times for the Cl^- channels after receptor activation by the two amino acids. Kinetic studies, however, show that the rate of onset of the GABA action was significantly faster than that for glycine while the rate of offset was significantly slower. It would thus appear likely that the density of glycine-receptors is greater than that of GABA-receptors in the lateral dendrite of the Mauthner cell.

GABA INTERACTS ALLOSTERICALLY WITH THE GLYCINE-RECEPTOR IN THE MAUTHNER CELL

When the response to glycine was examined in the presence of a constant dose of GABA, the dose-response curve was markedly shifted to the left without any apparent increase in the maximum response (Fig. 4) or change in slope of the log-log curve of the response to

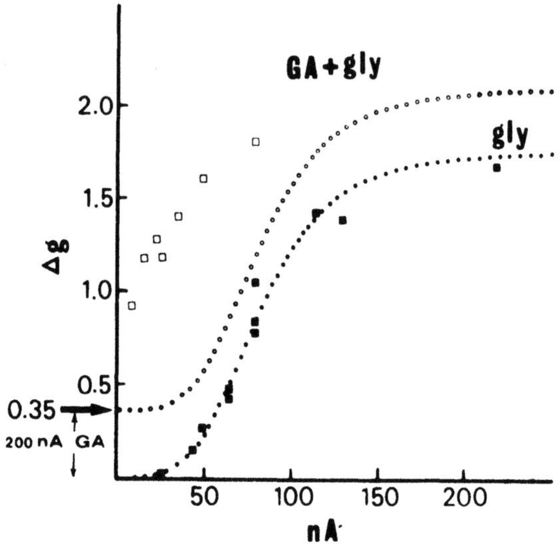

Fig. 4. Allosteric interaction of GABA (GA) with glycine (gly) receptor; glycine dose-response curve in presence and absence of GABA. The lower curve and black squares illustrate theoretical and experimental responses to glycine alone. In the presence of a constant dose of GABA (arrow), the algebraic sum of the responses is given by the intermediate curve. Actual experimental values are shown by white squares (from Werman, 12; Werman and Mazliah, 19).

glycine. This shift can be seen to be much greater than that predicted from an algebraic addition of the background GABA effect with the control glycine dose-response curve (Fig. 4), indicating that GABA is exerting an action on glycine-receptors. This conclusion was confirmed by double-reciprocal plotting of conductance change against the fourth power of glycine concentration: the maximal response was seen not to change but the slope was greatly reduced. From this finding it is clear that the GABA action is dual, both on its own set of receptors and an independent action on the glycine-receptor which is not expressed as a conductance change. The GABA action on the glycine-receptor is to increase the affinity of glycine for its receptor without decreasing the number of molecules (four) of glycine needed to activate the receptor. This occult action of GABA must thus be understood as a reaction on sites close to — but not with — the glycine-recognition sites which reduce the energy barriers for the glycine-receptor interactions; i.e., an allosteric interaction of GABA with the glycine-receptor. Since the glycine-receptor interaction is characterized by strong positive cooperativity, it is highly likely that the GABA action is to lower the energy barrier for the critical first step, the binding of the first glycine molecule to its receptor. The molecularity, n, of this GABA response has not been determined.

Since the GABA-induced conductance change was so much smaller than the corresponding glycine response, it was difficult to investigate the effect of glycine on the GABA-receptor interaction. In one cell, however, where GABA, even in large doses, failed to produce a measurable conductance increase, the presence of a small background conductance produced by a small constant dose of glycine was sufficient to reveal a GABA response. It is not clear from this experiment whether glycine acts allosterically on the GABA-receptor, or whether the glycine response itself was increased, or whether glycine acted agonistically with GABA on the GABA-receptor to reduce the molecularity of the GABA response.

Although these studies are by no means complete and are technically difficult, they hold much promise for the understanding of transmitter-receptor interactions in the vertebrate CNS. Among other implications they must serve as a basis for the interpretation of binding studies in isolated systems of CNS origin.

SUMMARY

1. Measurement of steady-state conductance changes provides a reliable method for determination of transmitter-receptor stoichiometry in general and for GABA and its receptor interactions in particular. Log-log plotting of steady-state conductance changes as

a function of decreasing transmitter concentration gives the molecularity of the interaction as a limiting slope.

2. Suitable measurements of GABA action in locust muscle show a molecularity of 3 with strong positive cooperativity. One molecule of picrotoxin is sufficient to block this reaction. Kinetic studies reveal the presence of occult desensitization. Thermodynamic studies reveal strong negative heats of interaction compatible with conformational changes in a multi-subunit receptor.

3. Measurements in the lateral dendrite of the goldfish Mauthner cell reveal that glycine has a more powerful action than GABA. Both interactions, however, utilize 4 molecules of amino acid with strong positive cooperativity to activate its receptors. The receptors are apparently distinct and there appears to be a higher glycine-receptor density.

4. In addition to its action on its own receptor, GABA allosterically modulates the glycine-receptor interaction in the Mauthner cell by lowering the energy barrier for the binding of the first glycine molecule, thereby increasing the affinity of glycine for its receptor.

ACKNOWLEDGEMENTS

This work was partly supported by Grant NS-13935-01 from the U.S. Public Health Service. I thank Mr. E. Armon, Mr. Y. Mazliah, Dr. N. Brookes and Dr. M. E. Spira for use of unpublished material.

REFERENCES

(1) Werman, R., Davidoff, R. A. and Aprison, M. H. (1968) J. Neurophysiol. 31, 81-95.
(2) Werman, R. (1969) Comp. Biochem. Physiol. 30, 997-1017.
(3) Werman, R. (1975) In: Stability and Origin of Biological Information (I. R. Miller, ed.) pp. 226-244, Wiley, New York.
(4) Werman, R; (1976) In: Electrobiology of Nerve, Synapse and Muscle (J. P. Reuben, D. P. Purpura, M. V. L. Bennett and E. R. Kandel, eds.) pp. 45-60, Raven Press, New York.
(5) Brookes, N. and Werman, R. (1973) Mol. Pharmacol. 9, 571-579.
(6) Werman, R. (1971) Comp. Gen. Pharmacol. 2, 129-137.
(7) Brookes, N. and Werman, R. (1970) Israel J. Med. Sci. 6, 322.
(8) Monod, J., Wyman, J. and Changeux, J.-P. (1965) J. Mol. Biol. 12, 88-118.

(9) Brookes, N., Blank, M. and Werman, R. (1973) Mol. Pharmacol. 9, 580-589.
(10) Armon, E., Spira, M. E. and Werman, R. (1975) Israel J. Med. Sci. 11, 856.
(11) Spira, M. E., Armon, E. and Werman, R. (1975) Israel J. Med. Sci. 11, 856.
(12) Werman, R. (1978) Proc. VII Int. Congr. Pharmacol. 8 pp. (in press).
(13) Hochner, B., Spira, M. E. and Werman, R. (1976) Brain Res. 107, 85-103.
(14) Krnjević, K., Puil, E. and Werman, R. (1977) Can. J. Physiol. Pharmacol. 55, 658-669.
(15) Krnjević, K., Puil, E. and Werman, R. (1977) Can. J. Physiol. Pharmacol. 55, 670-680.
(16) Werman, R. (1972) Res. Publ. A.R.N.M.D. 50, 147-180.
(17) Werman, R. (1972) Ann. Rev. Physiol. 34, 337-374.
(18) Werman, R. and Mazliah, J. (1974) Proc. Int. Union Physiol. Sci. 11, 149.
(19) Werman, R. and Mazliah, J. (1978) In: Iontophoresis and Transmitter Mechanisms in the Mammalian Central Nervous System (R. W. Ryall and J. S. Kelly, eds.) pp. 400-402, Elsevier/North Holland Biomedical Press, Amsterdam.
(20) Diamond, J. (1968) J. Physiol. (Lond.) 194, 669-723.

GABA RECEPTOR AGONISTS: RELATIONSHIP BETWEEN STRUCTURE AND BIOLOGICAL ACTIVITY *IN VIVO* AND *IN VITRO*

Povl Krogsgaard-Larsen and Jørn Arnt

Royal Danish School of Pharmacy, Departments of Chemistry BC and Pharmacology, 2 Universitetsparken, DK-2100 Copenhagen, Denmark

INTRODUCTION

Malfunctions of the central γ-aminobutyric acid (GABA) system contribute to the development of neurological and psychiatric disorders like Parkinson's disease (1-4) and Huntington's chorea (4-8). Furthermore, abnormalities in GABA-containing neuronal pathways may be factors of importance in epilepsy (9) and schizophrenia (10,11). Consequently, processes associated with GABA-ergic synaptic transmission are likely to be susceptible sites for pharmacological intervention in such diseases. The mechanisms underlying the processes involved in GABA-mediated neurotransmission, however, are still poorly understood, and pharmacological strategies for stimulation of the synaptic activity of GABA must be developed on a basis which includes a variety of unknown factors.

There are several potential sites of pharmacological attack at GABA-operated synapses. GABA-ergic neurotransmission might be facilitated in a flexible manner via inhibiting neuronal and glial GABA uptake processes (12) or the activity of the GABA-metabolizing enzyme GABA-transaminase (GABA-T) (13). Alternatively, GABA replacement therapies based on specific GABA receptor agonists may be relevant for the treatment of diseases in which impairments of the central GABA systems have been demonstrated.

GABA RECEPTOR AGONISTS: FUTURE PSYCHOTHERAPEUTIC DRUGS?

The development on a rational basis of therapeutically useful GABA receptor agonists involves a variety of problems. Fundamental aspects of the structure and mechanism of action of GABA-receptors are unknown (14), and the interactions of GABA-ergic neurones and neurones that involve other transmitter systems in different brain areas are poorly understood. These and many other aspects of the extremely complex functions of the brain make the full consequences of a GABA receptor agonist therapy unpredictable.

In diseases like Huntington's chorea and Parkinson's disease the degeneration of GABA-ergic neurones seems to be restricted to certain brain areas, suggesting therapies based on GABA-agonists with selective affinity for the GABA-receptors in the impaired brain region. Studies of the pharmacological characteristics of GABA-receptors in the human brain indicate that their affinities for GABA-agonists vary significantly among different brain areas, possibly due due to an inhomogeneous distribution of an endogenous receptor inhibitor in the brain (15). Furthermore, degeneration of GABA-ergic neurones might provoke supersensitivity of the corresponding post-synaptic receptors to GABA-agonists (16), although the physiological significance of these observations remains unknown. In any case, GABA replacement therapies based on selective manipulations of GABA-receptors is a realistic possibility. A prerequisite for progress in this potentially new area of psychopharmacology, however, is the development of potent and specific GABA receptor agonists with pharmacokinetic and toxicological properties acceptable for use in the human clinic.

GABA RECEPTOR AGONISTS: PHARMACOLOGICAL EVALUATION

Receptor Affinity Binding and Electrophysiological Studies

The process of developing GABA receptor agonists naturally involves studies of the compounds as inhibitors of the Na^{+}-independent binding of GABA to synaptic membranes. The characteristics of this binding strongly suggest that it represents an interaction of the ligands with GABA receptor sites (17-19), possibly primarily postsynaptic sites (20). This procedure of receptor affinity binding allows a quantitative comparison of the potencies of GABA receptor agonists and measurements of their binding constants. The scope of this biochemical determination of agonist-receptor interaction is, of course, limited. The binding sites apparently are receptors in more or less disintegrated synaptic environments measured under non-physiological conditions. Furthermore, the receptor binding procedure does not readily distinguish between receptor agonists and antagonists.

The present approach toward evaluating GABA receptor agonists also involves electrophysiological procedures. The compounds have been administered extracellularly on single neurones of the cat spinal cord by using microelectrophoretic techniques in order to determine their depressant actions on neuronal firing. The identification of this effect as a result of specific GABA-receptors has been accomplished by using selective neurotransmitter antagonists. The compounds under study have been administered on single neurones from multi-barrelled micropipettes with simultaneous application of strychnine or the selective GABA receptor antagonist, bicuculline-methochloride (BMC) (21). Thus, a GABA receptor agonist is defined as a bicuculline-sensitive, strychnine-insensitive depressant of neuronal firing.

This technique for *in vivo* studies of GABA receptor agonists also has inherent limitations. The exact concentration of a microelectrophoretically administered compound at the receptors is not known, and the method only allows an approximate determination of the relative potencies of receptor agonists. However, a combined use of this technique and the receptor affinity binding procedure permits identification of GABA receptor agonists and quantitative measurements of their potencies.

It must be understood that a direct comparison of potencies determined by using the two techniques discussed above is not possible. The substrate-specificities of GABA-receptors in brain and spinal neurones, the latter of which may be, at least partially, of presynaptic character (22), seem to be somewhat different. Muscimol, for example, is a much more potent depressant of the firing of spinal than of cortical neurones (23,24). However, with a few exceptions the present studies demonstrate a close correlation between the potencies of GABA receptor agonists as inhibitors of the binding of GABA to receptor sites *in vitro* and as BMC-sensitive neuronal depressants.

Unilateral Injection into Substantia Nigra

Unilateral injection of GABA into the pars reticulata of the substantia nigra in rats induces a contralateral turning behavior, and this effect is mimicked by muscimol, Lioresal, and imidazoleacetic acid (25). This turning response apparently is mediated via activation of GABA-receptors (26), but due to the complex and incompletely elucidated anatomy of this part of the brain an interpretation of this behavior, in terms of neuronal interactions, is not yet possible. In addition to the affinity of GABA agonists for GABA-receptors, a variety of pharmacokinetic factors may be involved in regulating the intensity and duration of the turning behavior after intranigral injection. Only compounds capable of penetrating membrane structures may easily reach the synaptic GABA-receptors. Furthermore, such factors must be involved in the spreading of the injected compound in the nervous tissue and in its distribution among various tissue elements.

Although a variety of factors of unknown significance must influence the described turning response, this behavior may represent a useful in vivo model for testing GABA receptor agonists. This model could provide a liaison between biochemical and electrophysiological experiments, on one hand, and behavioral studies based on systemic administration of GABA receptor agonists, on the other.

GABA RECEPTOR AGONISTS: DESIGN AND DEVELOPMENT

Muscimol as a GABA Receptor Agonist and Model Compound

Muscimol, which is a structural analogue of GABA (27) (Fig. 1), depresses the firing of central neurones (28) via activation of BMC-sensitive GABA-receptors (23,29). Muscimol also inhibits neuronal (30) and glial (31) GABA uptake processes, being a substrate for the neuronal transport carrier (G. A. R. Johnston, unpublished results).

It has been suggested that muscimol might be useful for the treatment of diseases caused by loss of GABA-ergic neurones (32). However, muscimol apparently is excessively metabolized after systemic administration to mice (33) and rats (S. J. Enna, unpublished results), and only a small proportion of peripherally administered muscimol appears to enter the brain (34). In addition, muscimol is a relatively toxic compound (35,36); this property is either inherent to muscimol, or due to its conversion to toxic metabolites in vivo. Therefore, muscimol does not appear to be a potentially useful drug. However, its dual effect on the GABA system makes it a possible model compound for the development of specific uptake inhibitors and GABA receptor agonists. Extensive structure-activity studies

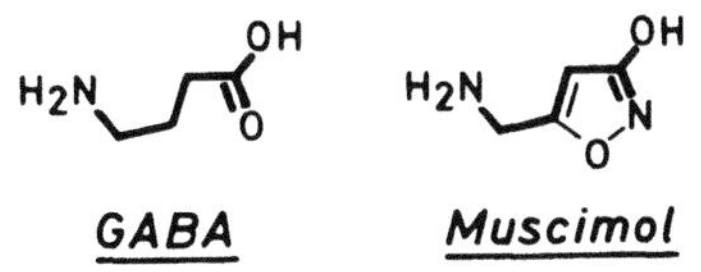

Fig. 1. Structural relationship between GABA and muscimol.

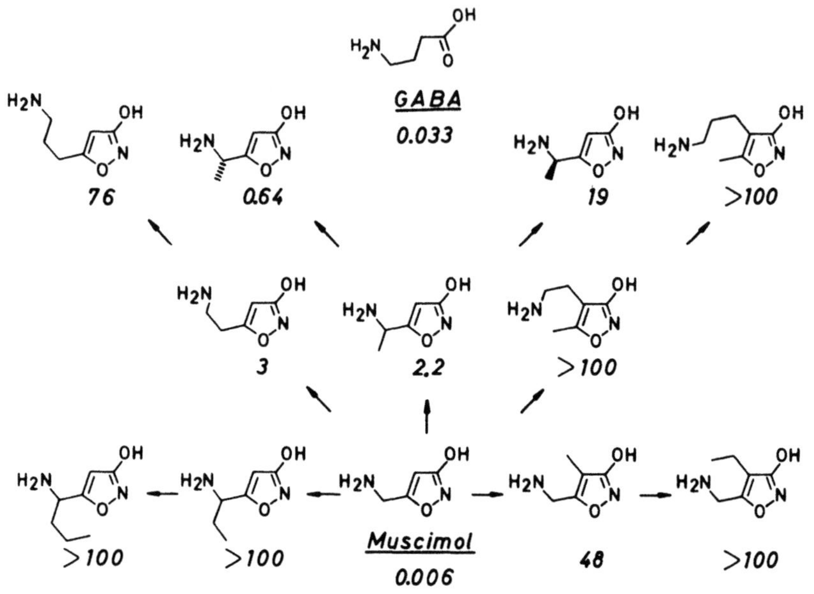

Fig. 2. Muscimol and related compounds, developed via systematic variations of the side chain of muscimol, as inhibitors of [^{3}H]GABA binding. IC_{50} values (μM) are indicated below the formulas. Receptor affinity binding experiments were performed according to the procedure of Krogsgaard-Larsen et al. (18).

on muscimol analogues already have led to the development of potent inhibitors of GABA uptake, like nipecotic acid, guvacine, and cis-4-hydroxynipecotic acid (12,37,38,39), and potent and specific GABA receptor agonists (29,40,41,42).

Mono- and Bicyclic Muscimol Analogues

The 3-isoxazolol nucleus of muscimol apparently is accepted as a masked carboxyl group by GABA-receptors. Furthermore, the aminomethyl side chain seems to be the most vulnerable part of muscimol with respect to metabolic conversions. Thus, preparation of 3-isoxazolol derivatives with various aminoalkyl groups in different positions of the ring might provide a rational approach for the development of GABA receptor agonists devoid of the undesirable side-effects of muscimol.

Various simple analogues of muscimol have been synthesized (43-46) and tested as inhibitors of GABA binding to receptor sites in vitro (18,41). It is evident that the structure of muscimol is almost optimum for binding to GABA-receptors (see Fig. 2). Minor changes of the side chain of muscimol and even introduction of small alkyl groups in the vicinity of the unaltered aminomethyl group result in compounds with dramatically reduced affinities for GABA-receptors. Introduction of an additional methylene group or a methyl group into the side chain of muscimol, however, leads to compounds with relatively high affinity for GABA-receptors, the S-(-)-form of 5'-methylmuscimol (5'-Me-muscimol) being more potent than the corresponding R-(+)-form (Fig. 2 and Table I). S-(-)-5'-Me-muscimol is of pharmacological interest, and the side chain of this compound may be much less susceptible to metabolism than the aminomethyl group of muscimol. In general, however, this pharmacochemical approach to the development of therapeutically useful GABA receptor agonists has been relatively unsuccessful.

As an extension of this work various bicyclic muscimol analogues have been synthesized (47-49) in which the amino function is located on, or incorporated into, an additional ring structure (Fig. 3). Structure-activity studies of these semi-rigid compounds further emphasize the remarkable structural specificity of GABA-receptors. Only one member of this series of muscimol analogues is a potent GABA receptor agonist, namely 4,5,6,7-tetrahydroisoxazolo[5,4-c]-pyridin-3-ol (THIP). THIP specifically activates GABA-receptors (Table I), being inactive on GABA uptake into rat brain slices and on the activities of GABA-metabolizing enzymes (40). Structure-activity considerations of THIP indicate that muscimol, and possibly GABA, interact with GABA-receptors in almost planar and partially folded conformations (39-41). The "receptor-active conformation"

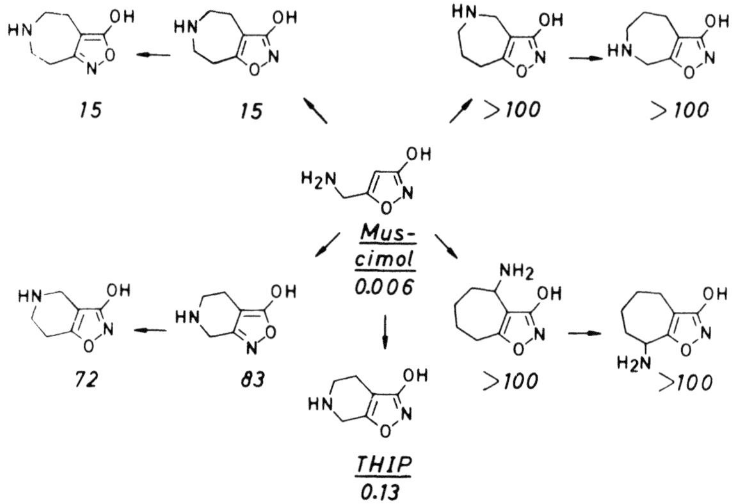

Fig. 3. Bicyclic muscimol-analogues as inhibitors of [^{3}H]GABA binding. IC_{50} values (μM) are indicated below the formulas. Receptor affinity binding experiments were performed according to the procedure of Krogsgaard-Larsen et al. (18).

of muscimol seems to be different from its low-energy conformation (27,50). This finding may explain why the introduction of additional structural elements into the molecule of muscimol (Fig. 2) impedes or even excludes contact between the muscimol analogues concerned and GABA-receptors (39,41,42).

The data shown in Table I indicate that a good correlation exists between the receptor binding data and the results of electrophysiological experiments, the only exception being RS-5'-ethylmuscimol (RS-5'-Et-muscimol). This compound, which is virtually inactive in receptor affinity binding experiments (41), is a moderately potent but BMC-insensitive depressant of neuronal firing (29). RS-5'-Et-muscimol may belong to a group of substances which interact with a sub-class of GABA-receptors insensitive to BMC (51). The potencies of the muscimol analogues listed in Table I with re-

Table I - Structures, Potencies as GABA-Agonists, and Turning Responses after Unilateral Intra-nigral Injection of Some Muscimol-Analogues

COMPOUND	STRUCTURE	INHIBITION OF GABA BINDING	ACTIVITY IN VIVO		TURNING RESPONSE DURATION	
		IC_{50} (µM)	Relative potency	Antagonism by BMC	µg	min.
Muscimol	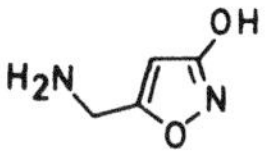	0.006	– – – –	+	0.01	75
THIP	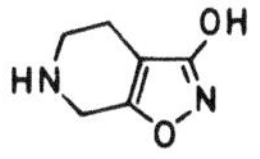	0.13	– – –(–)	+	0.3	55
RS-5'-Me-muscimol	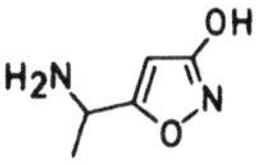	2.2	– – –	+	1.0	60
S-(-)- -"-	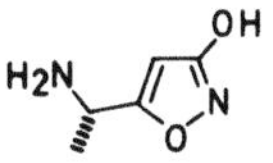	0.64	n.t.		1.0	70
R-(+)- -"-	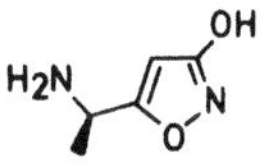	19	n.t.		25	90
Homomuscimol	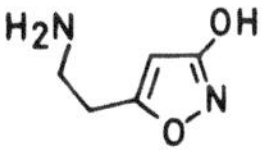	3	– – –	+	3.0	90
4-Me-muscimol	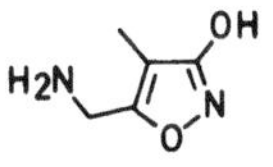	48[a]	–	0	15	60
RS-5'-Et-muscimol	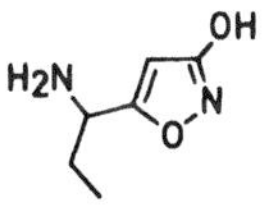	>100	– –	0	50	–

GABA receptor affinity binding was measured as described by Krogsgaard-Larsen et al. (18). GABA-agonist activities in vivo were studied using microelectrophoretic techniques, and the potencies of neuronal depressants are indicated relative to that of GABA (- - -). Unilateral intra-nigral injections were performed, and turning responses were measured as described by Scheel-Krüger et al. (25); n.t. indicates not tested.

[a] From Johnston et al. (50).

spect to induction of turning response after intra-nigral injections correlate reasonably well with the electrophysiological and receptor binding data. In this model RS-5'-Et-muscimol is almost inactive and THIP, which is the only bicyclic muscimol analogue studied, is relatively weak. The similar relative potencies of the compounds concerned in all three test systems may be a consequence of their closely related structures and probably similar physiochemical properties.

THIP is a relatively non-toxic compound in mice and dogs, and it seems to penetrate easily into the brain after peripheral administration (V. Christensen, unpublished results). Muscimol and THIP have been studied in various animal models assumed to reflect GABA receptor agonist activity. Of particular interest are determinations of their protection against isoniazid-induced convulsions (52), and their antagonism of morphine-induced hypermotility (54). In general, the effects of THIP are slightly weaker than, but qualitatively similar to those observed for muscimol (V. Christensen, unpublished results). As the aminoalkyl function of THIP is built into a ring structure, it might be relatively resistant to metabolic processes after peripheral administration; therefore, THIP might be relevant to study in the human clinic.

Heterocyclic Muscimol Analogues

The heterocyclic ring of muscimol has been subjected to systematic structural manipulations without altering the positions of the aminomethyl side chain and the acidic hydroxy group (46,55) (H. Hjeds and P. Krogsgaard-Larsen, unpublished results). The relative potencies of these muscimol-analogues with respect to displacement of [^{3}H]GABA from GABA receptor sites are provided in Fig. 4 and Table II. Despite an immediate structural similarity of the compounds concerned, their affinities for GABA-receptors is strictly dependent on the structure of the heterocyclic rings. Interchange of the ring oxygen and nitrogen atoms of muscimol and substitution

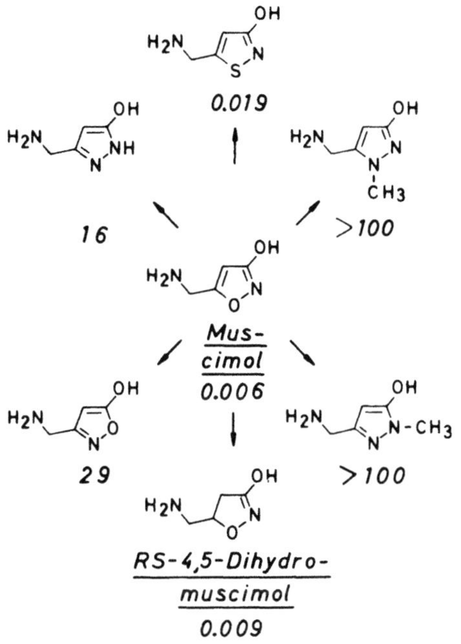

Fig. 4. Heterocyclic muscimol-analogues, developed via systematic variations of the ring of muscimol, as inhibitors of [^{3}H]GABA binding. IC_{50} values (μM) are indicated below the formulas. Receptor affinity binding experiments were performed according to the procedure of Krogsgaard-Larsen et al. (18).

of this oxygen atom for nitrogen result in dramatic decreases in, or almost complete loss of receptor affinity. However, the sulfur analogue of muscimol (thiomuscimol) is only slightly less potent (18,56). and the non-heteroaromatic derivative RS-4,5-dihydromuscimol is equipotent (18) with muscimol at inhibiting GABA receptor binding. Except for 1- and 2-methylazamuscimol, the compounds depicted in Fig. 4 have been studied as depressants of the firing of feline spinal neurones using microelectrophoretic techniques (Table II). The receptor affinity binding data reflect reasonably well the potencies

Table II - Structures, Potencies as GABA-Agonists, and Turning Responses after Unilateral Intra-nigral Injection of Some Muscimol-Analogues

COMPOUND	STRUCTURE	INHIBITION OF GABA BINDING IC_{50} (μM)	ACTIVITY IN VIVO Relative potency	Antagonism by BMC	TURNING RESPONSE μg	DURATION min.
Muscimol	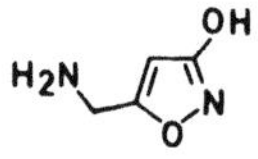	0.006	- - - -	+	0.01	75
RS-4,5-dihydro-muscimol	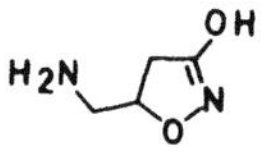	0.009	- - - -	+	0.04	65
Thio-muscimol	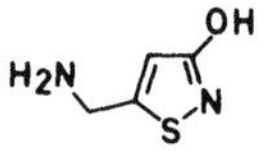	0.019	- - - -	+	5	7
Iso-muscimol	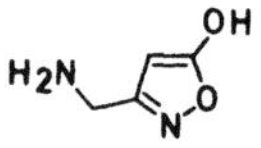	29	-	0	50	90
Aza-muscimol	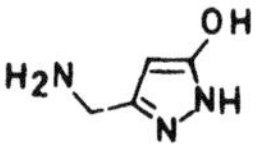	16	- -	0	6	45

GABA receptor affinity binding was measured as described by Krogsgaard-Larsen et al. (18). GABA-agonist activities in vivo were studied using microelectrophoretic techniques, and the potencies of neuronal depressants are indicated relative to that of GABA (- - -). Unilateral intra-nigral injections were performed, and turning responses were measured as described by Scheel-Krüger et al. (25).

of these compounds as GABA receptor agonists in vivo. The only exception is azamuscimol, which is a moderately potent, but BMC-insensitive, depressant of neuronal firing (29) with low affinity for GABA receptor sites in vitro (18). These effects of azamuscimol are similar to those described above for RS-5'-Et-muscimol (cf. Table I).

The potencies of some of these muscimol analogues with respect to induction of turning behavior after intra-nigral injection do not agree with their GABA-agonist activities. The turning response of thiomuscimol, for example, is remarkably weak and very short-lived. This effect of thiomuscimol might be explained by its rapid diffusion away from the injection site and possibly by its incorporation into fatty elements of nervous tissue. An alternative but rather unlikely explanation is that thiomuscimol undergoes very rapid chemical or metabolic degradation after its intra-nigral injection. The turning response of azamuscimol seems to correlate with its potency as a neuronal depressant and with its affinity for GABA receptor sites in vitro, in contrast to the findings for RS-5'-Et-muscimol (cf. Table I).

An interpretation of the relationship between structure and GABA receptor agonist activity of the heterocyclic GABA-analogues involves consideration of a variety of different structural factors. All of the compounds depicted in Table II are zwitterions, the hydroxy groups being the acidic parts of the molecules. However, the distribution of the negative charges of the compounds probably is very different, and this factor seems to be of critical importance for their affinities for GABA-receptors (18,42). The potency of RS-4,5-dihydromuscimol is remarkable. The non-heteroaromatic structure of this compound might reflect the "receptor active structure" of GABA almost perfectly, and possibly even better than muscimol, with respect to conformation of the carbon skeleton and delocalization of charges.

With the exception of muscimol the compounds considered here have only been subjected to preliminary pharmacological and toxicological studies. Thiomuscimol and RS-dihydromuscimol might be of pharmacological interest, although the aminomethyl side chains of these compounds, like that of muscimol, may be highly susceptible to metabolism.

Isoguvacine and Related Amino Acids

Since the 3-isoxazolol nucleus of muscimol can be regarded as a masked carboxyl group the amino carboxylic acid analogue of THIP, isoguvacine, has been synthesized (41) and studied as a GABA agonist

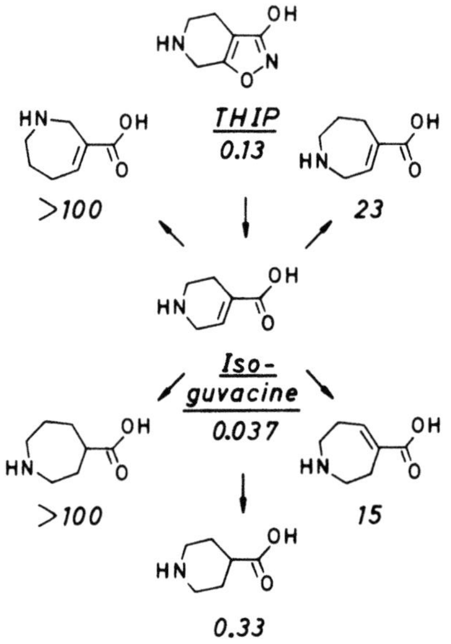

Fig. 5. Isoguvacine and related compounds as inhibitors of [^{3}H]GABA binding. IC_{50} values (μM) are indicated below the formulas. Receptor affinity binding experiments were performed according to the procedure of Krogsgaard-Larsen et al. (18).

in electrophysiological (40) and receptor affinity binding experiments (18,41; Table III). Isoguvacine does not interact with neuronal (40) or glial (31) GABA uptake processes and it is not an inhibitor of the activity of GABA-T (40). The relationships between structure and affinity for GABA receptor sites of isoguvacine and a number of related cyclic amino acids are illustrated in Fig. 5. Isoguvacine is a very potent and specific GABA receptor agonist, suggesting that the "GABA structure element" of this compound has an almost optimum conformation for activation of GABA-receptors.

In general, the turning responses observed after unilateral intra-nigral injections of the GABA receptor agonists listed

Table III - Structures, Potencies as GABA-Agonists, and Turning Responses after Unilateral Intra-nigral Injection of Some Amino Acids

COMPOUND	STRUCTURE	INHIBITION OF GABA BINDING IC_{50} (μM)	ACTIVITY IN VIVO Relative potency	ACTIVITY IN VIVO Antagonism by BMC	TURNING RESPONSE DURATION μg	TURNING RESPONSE DURATION min.
t-ACA	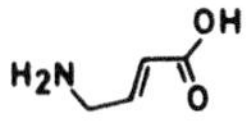	0.14[a]	- - -[b]	+	50	7
S-(-)-4-Me-t-ACA	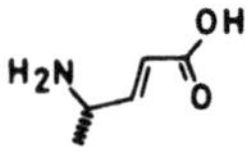	4.1	- -(-)[c]	+	10	>90
3-APS	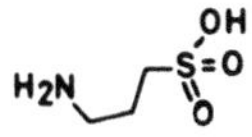	0.3[d]	- - -	+	50	120
Isoguvacine	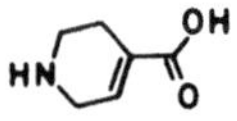	0.037	- - - -	+	3.0	>180
Isonipecotic acid	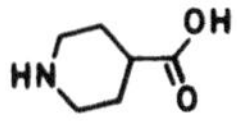	0.33	- - -	+	50	10

GABA receptor affinity binding was measured as described by Krogsgaard-Larsen et al. (18). GABA-agonist activities *in vivo* were studies using microelectrophoretic techniques, and the potencies of neuronal depressants are indicated relative to that of GABA (- - -). Unilateral intra-nigral injections were performed, and turning responses were measured as described by Scheel-Krüger *et al*. (25).

[a] From Johnston *et al*. (50).

[b] From Johnston *et al*. (57).

[c] D. R. Curtis (unpublished results).

[d] From Enna *et al*. (58).

in Table III are remarkably weak compared to their potencies in electrophysiological and receptor affinity binding experiments.

In the case of *trans*-4-aminocrotonic acid (*t*-ACA) the very weak and short-lived turning response observed may be the result of its rapid transamination, since *t*-ACA is a substrate for GABA-T *in vitro* (59). Furthermore, *t*-ACA is a potent inhibitor of neuronal GABA uptake (60) and it may be a substrate for the GABA transport carrier. Accordingly *S*-(-)-*trans*-4-methyl-4-aminocrotonic acid (*S*-(-)-4-Me-*t*-ACA), which is a specific GABA receptor agonist (42, 50), is more potent with respect to induction of the turning response. Isoguvacine was only moderately potent in this animal model, but markedly prolonged the duration of the turning response. The structure of isoguvacine suggests that this compound may interact irreversibly with GABA receptor sites. The compounds listed in Table III may be slow at penetrating membrane structures surrounding synapses, which could explain their relatively weak effects after intranigral injection.

Isoguvacine apparently does not easily penetrate the blood-brain barrier after systemic administration. Hence, a variety of its derivatives have been developed as part of a systematic search for "pro-drugs" of isoguvacine. The goal of this approach is to develop a non-toxic derivative of isoguvacine, which easily penetrates into the brain after systemic administration and which is subsequently hydrolyzed therein to give isoguvacine. A "pro-drug" of isoguvacine which meets these pharmacokinetic requirements may be of pharmacological and possibly of therapeutic interest.

ACKNOWLEDGEMENTS

This project has been supported by The Danish Medical Research Council. Electrophysiological experiments were performed by Professor D. R. Curtis. Professor D. R. Curtis and Dr. G. A. R. Johnston of The Australian National University, Canberra, Australia, and Dr. J. Scheel-Krüger, Sct. Hans Hospital, Roskilde, Denmark, are gratefully acknowledged for their collaboration and for their stimulating advice and discussions. Thanks are due to Mrs. B. Hare and Mr. S. Stilling for skillful secretarial and technical assistance.

REFERENCES

(1) Hornykiewicz, O., Lloyd, K. G. and Davidson, L. (1976) In: GABA in Nervous System Function (E. Roberts, T. N. Chase, and D. B. Tower, eds.) pp. 479-485, Raven Press, New York.
(2) Lloyd, K. G., Möhler, H., Heitz, Ph. and Bartholini, G. (1975) J. Neurochem. 25, 789-795.
(3) Lloyd, K. G., Shemen, L. and Hornykiewicz, O. (1977) Brain Res. 127, 269-278.
(4) McGeer, P. L. and McGeer, E. G. (1976) J. Neurochem. 26, 65-76.
(5) McGeer, P. L. and McGeer, E. G. (1976) In: GABA in Nervous System Function (E. Roberts, T. N. Chase, and D. B. Tower, eds.) pp. 487-495, Raven Press, New York.
(6) Enna, S. J., Stern, L. Z., Wastek, G. J. and Yamamura, H. I. (1977) Life Sci. 20, 205-211.
(7) Lloyd, K. G., Dreksler, S. and Bird, E. D. (1977) Life Sci. 21, 747-753.
(8) Crow, T. J. (1974) Nature (Lond.) 252, 634-635.
(9) Meldrum, B. S. (1975) In: International Review of Neurobiology (C. C. Pfeiffer and J. R. Smythies, eds.), Vol. 17, pp. 1-36, Academic Press, New York.
(10) Van Kammen, D. P. (1977) Amer. J. Psychiat. 134, 138-143.
(11) Roberts, E. (1974) Biochem. Pharmacol. 23, 2637-2649.
(12) Brehm, L., Krogsgaard-Larsen, P. and Jacobsen, P. (1979) In: GABA-Neurotransmitters (P. Krogsgaard-Larsen, J. Scheel-Krüger and H. Kofod, eds.) pp. 247-262. Munksgaard, Copenhagen.
(13) Chase, T. N. and Walters, J. R. (1976) In: GABA in Nervous System Function (E. Roberts, T. N. Chase and D. B. Tower, eds.) pp. 497-513, Raven Press, New York.
(14) DeFeudis, F. V. (1977) In: Progress in Neurobiology (G. A. Kerkut and J. W. Phillis, eds.) Vol. 9, 123-145.
(15) Enna, S. J., Ferkany, J. W. and Krogsgaard-Larsen, P. (1979) In: GABA-Neurotransmitters (P. Krogsgaard-Larsen, J. Scheel-Krüger and H. Kofod, eds.) pp. 191-200. Munksgaard, Copenhagen.
(16) Compochiaro, P., Schwarcz, R. and Coyle, J. T. (1977) Brain Res. 136, 501-511.
(17) Enna, S. J. and Snyder, S. H. (1975) Brain Res. 100, 81-97.
(18) Krogsgaard-Larsen, P., Hjeds, H., Curtis, D. R., Lodge, D. and Johnston, G. A. R. (1979) J. Neurochem. (in press).
(19) Snyder, S. H. and Bennett, J. P., Jr. (1976) Ann. Rev. Physiol. 38, 153-175.
(20) Enna, S. J. and Snyder, S. H. (1977) J. Neurochem. 28, 857-860.
(21) Johnston, G. A. R., Beart, P. M., Curtis, D. R., Game, C. J. A., McCulloch, R. M. and MacLachlan, R. M. (1972) Nature New Biology, 240, 219-220.
(22) Curtis, D. R. and Johnston, G. A. R. (1974) Ergebn. Physiol. 69, 97-188.
(23) Curtis, D. R., Duggan, A. W., Felix, D. and Johnston, G. A. R. (1971) Brain Res. 32, 69-96.

(24) Curtis, D. R., Duggan, A. W., Felix, D., Johnston, G. A. R. and McLennan, H. (1971) Brain Res. 33, 57-73.
(25) Scheel-Krüger, J., Arnt, J. and Magelund, G. (1977) Neurosci. Lett. 4, 351-356.
(26) Arnt, J., Scheel-Krüger, J., Magelund, G., Krogsgaard-Larsen, P. and Hjeds, H. (1978) In: Amino Acids as Neurotransmitters (F. Fonnum, ed.) pp. 323-326. Plenum Press, New York.
(27) Brehm, L., Hjeds, H. and Krogsgaard-Larsen, P. (1972) Acta Chem. Scand. 26, 1298-1299.
(28) Johnston, G. A. R., Curtis, D. R., DeGroat, W. C. and Duggan, A. W. (1969) Biochem. Pharmacol. 17, 2488-2489.
(29) Krogsgaard-Larsen, P., Johnston, G. A. R., Curtis, D. R., Game, C. J. A. and McCulloch, R. M. (1975) J. Neurochem. 25, 803-809.
(30) Johnston, G. A. R. (1971) Psychopharmacologia, 22, 230-233.
(31) Schousboe, A., Krogsgaard-Larsen, P., Svenneby, G. and Hertz, L. (1978) Brain Res. 153, 623-626.
(32) Enna, S. J., Collins, J. F. and Snyder, S. H. (1977) Brain Res. 124, 185-190.
(33) Ott, J., Wheaton, P. S. and Chilton, W. S. (1975) Physiol. Chem. & Physics, 7, 381-384.
(34) Snodgrass, S. R. (1978) Nature (Lond.) 273, 392-394.
(35) Theobald, W., Büch, O., Kunz, H. A., Krupp, P., Stenger, E. G. and Heimann, H. (1968) Arzneimittelforsch. 18, 311-315.
(36) Biggio, G., Casu, M., Corda, M. G., Vernaleone, F. and Gessa, G. L. (1977) Life Sci. 21, 525-531.
(37) Krogsgaard-Larsen, P. and Johnston, G. A. R. (1975) J. Neurochem. 25, 797-802.
(38) Johnston, G. A. R., Krogsgaard-Larsen, P. and Stephanson, A. (1975) Nature (Lond.) 258, 627-628.
(39) Krogsgaard-Larsen, P. (1978) In: Amino Acids as Neurotransmitters (F. Fonnum, ed.) pp. 305-321. Plenum Press, New York.
(40) Krogsgaard-Larsen, P., Johnston, G. A. R., Lodge, D. and Curtis, D. R. (1977) Nature (Lond.) 268, 53-55
(41) Krogsgaard-Larsen, P. and Johnston, G. A. R. (1978) J. Neurochem. 30, 1377-1382.
(42) Krogsgaard-Larsen, P., Honoré, T. and Thyssen, K. (1978) In: GABA-Neurotransmitters (H. Kofod, P. Krogsgaard-Larsen and J. Scheel-Krüger, eds.) Munksgaard, Copenhagen (in press).
(43) Krogsgaard-Larsen, P. and Christensen, S. B. (1974) Acta Chem. Scand. B28, 636-640.
(44) Brehm, L., Krogsgaard-Larsen, P. and Hjeds, H. (1974) Acta Chem. Scand. B28, 308-316.
(45) Hjeds, H. and Krogsgaard-Larsen, P. (1976) Acta Chem. Scand. B30, 567-573.
(46) Krogsgaard-Larsen, P., Larsen, A. L. N. and Thyssen, K. (1978) Acta Chem. Scand. B32, 469-477.
(47) Krogsgaard-Larsen, P. and Hjeds, H. (1974) Acta Chem. Scand. B28, 533-538.
(48) Krogsgaard-Larsen, P. (1977) Acta Chem. Scand. B31, 584-588.

(49) Krogsgaard-Larsen, P., Natova, L. and Christensen, S. B. (1977) Acta Chem. Scand. B31, 577-583.
(50) Johnston, G. A. R., Allan, R. D., Kennedy, S. M. E. and Twitchin, B. (1979) In: GABA-Neurotransmitters (P. Krogsgaard-Larsen, J. Scheel-Kruger and H. Kofod, eds.) pp. 149-164. Munksgaard, Copenhagen.
(51) Johnston, G. A. R. (1976) In: GABA in Nervous System Function (E. Roberts, T. N. Chase and D. B. Tower, eds.) pp. 395-411, Raven Press, New York.
(52) Mao, C. C., Guidotti, A. and Costa, E. (1975) Naunyn-Schmiedeberg's Arch. Pharmacol. 289, 369-378.
(53) Scheel-Krüger, J., Christensen, A. V. and Arnt, J. (1978) Life Sci. 22, 75-84.
(54) Christensen, A. V., Arnt, J. and Scheel-Krüger, J. (1978) Eur. J. Pharmacol. 48, 459-462.
(55) Lykkeberg, J. and Krogsgaard-Larsen, P. (1976) Acta Chem. Scand. B30, 781-785.
(56) Krogsgaard-Larsen, P., Hjeds, H., Curtis, D. R., Lodge, D. and Johnston, G. A. R. (1977) Proc. Int. Soc. Neurochem. 6, 626.
(57) Johnston, G. A. R., Curtis, D. R., Beart, P. M., Game, C. J. A., McCulloch, R. M. and Twitchin, B. (1975) J. Neurochem. 24, 157-160.
(58) Enna, S. J. and Snyder, S. H. (1977) Mol. Pharmacol. 13, 442-453.
(59) Beart, P. M. and Johnston, G. A. R. (1973) Brain Res. 49, 459-462.
(60) Beart, P. M., Johnston, G. A. R. and Uhr, M. L. (1972) J. Neurochem. 19, 1855-1861.

REGIONAL VARIATION AND CHARACTERISTICS OF GABA-RECEPTORS IN THE MAMMALIAN CNS

S. J. Enna

Departments of Pharmacology and of Neurobiology & Anatomy, The University of Texas Medical School at Houston, Houston, Texas 77025, U.S.A.

INTRODUCTION

Historically, the regional distribution of neurotransmitter systems in the brain has been the subject of considerable investigation(17). Regional distribution studies are of interest since a heterogeneous distribution of a putative neurotransmitter substance, its precursors, and synthesizing enzymes is one of the generally accepted criteria for a transmitter agent. Furthermore, knowledge of transmitter distribution in brain may yield insights into the functional activity regulated by a particular system. Thus, in the past, the high concentration of dopamine in the basal ganglia led to the belief that dopamine is important in regulating movement and muscle tone. While this premise undoubtedly has some validity, most investigators would now agree that other parameters (e.g., turnover rate) are of at least equal importance. In addition to these presynaptic factors, the characteristics of postsynaptic sites must also be analyzed to determine the level of activity of a transmitter system. For example, norepinephrine-sensitive adenylate cyclase is most enriched in cerebellum, an area which has relatively little noradrenergic innervation (29), suggesting that this transmitter plays a definite role in cerebellar function, a role which would not have been predicted if the judgement was based only upon cerebellar norepinephrine content. Another parameter which must be considered in this regard is the number and affinity of receptor recognition sites available to the neurotransmitter in any given brain area. Until recently it was virtually impossible to accurately measure this factor. However, with the development of ligand binding assays to study neurotransmitter recognition sites (33) it is now possible to quantify these sites in discrete brain regions. In general, these studies have indicated that no positive relationship exists between neurotransmitter content and receptor recognition sites in various brain areas (10,11).

The present communication is a review of studies concerned with the regional distribution of GABA-receptors in the mammalian central nervous system (CNS). The results of these investigations have indicated that GABA-receptors are present in virtually all areas of the CNS and that the postsynaptic response to GABA may be regulated not only by receptor density, but also by receptor affinity for GABA.

REGIONAL DISTRIBUTION OF GABA-RECEPTORS

Early neurophysiological studies revealed that virtually all CNS neurones respond to iontophoretically applied GABA, in contrast to the regional variation in response to transmitters such as acetylcholine, leading naturally to the conclusion that the GABA response was non-specific (7). However, intracellular recording studies combined with iontophoresis have revealed that GABA, as well as β-alanine and glycine, can hyperpolarize neurones and that this hyperpolarization could be reversed by intracellular injections of Cl^-; since the action of the natural inhibitory neurotransmitter on these cells could also be reversed by intracellular Cl^- injections, this suggested a physiological role for GABA, though its specificity was still unproven (21). Proof of specificity and the determination of a regional variation in response to GABA was demonstrated pharmacologically when it was shown that the convulsants picrotoxin, and especially bicuculline, could selectively and potently block responses to GABA with little or no effect on the inhibition induced by other amino acids (8). With these pharmacological tools, several investigators, especially Curtis and Johnston and their colleagues, demonstrated neurophysiologically and pharmacologically specific GABA-receptors in cerebral cortex, vestibular cortex, hippocampus, cerebellar cortex, ventrobasal thalamus, olfactory bulb and spinal cord (3).

Neurochemical studies on the regional distribution of GABA in the mammalian CNS indicated that significant quantities of this amino acid could be found in virtually all areas studied with only slight, but significant, regional variation (17). Thus, GABA levels are highest in the substantia nigra, globus pallidus and hypothalamus, intermediate in regions such as the dentate nucleus, putamen, caudate nucleus and colliculi, and low in regions such as the cerebral and cerebellar cortices, hippocampus and spinal cord. However, there exists only a 3- to 4-fold difference in content between the lowest and highest regions, which contrasts with other neurotransmitter systems for which the variation can be from 10- to 100-fold. The regional distributions of glutamate decarboxylase (GAD), the enzyme which catalyzes the conversion of glutamate to GABA, and high-affinity GABA uptake, both of which are markers for GABA-ergic terminals, are similar to that observed for GABA content (13,19,23).

More recently, the anatomical distribution of GABA-ergic neurones has been studied immunocytochemically using antibodies to GAD (1,24, 32). Results obtained with this technique have led to the suggestion that cerebellar Purkinje cells, Golgi II cells, basket cells, and possibly stellate cells all utilize GABA as their transmitter agent (32). Thus, though cerebellar GABA levels are among the lowest found in the brain, it is possible that four of the five intrinsic cell types of this brain region utilize GABA as a transmitter substance, further revealing the inadequacy of testing only neurotransmitter content as an indication of functional importance.

Table I - Regional Distributions of GABA-Receptor Binding in Human and Monkey CNS and of GABA Uptake and Glutamate Decarboxylase (GAD) Activity in Monkey CNS[a]

Region	Human Receptor binding (b)	(c)	(d)	Monkey Uptake (d)	GAD (d)
Cerebrum					
Frontal cortex	175	202	61	179	69
Occipital cortex	-	211	81	157	110
Temporal cortex	-	284	27	218	59
Parietal cortex	-	255	-	-	-
Olfactory cortex	-	186	-	-	-
Cingulate cortex	-	101	60	225	75
Insular cortex	-	306	-	-	-
Pre- and post-central gyrus	217	-	34	163	86
Superior temporal gyrus	-	-	87	240	66
Medial temporal gyrus	-	-	63	220	72
Limbic cortex					
Amygdala	83	113	56	250	69
Hippocampus	211	309	29	115	37
Septum	-	-	48	145	65
Midbrain tegmentum	44	8	28	119	109
Hypothalamus	-	-	16	232	151
Thalamus	94	38	40	66	57
Extrapyramidal areas					
Caudate	100	100	100	100	100
Putamen	44	103	94	88	98
Globus pallidus	39	24	79	251	362
Substantia nigra	28	47	45	173	206
Nucleus accumbens	-	87	-	-	-
Cerebellum; lower brain stem					
Cerebellar cortex	864	495	68	68	61
Anterior cerebellar vermis	369	-	-	-	-
Posterior cerebellar vermis	361	-	-	-	-
Nodule and uvula	-	-	22	84	66
Dentate nucleus	31	23	12	44	66
Inferior olive	64	-	-	-	-
Pons					
Basis	69	-	8	47	9
Tegmentum	33	-	-	-	-
Medulla	25	-	2	80	35
Spinal cord					
Cervical	-	-	3	45	9
Thoracic	-	-	4	30	6
Lumbar	-	-	3	58	20

[a] All values represent percentages of those for caudate nucleus.
[b] Adapted from ref. 11. Caudate value is 36 f-moles/mg protein.
[c] Adapted from ref. 22. Caudate value is 66 f-moles/mg protein.
[d] Adapted from ref. 13. Caudate values are 2.52 f-moles/mg wet wt for receptor binding, 339 f-moles/mg wet wt for uptake and 13 n-moles/mg wet wt/hr for GAD activity.

Studies aimed at determining the CNS regional distribution of GABA-receptor recognition sites were made possible by the development of *in vitro* GABA-receptor binding assays (2,14,25,34). For these assays, a sample of washed CNS membrane fragments is incubated in buffer containing a small amount of $[^3H]$GABA, $[^3H]$muscimol or $[^3H]$bicuculline, in the presence or absence of relatively high concentrations of unlabelled GABA. By subtracting the amount of tritium bound in the presence of the unlabelled amino acid from the total amount bound in the absence of unlabelled amino acid, the amount of isotope which specifically labels the GABA-receptor recognition site (specifically bound GABA) can be determined. Numerous neurochemical and pharmacological studies have indicated that this procedure provides a valid measure of the neurophysiologically relevant GABA-receptor site (14,15,27). Using this assay, the regional distribution of GABA-receptors in some 25 areas of the human and monkey CNS has now been determined (Table I). The results of these studies indicate that in human the highest density of GABA-receptors is found in the cerebellar cortex, followed by cerebral cortex, hippocampus, and cerebellar vermis, the cerebellar vermis displaying one-half to one-third the binding capacity of cerebellar cortex. Areas such as the caudate nucleus, amygdala, thalamus, inferior olive and pons tend to have intermediate values (i.e., about 10 - 20% those of cerebellum). Other areas, including the substantia nigra, globus pallidus, dentate nucleus and medulla, have only 5% or less of the binding capacity of cerebellum. This regional distribution of GABA-receptors in human does not correlate significantly with the regional distributions of presynaptic markers for GABA in monkey CNS, such as GAD activity and high-affinity GABA uptake (Table I), which tend to be highest in the basal ganglia and hypothalamus and lowest in the cerebellum, hippocampus and certain areas of the cerebral cortex. These results on GAD activity and GABA uptake of monkey brain are similar to those reported for human brain (17). Interestingly, the regional distribution of GABA-receptor binding in monkey brain correlated much better with the presynaptic markers for this neurotransmitter, with binding being highest in the basal ganglia and only low to intermediate in the cerebellum (Table I). This suggested that, unlike other neurochemical markers for the GABA system, receptor binding shows some species variation.

In rat CNS, the regional distribution of GABA-receptor binding, using [^{3}H]GABA as a marker, was very similar to that observed in human CNS (Table II). Thus, binding was highest in cerebellum, followed by cerebral cortex, hippocampus and midbrain, with the corpus striatum, medulla and hypothalamus being the next highest areas (about one-fifth of that seen in the cerebellum), and with the spinal cord having the lowest binding (only about one-twentieth the amount of [^{3}H]GABA bound per mg protein of cerebellum).

Recently, Möhler and Okada were able to demonstrate GABA-receptor binding using the labelled antagonist, [^{3}H]bicuculline, as the ligand (25; see also Möhler, this volume). It was found that the regional distribution of [^{3}H]bicuculline binding sites was very similar to that observed using the agonist, [^{3}H]GABA (Table II), with one exception; i.e., cerebral cortical membranes bound at least as much [^{3}H]bicuculline as cerebellum, whereas with [^{3}H]GABA binding in the rat, cerebellar membranes bound significantly more of the isotope than those prepared from cerebral cortex. These findings tend to indicate that, as has been suggested for other neurotransmitter receptors, the GABA-receptor exists in two distinct states; i.e., in agonist and antagonist states (30). This finding is not surprising since the bicuculline binding affinity for the receptor is enhanced by chaotropic agents and unaffected or decreased by Triton treatment, whereas GABA binding is enhanced by Triton and unaffected or decreased by chaotropic agents (15). Thus, these findings from regional distribution studies, coupled with the differences in the chemical characteristics of agonist versus antagonist ligand binding, suggest that the attachment sites for these two ligands are either distinct, but complementary, or that the receptor exists in two interconvertible states.

Table II - Regional Distributions of $[^3H]$GABA and $[^3H]$Bicuculline Binding in Rat CNS

Region	Receptor binding[a]	
	$[^3H]$GABA[b]	$[^3H]$Bicuculline[c]
Cerebral cortex	216	211
Hippocampus	121	146
Corpus striatum	100	100
Midbrain	133	138
Hypothalamus	85	126
Cerebellum	461	172
Medulla oblongata - pons	87	41
Spinal cord	12	22

[a] All values represent percentages of those for corpus striatum. [b] Adapted from ref. 14. Corpus striatum value is 33 f-moles/mg protein. [c] Adapted from ref. 26. Corpus striatum value is 39 f-moles/mg protein.

PHARMACOLOGICAL AND KINETIC CHARACTERISTICS OF THE GABA-RECEPTOR IN HUMAN BRAIN REGIONS

Neurophysiological and biochemical findings which have revealed that GABA-receptors are present in virtually all areas of the mammalian CNS suggest that pharmacological manipulation of this neurotransmitter system will be relatively non-specific. Thus, for example, in Huntington's disease there appears to be a relatively selective loss of GABA neurones in the basal ganglia which is undoubtedly related to the choreiform movements seen in this disorder (16). Based on the foregoing regional distribution studies, it could be argued that administration of a GABA-receptor agonist to a patient with Huntington's disease may not be beneficial since, in addition to striatal GABA-receptors, those GABA-receptors in other unaffected brain areas would also be activated and this might induce side-effects that could override any beneficial response to the drug. However, this may not be the case if, as for the cholinergic and adrenergic systems, there exist two or more pharmacologically distinct types of GABA-receptor. If this were the case, and if the regional distributions of the different receptor types were known, then it may be possible to activate, more specifically, the GABA-receptors of selected brain areas.

To explore this possibility, the pharmacological specificity of [^{3}H]GABA receptor binding was investigated in seven regions of the human brain by determining the concentrations of a variety of agonists required to inhibit [^{3}H]GABA binding by 50% (IC_{50}) (12). If the pharmacological specificity of the GABA-receptor varies among different brain regions, then the relative potencies of a series of agonists should also be expected to vary, assuming that some of these agents are more potent on one type of GABA-receptor than on the other. Results of this investigation have indicated that, with the compounds studied, the pharmacological characteristics of the GABA-receptor are similar in these seven brain areas. Thus, the order of potency of the fourteen agents studied was the same in all brain areas.

As mentioned above, pre-treating membranes with Triton X-100 enhances [^{3}H]GABA binding to its receptor, possibly by removing some endogenous inhibitor of binding (18). Since the foregoing study on pharmacological specificity was conducted on tissue treated with Triton, it was of interest to determine whether a similar profile would be obtained using tissue which had not been treated with the detergent. Although the results of this study indicated that, as observed with Triton-treated tissue, the relative potencies of the agonists remained the same throughout the brain, it was noted that the effect of Triton of enhancing affinity of the agonist for the receptor varied among different brain regions (12). Since the IC_{50} value for a compound represents an approximation of the receptor affinity for that agent, this point can be illustrated by calculating the ratios of the IC_{50} determined in the absence of Triton treatment to that found in membranes pre-treated with Triton (Table III). Thus, in caudate nucleus, frontal cortex, hippocampus and cerebellum, Triton enhanced agonist binding 12- to 50-fold, whereas in midbrain, substantia nigra and dentate nucleus the enhancement was generally less pronounced. This difference was especially evident in the dentate nucleus where Triton treatment either had no effect on receptor affinity or increased it by only 2- to 5-fold (Table III). It is noteworthy that this difference between the dentate nucleus and other brain regions is due to the fact that the IC_{50} for the various agonists in the absence of Triton is similar to the IC_{50} for these compounds in other brain areas after treatment with the detergent. Thus, the IC_{50} for GABA in the dentate nucleus is about 20 nM, with or without Triton treatment, whereas that for GABA in the caudate nucleus is about 200 nM without Triton treatment, but only 17 nM after treatment with the detergent (12). If the enhanced ligand affinity seen after Triton treatment is due to the removal of an endogenous inhibitor of GABA receptor binding, then these results would suggest that the midbrain, substantia nigra and, in particular, the dentate nucleus, contain significantly less of this inhibitor; i.e., under normal circumstances, the affinity of the GABA-receptor in this brain area would be significantly higher than that of other brain areas. If this is the case, then administration of a GABA-agonist would have a somewhat selective effect since those brain regions with higher affinity for the ligand would be activated at lower doses than those regions possessing a lower affinity.

RELATIONSHIP BETWEEN BENZODIAZEPINES AND GABA RECEPTORS IN HUMAN BRAIN

Much evidence suggests that the mechanism of action of benzodiazepines is related to their action on the GABA-ergic system (5,6, 9,20,28). Some of the most convincing evidence has been provided by Polc _et al_. (28), who demonstrated that the enhancement of spinal presynaptic inhibition evoked by benzodiazepines could be abolished

Table III - Human Brain: Regional Distribution of Non-Triton/Triton IC_{50} Ratio for Various GABA Receptor Agonists[a]

Compound	Brain region; IC_{50} ratio						
	Caudate nucleus	Frontal cortex	Hippo-campus	Cerebellar cortex	Mid-brain	Substantia nigra	Dentate nucleus
GABA	12	7	8	4	10	5	1
Muscimol	13	15	16	27	4	4	2
Thiomuscimol	15	25	23	45	6	2	1
Isoguvacine	33	19	18	26	25	14	3
Isoxazole-16	52	16	28	18	40	15	5

[a] Adapted from ref. 12. Each value represents a ratio determined by dividing the IC_{50} value for each compound for displacement of specifically bound [^{3}H]GABA from untreated tissue by the IC_{50} value determined using tissue pre-incubated with 0.05% Triton.

by reducing the content of GABA in the spinal cord. This finding suggested that benzodiazepines require an intact GABA system to be effective, at least with regard to presynaptic inhibition; i.e., that these drugs do not directly activate the GABA-receptor. Further support for the contention that the benzodiazepines do not interact directly with the GABA-receptor recognition site was provided by the findings that these agents do not inhibit [^{3}H]GABA or [^{3}H]bicuculline binding at reasonable concentrations (24,31), and that GABA does not interfere with [^{3}H]diazepam binding to specific benzodiazepine receptor sites in brain (26). Since the concentration of benzodiazepines required to inhibit GABA uptake and metabolism or to enhance neuronal release of GABA are quite high relative to the affinity constant of this drug for its receptor, it is unlikely that the primary action of these agents is mediated by one of these mechanisms. Recently, Guidotti and Costa (18) have proposed a novel mechanism by which the benzodiazepines might facilitate the action of GABA. Briefly, these investigators have isolated an endogenous inhibitor of high-affinity [^{3}H]GABA binding. This inhibitor can be removed from brain membranes by Triton X-100, which would explain the enhanced membrane binding of GABA observed after treatment with this detergent. Furthermore, they have shown that pre-treating brain membranes with diazepam can also remove inhibitor and lead to an enhanced affinity of GABA for its receptor site. These exciting findings suggest that the benzodiazepines might enhance GABA-ergic transmission by removing the endogenous inhibitor of GABA binding, thereby increasing the sensitivity of the receptor for the neurotransmitter.

If this hypothesis is correct, then the regional distribution of [^{3}H]diazepam binding in the brain should parallel the distribution of [^{3}H]GABA receptor binding. In fact, the regional distributions of the two ligands in human brain have been shown to be quite similar (Table IV), having a Spearman correlation coefficient of 0.79 ($P < 0.01$). In contrast, no significant correlation existed between the regional distribution of [^{3}H]diazepam binding in human

Table IV - Comparison of the Regional Distributions of [^{3}H]GABA and [^{3}H]Diazepam Receptor Binding in Human Brain

Region	Receptor binding (% of caudate)			
	[^{3}H]GABA[a]	5-HT[a]	Muscarinic cholinergic[a]	[^{3}H]Diazepam[b]
Frontal cortex	175	230	61	184
Amygdala	83	156	62	116
Hippocampus	211	162	41	173
Thalamus	94	47	38	75
Caudate nucleus	100	100	100	100
Putamen; globus pallidus	42	120	56	82
Cerebellar cortex	864	23	0	132
Dentate nucleus	31	-	-	36
Pons	51	58	3	36
Medulla oblongata	25	26	-	45

[a] Adapted from ref. 11. Caudate values were 36 f-moles/mg protein for GABA, 66 f-moles/mg protein for 5-HT and 480 f-moles/mg protein for muscarinic cholinergic receptor binding. [b] Adapted from ref. 4. Caudate value was 440 f-moles/mg protein.

brain and the regional distribution of serotonin-receptors or muscarinic cholinergic-receptors for which the respective correlation coefficients were 0.59 and 0.24. Although correlations cannot prove causality, these comparisons of the regional distributions of benzodiazepine and neurotransmitter receptors do support the notion that benzodiazepines may act, in some way, at the GABA receptor site.

SUMMARY AND CONCLUSIONS

Neurophysiological, biochemical and histochemical analyses have indicated a heterogeneous distribution of the GABA-ergic system in mammalian brain. The poor correlation between the regional distributions of presynaptic markers and synaptic receptors for GABA suggests that it is impossible to predict the functional importance of this neurotransmitter system in any particular brain region using either pre- or post-synaptic determinations alone. The possibility that the receptor affinity for GABA may vary significantly among different brain regions suggests that this may be another mechanism whereby the brain regulates the activity level of this transmitter system. Thus, in any given brain region, the storage and uptake capacities for GABA, the rates of its synthesis and release, and the number and affinity of GABA receptor recognition sites can all play a role in regulating GABA-ergic tone. Subtle alterations in any one or more of these parameters can lead to dysfunction, whose manifestation will depend upon the brain region involved.

Regional distribution studies can also yield insights into the possible mechanisms of action of centrally-active drugs. Thus, a significant correlation between the regional distribution of a neurotransmitter receptor and a particular class of drug provides circumstantial evidence that the drug interacts with this receptor site. Such a situation appears to exist for the benzodiazepines and the GABA-receptor.

ACKNOWLEDGEMENTS

Preparation of this manuscript was facilitated by grants from the Pharmaceutical Manufacturers Association, the Huntington's Chorea Foundation, Merck Sharpe & Dohme and a U.S.P.H.S. Research Career Development Award (NS-00335).

REFERENCES

(1) Barber, R. P. and Saito, K. (1976) In: GABA in Nervous System Function (E. Roberts, T. N. Chase and D. B. Tower, eds.) pp. 113-132, Raven Press, New York.

(2) Beaumont, K., Chilton, W. S., Yamamura, H. I. and Enna, S. J. (1978) Brain Res. 148, 153-162.

(3) Bloom, F. E. (1972) (ed.) Neurosci. Res. Bull. 10, 121-251.

(4) Braestrup, C., Albrechtsen, R. and Squires, R. (1977) Nature (Lond.) 269, 702-704.

(5) Choi, D. W., Farb, D. H. and Fischbach, G. D. (1977) Nature (Lond.) 269, 342-344.

(6) Costa, E. and Greengard, P. (1975) (eds.) Mechanism of Action of Benzodiazepines, Raven Press, New York.

(7) Crawford, J. M. and Curtis, D. R. (1964) Brit. J. Pharmacol. 23, 313-329.

(8) Curtis, D. R., Duggan, A., Felix, D. and Johnston, G. A. R. (1971) Brain Res. 32, 69-96.

(9) Curtis, D. R., Game, C. J. A. and Lodge, D. (1976) Exp. Brain Res. 25, 413-428.

(10) Enna, S. J. (1978) In: Amino Acids as Chemical Transmitters (F. Fonnum, ed.) Plenum Press, New York, pp. 445-456.

(11) Enna, S. J., Bennett, J. P., Bylund, D. B., Creese, I., Burt, D., Charness, M., Yamamura, H. I., Simantov, R. and Snyder, S. H. (1977) J. Neurochem. 28, 233-236.

(12) Enna, S. J., Ferkany, J. W. and Krogsgaard-Larsen, P. (1979) In: GABA – Neurotransmitters (H. Kofod, P. Krogsgaard-Larsen and J. Scheel-Krüger, eds.) Munksgaard Press, Copenhagen (in press).

(13) Enna, S. J., Kuhar, M. J. and Snyder, S. H. (1975) Brain Res. 93, 168-174.

(14) Enna, S. J. and Snyder, S. H. (1975) Brain Res. 100, 81-97.

(15) Enna, S. J. and Snyder, S. H. (1977) Mol. Pharmacol. 13, 442-453.

(16) Enna, S. J., Stern, L., Wastek, G. and Yamamura, H. I. (1977) Life Sci. 20, 206-212.

(17) Fahn, S. (1976) In: GABA in Nervous System Function (E. Roberts, T. N. Chase and D. B. Tower, eds.) pp. 169-186, Raven Press, New York.

(18) Guidotti, A., Toffano, G. and Costa, E. (1979) In: GABA – Neurotransmitters (H. Kofod, P. Krogsgaard-Larsen and J. Scheel-Krüger, eds.) Munksgaard Press, Copenhagen (in press).

(19) Hökfelt, T., Jonsson, G. and Ljungdahl, A. (1970) Life Sci. 9, 203-212.

(20) Kozhechkin, S. N. and Ostrovskaya, R. U. (1977) Nature (Lond.) 269, 72-73.

(21) Krnjević, K. and Schwartz, S. (1967) Exp. Brain Res. 3, 320-336.

(22) Lloyd, K. G., Shemen, L. and Hornykiewicz, O. (1974) Brain Res. 127, 269-274.

(23) Lowe, I. P., Robins, E. and Eyerman, G. S. (1958) J. Neurochem. 3, 8-18.

(24) McLaughlin, B. J., Wood, J. G., Saito, K., Barber, R., Vaughn, J., Roberts, E. and Wu, J. Y. (1974) Brain Res. 76, 377-391.
(25) Möhler, H. and Okada, T. (1977) Nature (Lond.) 267, 65-67.
(26) Möhler, H. and Okada, T. (1977) Science, 298, 849-851.
(27) Olsen, R. W., Ticku, M. K., Van Ness, P. C. and Greenlee, D. (1978) Brain Res. 139, 277-294.
(28) Polc, P., Möhler, H. and Haefely, W. (1974) Naunyn-Schmiedeberg's Arch. Pharmacol. 284, 319-337.
(29) Rall, T. and Gilman, A. (1970) Neurosci. Res. Bull. 8, 223-232.
(30) Snyder, S. H. (1977) New Engl. J. Med. 296, 266-271.
(31) Snyder, S. H. and Enna, S. J. (1975) In: Mechanism of Action of Benzodiazepines (E. Costa and P. Greengard, eds.) pp. 81-91, Raven Press, New York.
(32) Wood, J. G., McLaughlin, B. J. and Vaughn, J. (1976) In: GABA in Nervous System Function (E. Roberts, T. N. Chase and D. B. Tower, eds.) pp. 133-148, Raven Press, New York.
(33) Yamamura, H. I., Enna, S. J. and Kuhar, M. J. (1978) (eds.) Neurotransmitter Receptor Binding, Raven Press, New York.
(34) Zukin, S. R., Young, A. B. and Snyder, S. H. (1974) Proc. Nat. Acad. Sci., U.S.A. 71, 4802-4807.

THE GABA-RECEPTOR: STEREOSPECIFICITY AND STRUCTURE-ACTIVITY STUDIES

N. G. Bowery, J. F. Collins, G. Cryer, T. D. Inch and N. J. McLaughlin

Department of Pharmacology, St. Thomas's Hospital Medical School, London SE1 7EH; Department of Chemistry, City of London Polytechnic, 31 Jewry Street, London EC3N 2EY; and Chemical Defence Establishment, Porton Down, Salisbury, Wiltshire, England

INTRODUCTION

The amino acid γ-aminobutyric acid (GABA) has now been shown to be an important inhibitory neurotransmitter in the mammalian CNS (1). In order to study and further characterize the GABA-receptor it is important to develop new potent agonists and antagonists. In addition, potent agonists may be potentially useful therapeutic agents (2). To this end we have studied a number of novel compounds using two *in vitro* assay systems, viz: the rat superior cervical ganglion, and a binding assay employing a synaptosomal preparation derived from a P_2 pellet with $[^3H]$muscimol serving as the radio-ligand.

Neurones in the rat superior cervical ganglion have been shown to possess receptors for GABA (3) which when activated produce an increase in membrane conductance (4). The magnitude of the resulting depolarization is concentration-dependent and can be detected at the ganglion surface. Characteristics comparable with those of central GABA-receptors have been shown to exist for this tissue (3),

(2) R=H, Isoguvacine

(2) R=Me, N methyl isoguvacine

(3) isonipecotic acid

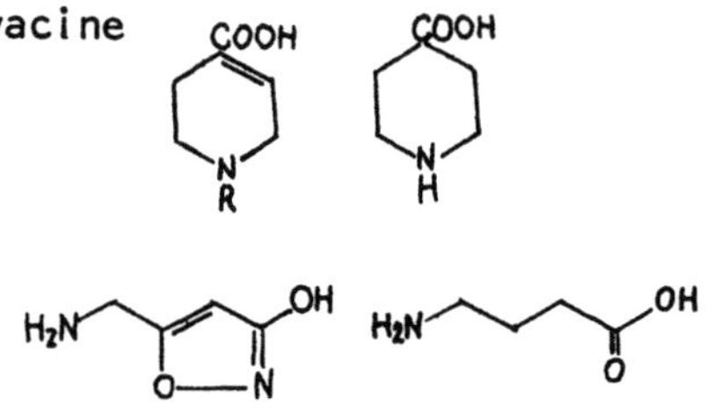

(1) GABA

Fig. 1. Agonist structures

thus providing a simple system in which the quantitative potencies of potential GABA-agonists and GABA-antagonists can be determined. The ganglion also contains nicotinic cholinergic receptors, and the effects of drugs on these can also be monitored.

Muscimol, a naturally occurring isoxazole (Fig. 1; (1)) is a more potent agonist than GABA (5) which has minimal affinity for the GABA uptake system (6). We have found specific [^{3}H]muscimol binding to be saturable and to possess characteristics comparable with the properties of the GABA-receptor derived from electrophysiological experiments. In these respects, our results are broadly similar to those obtained by Enna (7) and Snodgrass (8).

Krogsgaard-Larsen et al. (9) have recently reported that a conformationally-restricted analogue of GABA, isoguvacine (Fig. 1; (2) R = H), is a potent agonist on feline spinal interneurones. They suggested that its activity was comparable with that of muscimol. However, the relative molar potencies of both agonists and antagonists can only be determined when their concentrations in the vicinity of the receptor are known. In iontophoretic studies this is difficult to determine, and comparisons are usually based on currents required to expel sufficient compound to depress or antag-

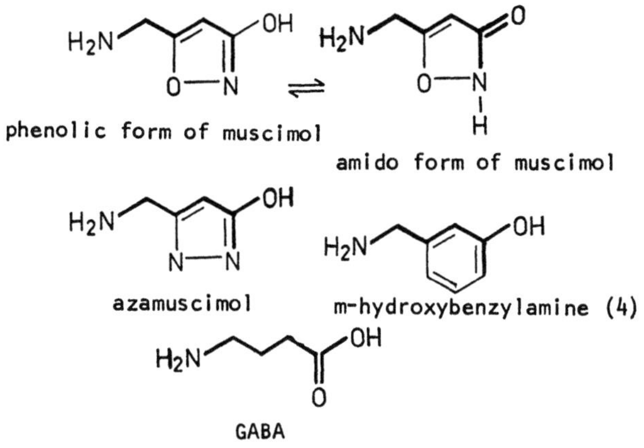

Fig. 2. Muscimol and potential analogues

onize the depression of the cell firing rate by the same amount as GABA or bicuculline (a GABA-antagonist), respectively.

In the present work we have therefore compared the potencies of muscimol, isoguvacine, the saturated analogue of isoguvacine, isonipecotic acid (Fig. 1; (3)) and N-methyl-isoguvacine (Fig. 1; (2) R = Me) in the rat superior cervical ganglion. The relative potencies of muscimol, GABA and isoguvacine have also been determined using the binding assay.

The structure of muscimol is normally formulated as 3-hydroxy-isoxazole, implying that the hydroxyl group is phenolic. Structural comparisons (10) have been drawn between muscimol, muscimol analogues and GABA (Fig. 2), and it has been suggested that the

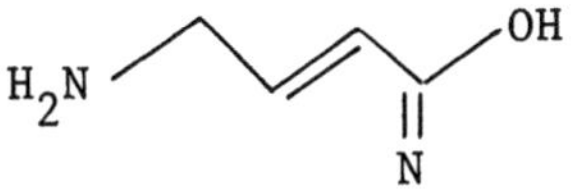 grouping in muscimol broadly corresponds to a similar grouping — 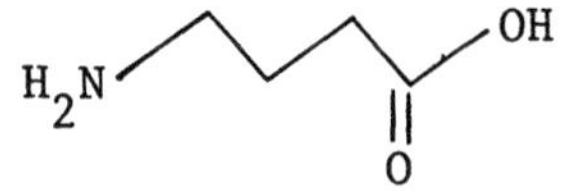 — in GABA and that muscimol and its analogues are restricted analogues of GABA. We have synthesized m-hydroxybenzylamine (Fig. 2; (4)), a restricted analogue of GABA containing a 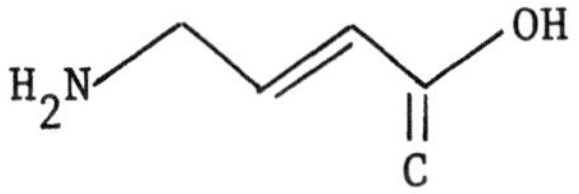 group, and

have evaluated it as a GABA-agonist in the sympathetic ganglion.

Bicyclophosphate esters of the general structure shown in Figure 7 have been shown to be potent GABA-antagonists in both the ganglion and the frog spinal cord (11). Their relative potencies have now been determined in the binding assay using [^{3}H]muscimol as radio-ligand.

Methods and Materials

Ganglia excised from rats anesthetized with urethane (1.4 g/kg) were desheathed and superfused with Krebs-Henseleit solution in the manner described by Brown and Marsh (12). The ganglion surface potential, measured with respect to the post-ganglionic trunk, was continuously monitored using a Smiths Servoscribe (1s) potentiometric recorder. Agonists to be evaluated were dissolved in the superfusion fluid and applied to the tissue for 1-minute periods at intervals of 15 minutes.

For the binding assay, membranes were prepared by a modification of the method of Zukin et al. (13). Briefly, whole rat brains (from three rats) were homogenized in 15 volumes of ice-cold 0.32 M sucrose solution using a Potter-Elvehjem glass homogenizer with a Teflon pestle. The P_1 pellet, collected after centrifugation at 1000 g for 10 minutes, was discarded and the supernatant was centrifuged at 20,000 g for 20 minutes. The pellet obtained was dispersed in cold distilled water using a high-speed stirrer. The suspension was centrifuged at 8000 g for 20 minutes. The supernatant was collected and the pellet, a bilayer with a soft upper coat, was rinsed carefully with the supernatant to collect the upper layer. The combined supernatant was then centrifuged at 48,000 g for 20 minutes. The final crude synaptic membranes (P_2 fraction) were suspended in water and centrifuged at 48,000 g. The washing process was repeated, and the membranes were stored at -30°C for at least 24 hours before being used.

The Assay

The frozen pellet was allowed to thaw and was then suspended in 25 ml distilled water and left for 30 minutes at room temperature. A sample was removed for protein determination by the method of Lowry et al. (14). The suspension was adjusted in volume to provide a concentration of 1 mg protein/ml. This suspension was then transferred in 1-ml aliquots to the appropriate number of 1.5-ml Eppendorf tubes and centrifuged at 9000 g for 2 minutes using a Beckman Microfuge B. The supernatant was removed and discarded. Each pellet was then resuspended in 0.05 M Tris-citric acid buffer (pH 7.1) and incubated with the appropriate drug for 10 minutes at 4°C, followed by incubation for a further 10 minutes in the presence of [^{3}H]muscimol (5 nM). Incubation was terminated by centrifugation at 9500 g for 5 minutes. The supernatant was carefully, but rapidly, removed and the tubes were carefully dried. The pellets were dissolved overnight using Soluene tissue solubilizer. After neutralization of the Soluene by 1.5 M hydrochloric acid, the samples were made up to 11 ml with Toluene-Triton X scintillant containing PPO (0.5%) POPOP (0.05%) and counted in a liquid scintillation spectrometer.

Methylamine-[^{3}H]muscimol (specific activity, 12.5 Ci/m-mole) was obtained from the Radiochemical Centre, Amersham. Its radiochemical purity was checked by thin-layer chromatography on silica gel in an isopropanol:methyl ethyl ketone:ammonia (65:15:25) system and it was found to be > 97% radiochemically pure. The R_f value corresponded to that of inactive muscimol. The lability of the label was tested by boiling [^{3}H]muscimol (1 mCi) in water for 48 hours. Less than 5% of the radioactivity associated with the [^{3}H]muscimol was lost (D. C. Warrell, the Radiochemical Centre; private communication).

Isoguvacine-hydrochloride and N-methyl-isoguvacine were prepared by a modification of the method of Oediger and Joop (15). m-Hydroxybenzylamine was prepared from m-methoxybenzylamine by demethylation with hydrobromic acid. Muscimol-monohydrate was obtained from Research Organics Inc., Cleveland, Ohio. The bicyclophosphate esters were prepared from the corresponding triols by cyclization with phosphorous oxychloride. (-)Bicuculline-methobromide was prepared by the method of Tietel _et al._ (16). Isonipecotic acid was obtained from Dr. C. Cooksey, Department of Chemistry, University College.

RESULTS AND DISCUSSION

Agonists

All five agonist analogues, muscimol, isoguvacine, N-methylisoguvacine, isonipecotic acid and m-hydroxybenzylamine, depolarized the rat superior cervical ganglion in a manner consistent with an action at GABA-receptors. They produced rapid short-lasting depolarizations without any subsequent hyperpolarizations (cf., the action of the cholinomimetic, carbachol, in this tissue, Fig. 3). Responses to each agonist could be blocked by the GABA-antagonists bicuculline-methochloride (BMC) and isopropyl-bicyclophosphate (IPTBO), as illustrated by the results shown in Figure 3. Submaximal matched responses were obtained to GABA (10 μM), muscimol (1.8 μM), isoguvacine (62 μM), N-methylisoguvacine (5.7 mM), isonipecotic acid (0.24 mM) and carbachol (14 μM). Addition of either BMC (20 μM) or IPTBO (24 μM) completely blocked the responses to the analogues without significantly suppressing that to carbachol. Recovery of the responses to all the analogues was obtained 20 minutes after removal of either the BMC or IPTBO. Log dose-response curves for muscimol, isoguvacine and isonipecotic acid were found to be parallel to that for GABA (Fig. 4). Muscimol was the only agonist which was more potent than GABA (5.08 $\pm$ 0.707; mean $\pm$ S.E.M.; 4 experiments). Isoguvacine was found to be 0.23 $\pm$ 0.026 (mean $\pm$ S.E.M.; 4 experiments) as potent as GABA, while isonipecotic acid

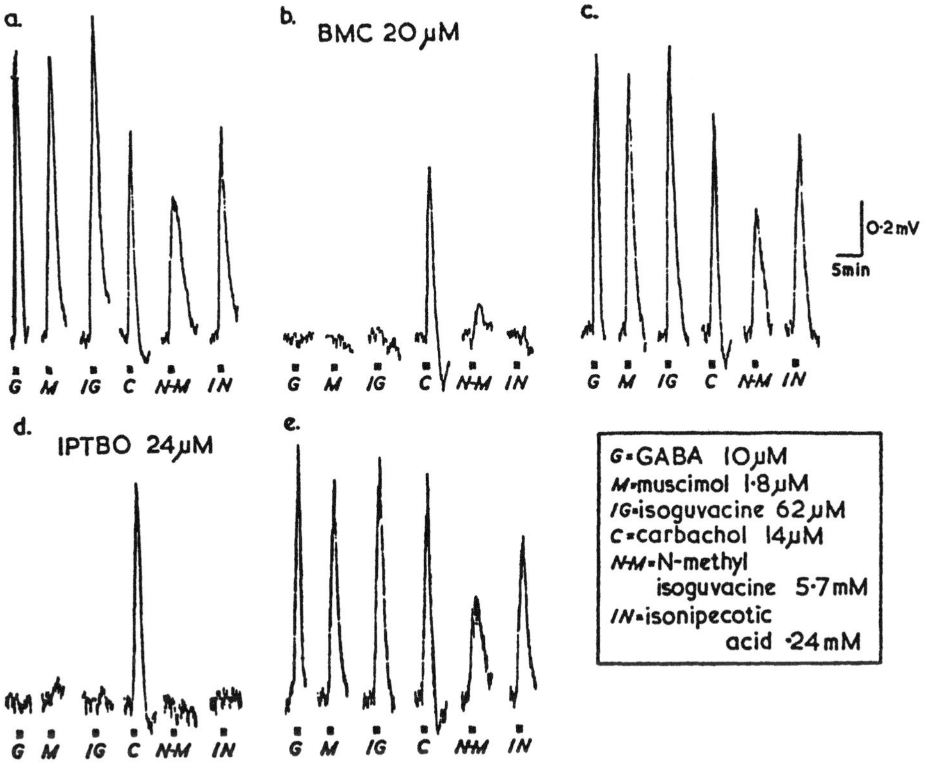

Fig. 3. Antagonism of the responses to the analogues produced by bicuculline-methochloride (BMC) and isopropyl-bicyclophosphate (IPTBO). Submaximal responses were obtained consecutively to GABA (G, 10 μM), muscimol (M, 1.8 μM), isoguvacine (IG, 62 μM), carbachol (C, 14 μM), N-methyliso-guvacine (NM, 5.7 nM) and isonipecotic acid (IN, 240 μM). Each substance was applied for 1 minute at intervals of at least 15 minutes indicated by the gaps between responses. Panel a: control responses; b: during continuous superfusion with BMC (20 μM); c: recovery; d: during IPTBO (24 μM); e: recovery. Hyoscine-hydrobromide (2.6 μM) was present throughout to restrict the action of carbachol to nicotinic receptors. All the responses were obtained in a single ganglion. Calibration 0.2 mV and 5 minutes.

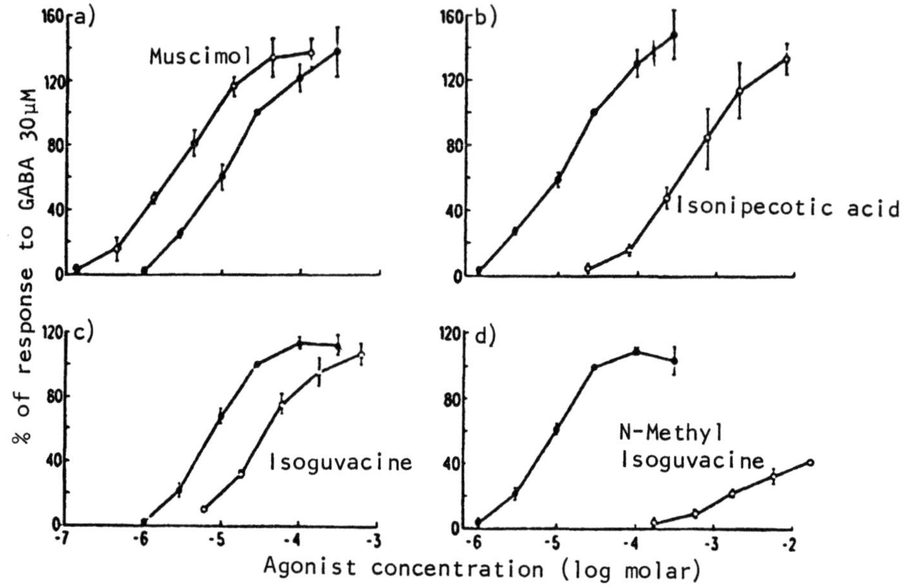

Fig. 4. Log dose-response curves for: a) muscimol; b) isonipecotic acid; c) isoguvacine; d) N-methyl-isoguvacine and GABA. (a) Induced depolarization of the isolated rat superior cervical ganglion. Data were obtained from three (b and d) or four ganglia (a and c) for each compound. In each experiment comparison was made with GABA and responses were normalized with respect to that produced by 30 μM GABA to facilitate comparison among experiments. Each point is the mean value (vertical bars = S.E.M.). Abscissa: log molar agonist concentration.

was found to be 0.011 + 0.0028 (mean ± S.E.M.; 3 experiments). Both N-methyl-isoguvacine (0.001 GABA) and m-hydroxybenzylamine (0.00015 GABA) were found to be very weak agonists.

The relative potencies of muscimol, GABA and isoguvacine were also measured using the [^{3}H]muscimol binding assay (Fig. 5). Again, muscimol was found to be more potent than GABA (4.5 x GABA), while isoguvacine was less potent (0.05 x GABA). Minimal displacement of the [^{3}H]muscimol binding was observed even at very high concen-

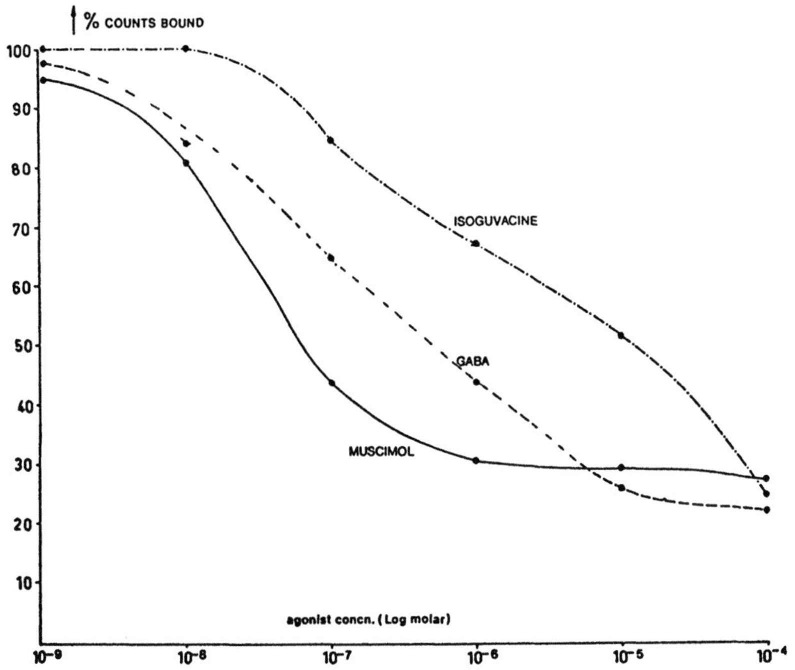

Fig. 5. Competition for the [^{3}H]muscimol binding sites by GABA and isoguvacine. The points represent the means of at least three determinations. The relative potencies were estimated at 60% of the total counts bound level.

trations (1 mM) of m-hydroxybenzylamine.

Although the potency of muscimol in both the ganglion and the binding assay compares favourably with the values obtained iontophoretically and in binding studies using other radio-ligands (17, 18), the potencies of isoguvacine and isonipecotic acid appear to be at least an order of magnitude weaker in the *in vitro* assays than when applied iontophoretically to spinal interneurones.

It has been shown (19) that isoguvacine and isonipecotic acid have little or no effect on neuronal [^{3}H]GABA uptake in rat cortical slices and also that they are without effect on glial [^{3}H]GABA uptake processes in sympathetic ganglia and rat retina. It is therefore unlikely that the weaker activity of these analogues can be attributed to any difference in their interaction with GABA transport processes. Isoguvacine, administered intra-cerebroventricularly, has been shown to be a weaker anticonvulsant than muscimol (20).

It is interesting to note that m-hydroxybenzylamine is a very much less potent agonist than muscimol. Muscimol, although normally formulated as an isoxazole (Fig. 2), is in fact a tautomeric mixture of the amido isoxazolone form and the phenolic isoxazole form. m-Hydroxybenzylamine is only capable of existing in the phenolic form. It may well be that muscimol acts as an agonist primarily in the isoxazolone tautomer. Also, the existence of the ring oxygen atom (O-1) in muscimol may be an important factor in the binding of muscimol to the receptor. As the unshared electron pairs on the oxygen are not required to make up the aromatic sextet, they are available for hydrogen bonding to the receptor. Azamuscimol (Fig. 2) has also been found to be a much weaker agonist than muscimol itself (17). This is consistent with the hydrogen-bonding theory.

Antagonists

The relative potencies of the GABA-antagonist (+)bicuculline-methobromide and its enantiomer (-)bicuculline-methobromide have been determined using the binding assay (Fig. 6). (-)Bicuculline-methobromide is very much weaker (by at least 2 - 3 orders of magnitude) than (+)bicuculline-methobromide in competing for [^{3}H]muscimol binding. This stereospecificity has been observed in other binding assays (8,18,26) and has also been demonstrated iontophoretically (21).

Bicyclophosphates having the general structure shown in Figure 7 have been found to be potent convulsants and GABA-antagonists in the ganglion and frog spinal cord (11). The t-butyl-bicyclophosphate, for example, has been found in the ganglion to be 5 times as potent as bicuculline as a GABA-antagonist, whereas the neopentyl-bicyclophosphate appears to be about twice as potent as bicuculline.

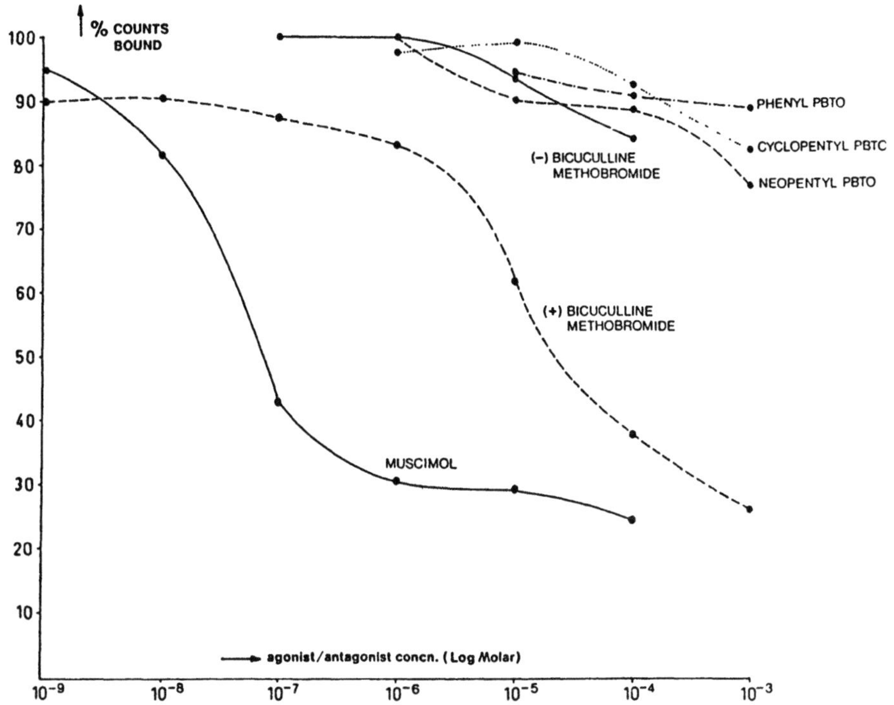

Fig. 6. Competition for the [^{3}H]muscimol binding by GABA antagonists. The points represent the means of at least three determinations. The relative potencies were estimated at 60% of the total counts bound level.

However, we have found, using the [^{3}H]muscimol binding assay, that even at high concentrations (10^{-4} M) of bicyclophosphates only small amounts of the ligand were displaced (Fig. 6). In experiments employing [^{3}H]GABA as radio-ligand it had been previously shown that some displacement of the [^{3}H]ligand could be observed with the bicyclophosphates (27). Olsen (22) has recently suggested, as a result of experiments with a binding assay employing [^{3}H]α-dihydropicrotoxinin, that picrotoxinin, tetramethylenedisulphotetramine (23) and the bicyclophosphates act, not at the GABA-receptor but at the associated Cl^- ionophore. The results obtained with the

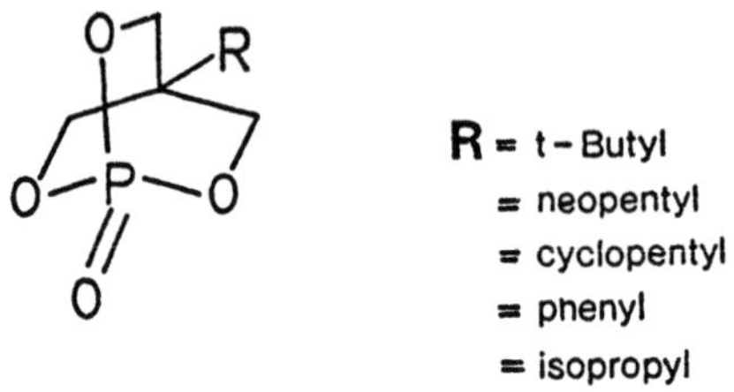

Fig. 7. Structures of potent bicyclophosphates.

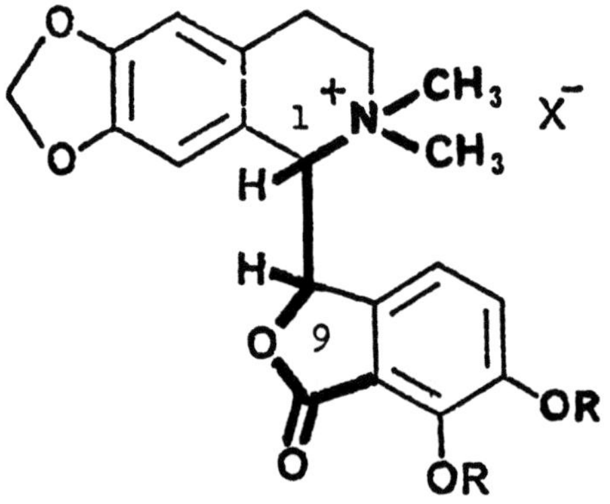

Phthaleidoisoquinoline structure

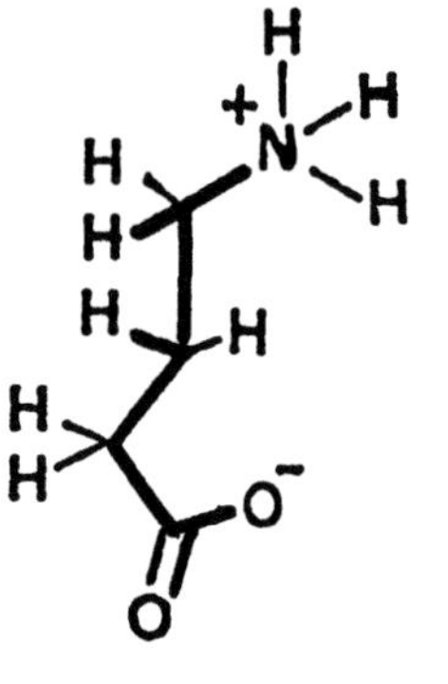

GABA

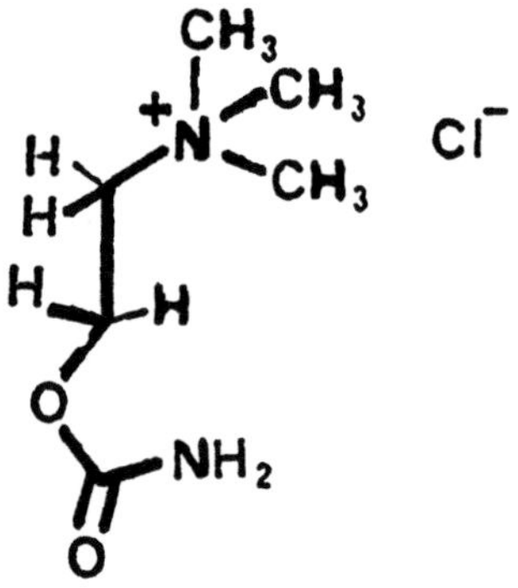

Carbachol

Fig. 8. Comparison of the phthaleidoisoquinoline structure with GABA and carbachol.

Table I - The Relative Potencies of the Phthaleidoisoquinoline Alkaloids as GABA and Carbachol Antagonists ((+)-Bicuculline-Methohalide = 1)

Compound (as methohalide)	vs GABA	vs Carbachol
Bicuculline		
1S,9R	1	1
1R,9S	0.0001	2
Narcotine		
1R,9S	0.0001	0.5
1S,9R	0.1	1 - 1.5
1S,9S	0.0001	1
Hydrastine		
1R,9S	0.0001	1
1R,9R	0.04	1
1S,9S	0.0001	5-10

Relative potencies were obtained at the 50% level of antagonism of carbachol (33 μM) and GABA (30 μM) to matched submaximal responses. At 9.6 μM (+)-bicuculline-methochloride caused a 31% diminution in the response to GABA and only a 5.5% diminution in that to carbachol.

bicyclophosphates in the binding assay with [^{3}H]muscimol demonstrate that they do not act at the postsynaptic binding site for muscimol, and support Olsen's suggestion.

Both bicuculline and bicuculline-methochloride have been shown to be competitive inhibitors of brain acetylcholinesterase in an *in vitro* bio-assay (24). However it has been suggested (25), as a result of an *in vivo* study, that these results are not of great pharmacological significance.

We have employed the superior cervical ganglion to evaluate a series of closely related phthaleidoisoquinoline alkaloids principally as GABA-antagonists, but also as nicotinic receptor antagonists. These alkaloids possess two chiral centers (Fig. 8; C-1 and 9). (+)Bicuculline-methochloride, which has the 1S 9R configuration, is the most potent GABA-antagonist in the series but it also causes some reduction in responses to carbachol (at 9.6 μM this was 5.5 ± 2.6%; mean ± S.E.M.; 18 experiments). (-)Bicuculline-methohalide, while ineffective as a GABA-antagonist (0.0001 times (+)bicuculline), is twice as potent as (+)bicuculline-methohalide as a carbachol-antagonist.

Results obtained with the other compounds are shown in Table I. From the table it can be readily seen that while a very clear stereo-specificity is required for GABA antagonism, all the phthaleido-isoquinoline alkaloids examined were carbachol-antagonists of approximately equal low potency. Obviously, the stereochemical configuration of the alkaloids, while important for GABA antagonism, is not so for carbachol antagonism. As can be seen from Figure 8, carbachol, like GABA, has a reasonable structural resemblance to important structural elements of the phthaleidoisoquinoline alkaloids.

In conclusion, it emerges from this study that the GABA-receptor is both regio- and stereo-specific.

REFERENCES

(1) Curtis, D. R. and Johnston, G. A. R. (1974) Ergebn. Physiol. 69, 98.

(2) Chase, T. N. and Walters, J. R. (1967) In: GABA in Nervous System Function (E. Roberts, T. N. Chase and D. B. Tower, eds.) pp. 497-513, Raven Press, New York.

(3) Bowery, N. G. and Brown, D. A. (1974) Brit. J. Pharmacol. 50, 205.

(4) Adams, P. R. and Brown, D. A. (1975) J. Physiol. (Lond.) 250, 85.

(5) Curtis, D. R., Duggan, A. W., Felix, D. and Johnston, G. A. R. (1971) Brain Res. 32, 69.

(6) Johnston, G. A. R. (1971) Psychopharmacologia, 22, 230.

(7) Enna, S. J. (1978) In: Amino Acids as Chemical Transmitters (F. Fonnum, ed.) Plenum Press, New York (in press).

(8) Snodgrass, S. R. (1978) Nature (Lond.) 273, 392.
(9) Krogsgaard-Larsen, P., Johnston, G. A. R., Lodge, D. and Curtis, D. R. (1977) Nature (Lond.) 268, 53.
(10) Krogsgaard, Larsen, P., Johnston, G. A. R., Curtis, D. R., Game, C. J. A. and McCulloch, R. M. (1975) J. Neurochem. 25, 83.
(11) Bowery, N. G., Collins, J. F., Hill, R. G. and Pearson, S. (1977) Brit. J. Pharmacol. 60, 275 P.
(12) Brown, D. A. and Marsh, S. (1975) J. Physiol. (Lond.) 246, 24 P.
(13) Zukin, S. R., Young, A. B. and Snyder, S. H. (1974) Proc. Nat. Acad. Sci., U.S.A. 71, 4802.
(14) Lowry, O. H., Rosebrough, N. J., Farr, A. L. and Randall, R. J. (1951) J. Biol. Chem. 193, 265.
(15) Oediger, H. and Joop, N. (1972) Ann. Chem. 764, 21.
(16) Teitel, S., O'Brien, J. and Brossi, A. (1972) J. Org. Chem. 37, 1879.
(17) Olsen, R. W., Greenlee, D., Van Ness, P. and Ticku, M. K. (1978) In: Amino Acids as Chemical Transmitters (F. Fonnum, ed.) Plenum Press, New York (in press).
(18) Möhler, H. and Okada, T. (1977) Nature (Lond.) 267, 65.
(19) Bowery, N. G., Collins, J. F., Hudson, A. L. and Neal, M. J. (1978) Experientia (in press).
(20) Anlezark, G., Collins, J. F. and Meldrum, B. (1977) Neurosci. Lett. 7, 397.
(21) Collins, J. F. and Hill, R. G. (1974) Nature (Lond.) 249, 845.
(22) Ticku, M. K., Ban, M. and Olsen, R. W. (1978) Mol. Pharmacol. 14, 391.
(23) Bowery, N. G., Brown, D. A. and Collins, J. F. (1975) Brit. J. Pharmacol. 53, 422.
(24) Svenneby, G. and Roberts, E. (1973) J. Neurochem. 21, 1025.
(25) Curtis, D. R., Johnston, G. A. R., Game, C. J. A. and McCulloch, R. P. (1974) J. Neurochem. 23, 605.
(26) Enna, S. and Snyder, S. H. (1976) Brain Res. 115, 174.
(27) Snyder, S. H., Enna, S. J. and Collins, J. F. (1977) Brain Res. 124, 185.

GABA RECEPTOR BINDING WITH [^{3}H]GABA AND [^{3}H](+)BICUCULLINE-METHIODIDE: AN IMPROVED METHOD

H. Möhler

Pharmaceutical Research Department, F. Hoffman-LaRoche & Co. Ltd., 4002 Basel, Switzerland

INTRODUCTION

In vitro GABA receptor binding studies have so far been performed using either GABA-agonists (GABA, muscimol) or a GABA-antagonist ((+)bicuculline-methiodide; [^{3}H](+)BM) as labelled ligands (1,2,3,4, 10). These two approaches for investigating GABA-receptors, however, differ not only in the assay conditions, but also in the membrane material used, since the treatment of the synaptic membrane fraction prior to the binding assay is different when either [^{3}H]GABA or [^{3}H](+)BM is used as labelled ligand. For example, for assays of high-affinity [^{3}H]GABA binding, membrane fractions are used after having been treated with Triton X-100 (1). However, in a membrane preparation treated in this way [^{3}H](+)BM binding is not detectable (4). It is difficult, therefore, to relate GABA-agonist and GABA-antagonist binding data to the same population of GABA-receptors when the membrane material used in both assays is not identical.

In this paper a method is described which allows [^{3}H]GABA and [^{3}H](+)BM binding studies to be performed under identical assay conditions using the same synaptic membrane fraction for both ligands. Thus, the GABA-receptor can be studied in the same structural and functional integrity and in the same membrane environment, using either a receptor agonist ([^{3}H]GABA) or an antagonist ([^{3}H](+)BM) as labelled ligand. This approach should help to clarify the structure of GABA-receptors and to resolve possible heterogeneous populations of GABA-receptors.

METHODS

Tissue Preparation

Rat cerebellar membranes were prepared according to Zukin et al. (5). Briefly, the tissue was homogenized in 0.32 M sucrose solution (15 vol) in a glass-Teflon homogenizer and centrifuged at 1000 g for 10 minutes. The supernatant was centrifuged at 20,000 g for 20 minutes and the pellet, including the soft white upper layer, was resuspended in water (4°C; 20 times the amount of original wet weight of tissue). The suspension was treated with a Polytron PT-20 homogenizer (setting 6; 30 sec), and centrifuged at 8000 g for 20 minutes; the supernatant plus the soft buffy upper coat of the pellet were suspended in H_2O and distributed into various tubes in amounts sufficient for the binding assays performed in one day. The suspensions were centrifuged at 48,000 g for 20 minutes, and the pellets were kept frozen at -30°. Binding assays were performed after resuspending the thawed pellet in 50 mM Tris-HCl buffer (pH 7.4) at room temperature or at 0°C, depending on whether the binding assay was to be performed at room temperature or at 4°C. About 20 mg protein of cerebellar synaptic membrane suspension were obtained from 1 g wet weight of original cerebellar tissue.

[^{3}H](+)BM Binding

The assay mixture contained, in a final volume of 2 ml, 150 mM $NaClO_4$, 50 mM Tris-HCl buffer (pH 7.4), [^{3}H](+)BM (generally 5 nM) and 2 mg protein of cerebellar synaptic membrane fraction in the presence or absence of 1 mM GABA. After incubation for 15 minutes at 0°C, the assay was terminated by centrifugation at 48,000 g for

10 minutes, the pellet was rinsed with 5 ml and then 10 ml of ice-cold 50 mM Tris-HCl-buffer (pH 7.4). After drying the walls of the tube, the pellet was solubilized in 2 mg Protosol and then rinsed with scintillant into a counting vial. At 5 nM [^{3}H](+)BM, total binding amounted to about 6,300 dpm/mg protein and non-specific binding amounted to about 2,500 dpm/mg protein; specific binding ranged between 50 - 60% of total binding. The [^{3}H](+)BM binding assay was performed at 0°C, i.e., at the same temperature as that generally used for the [^{3}H]GABA binding assay (see below). However, [^{3}H](+)BM binding can also be performed at room temperature with slightly higher amounts of specific binding but with IC_{50} values of inhibitors almost identical to those found at 0°C.

[^{3}H]GABA Binding

The assay was performed at 0°C in exactly the same way as that for [^{3}H](+)BM, except that [^{3}H]GABA (generally 8 nM) was used as labelled ligand with 1 mg of cerebellar synaptic membrane fraction. At 8 nM [^{3}H]GABA, total binding amounted to about 46,000 dpm/mg protein, non-specific binding amounted to about 7,000 dpm/mg protein, and specific binding ranged between 85 - 90% of total binding.

Materials

[^{3}H](+)BM (specific radioactivity, 9.3 Ci/m-mole) was prepared as previously described (6). [^{3}H]GABA (specific radioactivity, 34.5 Ci/m-mole) was purchased from New England Nuclear.

RESULTS AND DISCUSSION

Chaotropic Ions and GABA Receptor Binding

Chaotropic ions, like SCN^- or ClO_4^-, with their high partial molar entropies (of the aqueous ion) can alter the structure of water in the direction of more disorder and can make water more "lipophilic" (7). Both effects provide the basis for their use e.g. for solubilizing particulate proteins (7,8). This chaotropic action also appeared promising for in vitro binding studies of neurotransmitter-receptor systems, i.e., for either removing a component which might block the accessibility of a binding site to the ligand, or for removing a physiological regulatory component which could alter the receptor properties. It should be kept in mind that chaotropic ions are not included in the binding assays to mimic the effect of a physiological ion, but to alter the membrane structure in such a way as to render it suitable for binding studies.

By including the chaotropic ion ClO_4^- (150 mM) it was possible to perform studies on GABA receptor binding with $[^3H]$GABA or $[^3H]$(+) BM as labelled ligand under identical assay conditions; i.e., using the same membrane preparation. GABA receptor binding in the presence of ClO_4^- offers two main advantages over previously published procedures (1,3): 1) Only by using an identical membrane preparation can $[^3H]$GABA and $[^3H]$(+)BM binding reflect the properties of the GABA-receptor in the same structural and functional integrity and in the same membrane environment; 2) The concomitant use of GABA-agonists and GABA-antagonist as labelled ligands under the same assay conditions offers the most promising approach in searching for heterogeneous populations of GABA-receptors in vitro.

GABA Receptor Binding with $[^3H]$(+)BM as Labelled Ligand in the Presence of ClO_4^-

Use of the chaotropic ion, ClO_4^- (150 mM), improved $[^3H]$(+)BM specific binding considerably in comparison to previously published results (3,4) that were obtained using SCN^- as the chaotropic ion. Specific binding of $[^3H]$(+)BM in the presence of ClO_4^- amounted to about 60% of total binding as compared to the previous value of about 30% (at 5 nM $[^3H]$(+)BM). Also, a high-affinity binding site for (+)BM could be detected in addition to the previously described low-affinity binding site (Table I).

Table I - Apparent Dissociation Constants (K_D) and Maximal Binding Capacities (B_{max}) of Specific [^{3}H]GABA and [^{3}H](+)BM Binding Using the Same Membrane Preparation and the Same Assay Conditions (150 mM $NaClO_4$) for Both Labelled Ligands

	K_D (nM)	B_{max} (p-mole/mg protein)
Specific [^{3}H]GABA binding		
High affinity	38 ± 4	2.8 ± 0.1
Low affinity	171 ± 35	5.4 ± 0.3
Specific [^{3}H](+)BM binding		
High affinity	85 ± 12	3.3 ± 0.2
Low affinity	362 ± 34	7.0 ± 0.6

Binding assays were performed at 0°C with previously frozen synaptic membrane fractions of cerebellum, using 150 mM $NaClO_4$ buffer containing 5 - 600 nM [^{3}H]GABA or [^{3}H](+)BM (see "Methods").

Regarding the properties of the high-affinity binding site, analysis of saturation kinetics in previously-frozen synaptic membrane fractions provided an apparent dissociation constant K_D = 85 ± 12 nM and a maximum binding capacity of 3.3 ± 0.2 p-mole/mg protein for specific [^{3}H](+)BM binding. Similar values were obtained using freshly-prepared synaptic membrane fractions of cerebellum. The potencies of GABA-agonists and GABA-antagonists in inhibiting high-affinity specific [^{3}H](+)BM binding (Table II) were in line with their neurophysiological potencies. Thus, high-affinity specific [^{3}H](+)BM binding in the ClO_4^- assay appears to represent an interaction with the synaptic GABA-receptor. It is noteworthy that GABA-agonists showed a much lower potency at inhibiting [^{3}H](+)BM binding than [^{3}H]GABA binding, whereas the inverse relationship held for GABA-antagonists. This finding can be explained best by assuming a

Table II - Inhibition of the Specific Binding of $[^3H]$GABA and $[^3H]$(+)BM by Various Compounds Using the Same Membrane Preparation and the Same Assay Conditions for Both Labelled Ligands

Compound	IC_{50} (nM)	
	$[^3H]$GABA	$[^3H]$(+)BM
Muscimol	6.9 ± 0.5	58 ± 7
GABA	28.7 ± 6.2	480 ± 80
(+)Bicuculline	110 ± 14	10 ± 2
(+)Bicuculline-methiodide	570 ± 70	101 ± 16
(-)Bicuculline	> 100,000	72,500 ± 5,000

Binding studies were performed at 0°C with previously-frozen synaptic membrane fractions of cerebellum, using 150 mM $NaClO_4$ buffer containing 8 nM $[^3H]$GABA or 5 nM $[^3H]$(+)BM (see "Methods").

GABA-receptor in two conformations: an agonist conformation with a high affinity for GABA-agonists and a low affinity for GABA-antagonists; an antagonist conformation with reversed properties (4).

GABA Receptor Binding with $[^3H]$GABA as Labelled Ligand in the Presence of ClO_4^-

The results on specific $[^3H]$GABA binding in the presence of ClO_4^- agreed well with those found previously with Triton X-100 treated membrane fractions, except that GABA-antagonists of the bicuculline type had much higher potencies. Saturation kinetics of specific $[^3H]$GABA binding under the conditions presented here (150 mM ClO_4^-) indicated two binding sites, one with a high affinity (K_D = 38 ± 4 nM; B_{max} = 2.8 ± 0.1 p-moles/mg protein) and one with a low affinity (K_D = 171 ± 35 nM; B_{max} = 5.4 ± 0.3 p-mole/mg protein). The ligand specificity of the high-affinity binding site was in agreement with the rank orders of potency of GABA-agonists and GABA-antagonists in neurophysiological experiments, indicating that specific [^{3}H]GABA binding in the presence of ClO_4^- reflects an interaction with the synaptic GABA-receptor (Table II).

The GABA receptor binding sites demonstrated here for cerebellum appear to be similar to those described by Enna and Snyder (9) for Triton X-100 treated membranes of whole brain. 1) High- and a low-affinity GABA binding sites were observed under both conditions. 2) The potency of GABA-agonists in inhibiting high-affinity specific [^{3}H]GABA binding was nearly identical in the two systems. 3) Specific binding of [^{3}H]GABA (determined at 8 nM or 25 nM [^{3}H]GABA) accounted for about 90% of total binding in both systems. 4) The finding that the potencies of GABA-antagonists at inhibiting [^{3}H]GABA specific binding in the presence of ClO_4^- were about 10 times higher than that described by Enna and Snyder for Triton-treated membrane preparations is consistent with their observation that SCN^- (a chaotropic ion) enhances the potency of GABA-antagonists (9).

Thus for studies on high-affinity GABA binding, pre-treatment of synaptic membrane fractions with Triton X-100, with its drastic protein solubilizing action (4), has become obsolete and can be omitted in favor of including 150 mM $NaClO_4$ in the binding assay, which represents a milder, simpler, and less time-consuming procedure. The main advantage of this procedure, however, is that the binding of [^{3}H]GABA, as well as that of [^{3}H](+)BM, can be studied using the same membrane preparation. This method presents the basis for comparative studies on the GABA-receptor using both GABA-agonists and GABA-antagonists as labelled ligands.

REFERENCES

(1) Enna, S. J., Collins, J. F. and Snyder, S. H. (1977) Brain Res. 124, 185-190.
(2) Beaumont, K., Chitton, W., Yamamura, H. I. and Enna, S. J. (1978) Brain Res. 148, 153-162.
(3) Möhler, H. and Okada, T. (1977) Nature (Lond.) 267, 65-67.
(4) Möhler, H. and Okada, T. (1978) Mol. Pharmacol. 14, 256-265.

(5) Zukin, S. R., Young, A. B. and Snyder, S. H. (1974) Proc. Nat. Acad. Sci., U.S.A. 71, 4802-4807.
(6) Schwob, R. and Würsch, J. (1977) J. Labelled Comp. & Radiopharmaceuticals, 14, 355-360.
(7) Hatefi, Y. and Hanstein, W. G. (1969) Proc. Nat. Acad. Sci., U.S.A. 62, 1129-1136.
(8) Coleman, R. (1978) Biochim. Biophys. Acta, 300, 1-30.
(9) Enna, S. J. and Snyder, S. H. (1978) Mol. Pharmacol. 13, 442-453.
(10) Snodgrass, S. R. (1978) Nature (Lond.) 273, 392-394.

$[^3H]$MUSCIMOL BINDING TO RAT CEREBELLAR MEMBRANES

Richard Maurer

Biological and Medical Research Division,
Sandoz Ltd., Basel, Switzerland

INTRODUCTION

Binding studies with different labelled agonists or antagonists have been used in receptor pharmacology. Interactions with a GABA-ergic binding site have been studied with the transmitter itself as ligand (1,2), with $[^3H]$bicuculline (3) and $[^3H]$picrotoxinin (4), and more recently with $[^3H]$muscimol (5-7). Since an efficient Na^+-dependent uptake system (8,9), a high content of endogenous GABA in the CNS, and the metabolism of GABA interfere with receptor binding assays, membrane purification seems to be crucial.

The preparation of membrane suspensions for assaying Na^+-independent $[^3H]$GABA binding involves either an osmotic shock together with a freezing step, or the solubilization of membrane proteins with Triton X-100 (1,2,10,11). Fresh cerebellar membranes have been used only recently for studies with $[^3H]$muscimol (6). Tritiated muscimol of high specific activity is the ligand of choice for the GABA system not only for the binding studies (5-7), but also for receptor distribution and pharmacokinetics in the mammalian CNS (7,12). Muscimol was characterized and proposed as a GABA-agonist several years ago (13), but the inter-dependence of the two substances was never quite clear. Scatchard analyses of the Na^+-independent binding of GABA and muscimol to frozen or detergent-treated membranes of rat brain (11,14-16) revealed more than one population of binding sites. However, it is uncertain whether the multiplicity of sites is due to the preparative steps used, like freezing and detergent treatment, or if it really represents a true diversity. To circumvent the preparative problems, we used dialyzed membranes as an alternative treatment to get rid of endogenous substances (e.g., GABA, ions) hypothesized to be present.

METHODS

Regional distribution studies with $[^3H]$GABA and $[^3H]$muscimol showed that the cerebellum is the brain region which contains the highest density of binding sites (2,5). Therefore we used rat cerebellar tissue for our studies.

After a light CO_2 narcosis, Ivanovas rats (male or female, 200-300 g) were decapitated and their brains were removed and washed in physiological NaCl solution. The cerebellum was homogenized in 20 volumes of ice-cold 0.32 M sucrose solution in a polytron homogenizer for 15 seconds at position 3 (scale: 1 - 10), and then centrifuged at 1000 g for 10 minutes. The pellet and the white slime at the interface were discarded, and the supernatant was recentrifuged for 20 minutes at 20,000 g. The crude mitochondrial pellet was resuspended in the original volume of Tris-citrate buffer (0.05 M; pH, 6.7) with the polytron (15 sec, position 3). This suspension was either dialyzed against the same Tris-citrate buffer or used directly for binding studies. Specific binding of radioactive muscimol was allowed to occur in triplicate samples for 30 minutes at room temperature in 2 ml Tris-citrate buffer (0.05 M; pH, 6.7) containing 2 nM $[^3H]$muscimol (New England Nuclear; 12.9 Ci/m-mole) alone or in the presence of 10^{-6} M non-radioactive muscimol. The incubation was terminated by filtration through glass fiber filters (Gelman, Type A-E) under reduced pressure (100 mbar). Finally, each filter was washed twice with 5 ml of cold water, and then placed into a plastic scintillation vial containing 5 ml of Instagel and heated to 56°C

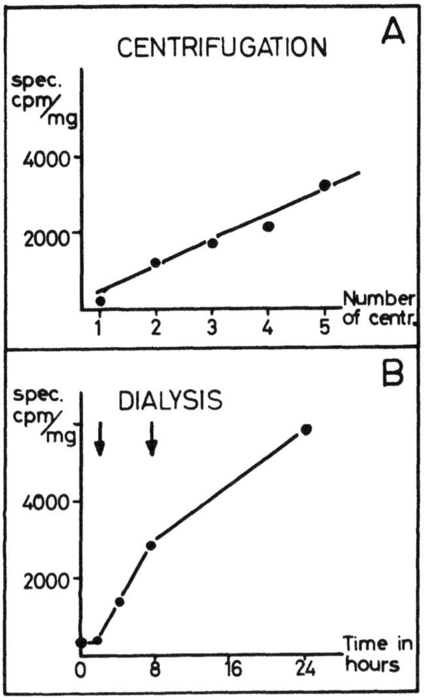

Fig. 1. Particulate (P_2) fractions of rat brain were either resuspended in Tris-citrate buffer (0.05 M; pH, 6.7) for varying times and centrifuged at 20,000 g for 20 min (A) or dialyzed against the same buffer (20 times excess volume) at 4°C (B) prior to assays for [^{3}H]muscimol binding as described in "Methods" (↓ : change of dialysis buffer).

for 30 minutes. Specifically bound ligand was calculated as the difference between the total amount of radioactive ligand trapped on the glass filter in the absence and in the presence of excess (10^{-6} M) non-radioactive ligand. Protein was measured according to Lowry _et al._ (17) using crystalline bovine serum albumin as standard.

Table I

Dialysis (hr)	Specific binding (cpm)	
	Tris-citrate	H_2O
24	1900	1000
48	3050	60
72	3350	0

A cerebellar membrane preparation (P_2 pellet) was suspended in Tris-citrate buffer (0.05 M; pH, 6.7) and either dialyzed against the same buffer or against water (20 times excess volume) for different periods of time at 4°C. Buffer was changed after 2, 6, 24, 30, 48 and 54 hours. Binding assays of aliquots of the suspension were performed as described in "Methods".

RESULTS

Centrifugation and Dialysis of Cerebellar Membranes

A preparation of cerebellar membranes was divided into two aliquots: one aliquot was resuspended several times with the polytron, being centrifuged each time for 20 minutes at 20,000 g, and then assayed; the other aliquot was dialyzed against the assay buffer at 4°C. Although the specific binding to fresh membranes is very low, the binding capacity was increased either by repeated washings or by dialysis (Fig. 1). These results strongly suggest that an endogenous inhibitor of muscimol binding, probably GABA, had been removed by both methods. With an increase in duration of dialysis there occurred a parallel increase in specific binding (Table I), which reached a plateau after about 48 hours. A complete loss of binding activity was observed when membrane suspensions were dialyzed for two days against water. Almost half of the specific binding was already lost after 24 hours.

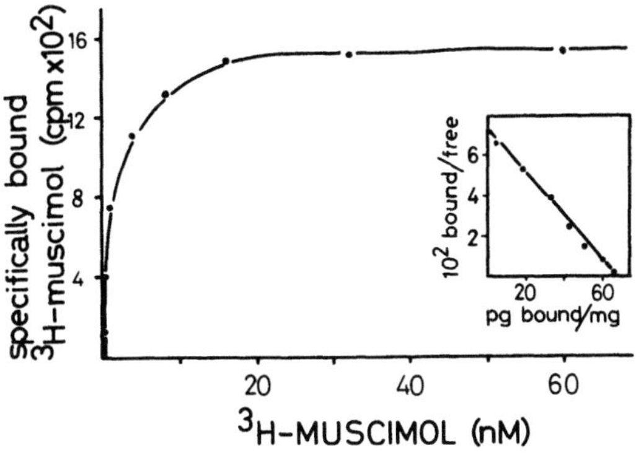

Fig. 2. Increasing concentrations of [^{3}H]muscimol were added to constant amounts of cerebellar membrane suspension (0.27 mg/assay), and specifically bound muscimol was determined. Insert: Scatchard plot of the same data.

Biochemical Properties of Dialyzed Membrane Suspensions

Whether muscimol and GABA bind to identical sites or if there exists any kind of cooperativity has so far not been unequivocally shown. With a classical pharmacological approach, we tried to resolve the above-mentioned problems. Adding different amounts of [^{3}H]muscimol in a wide concentration range (0.1 - 60 nM) to a constant amount of dialyzed cerebellar membrane suspension resulted in the expected saturation kinetics, with a single class of binding sites, when re-plotted in Scatchard form (Fig. 2).

Addition of a constant amount of non-radioactive GABA (2×10^{-8} or 6×10^{-8} M) under otherwise identical conditions led to a diminution of specifically bound [^{3}H]muscimol at most of the concentrations used, although the same plateau level was reached when using high concentrations of radioactive ligand (Fig. 3). Mathematical analyses* of the obtained curves indicated that GABA binds to the same site as muscimol. Apparent dissociation constants (K_D) of 1.3 ± 0.16 nM and 11 ± 1.9 nM have been calculated, respectively, for muscimol and GABA.

DISCUSSION

This paper describes an additional methodological approach using dialysis rather than centrifugation, freezing and thawing and/or Triton X-100 treatment (2,10,11) for the preparation of synaptosomal membranes. Prolonged dialysis facilitates the removal of endogenous low molecular weight compounds, like GABA and Na^+, that normally interfere with the assay system. The high content of endogenous GABA in particles of mammalian CNS is not as readily removed as expected when using the procedure of Zukin et al. (1). Even after an additional freezing step similar to that described by Greenlee et al. (16), GABA can be found in the supernatant (Maurer, unpublished results). Interference by additional endogenous substances with the binding of [^{3}H]GABA or [^{3}H]muscimol cannot, however, be ruled out.

Two GABA binding sites have been found in Triton-treated membranes of rat brain (2,11,16), but only one binding site for [^{3}H]-muscimol has been described using an identical membrane preparation (7). Our own results obtained with dialyzed crude cerebellar membrane preparations support the data obtained with [^{3}H]muscimol (7),

*For the calculations the SAAM program was used by Dr. M. Berman (NIH, Bethesda, U.S.A.):

$$B = \frac{B_M}{1 + \frac{K_{A*}}{[A*]}\left(1 + \frac{[B]}{K_B}\right)}$$

B_M = maximal binding capacity
K = dissociation constant
A* = ligand concentration (i.e., muscimol)
B = agonist or antagonist concentration (i.e., GABA)

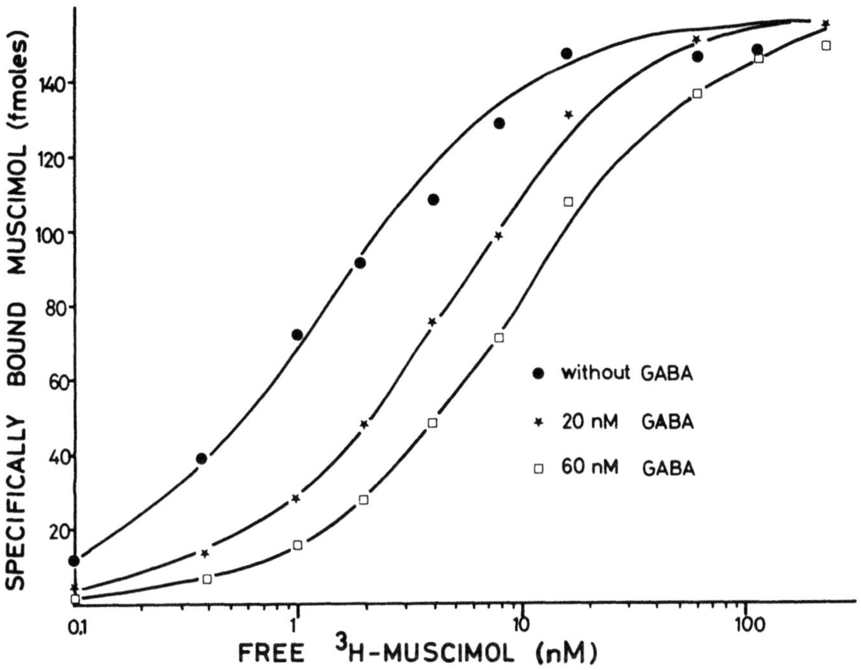

Fig. 3. Incubation of dialyzed cerebellar membrane suspension (0.27 mg protein/assay tube) with different concentrations of [^{3}H]muscimol in the absence (•) or presence *, 20 nM; □,60 nM) of non-radioactive GABA. Binding assays were performed as described in "Methods".

but differ from the findings with [^{3}H]GABA (2,11,16). Resuspension of the P_2 pellet in water, as used by Zukin et al. (1) to rupture synaptosomes, might be slightly detrimental to the GABA or muscimol binding sites (see Table I). Further purification steps involving freezing/thawing and Triton treatment (1,10,16) might also influence the binding sites and give rise to different sites in Scatchard plots. The gentle treatment of membrane suspensions by dialysis against assay buffer seems to be superior to the rather harsh conditions used by others, and might be useful for explaining the discrepancies in number of GABA and muscimol binding sites.

REFERENCES

(1) Zukin, S. R., Young, A. B. and Snyder, S. H. (1974) Proc. Nat. Acad. Sci., U.S.A. 71, 4802-4807.
(2) Enna, S. J. and Snyder, S. H. (1975) Brain Res. 100, 81-97.
(3) Möhler, H. and Okada, T. (1977) Nature (Lond.) 267, 65-67.
(4) Olsen, R. W., Greenlee, D., Van Ness, P. and Ticku, M. K. (1978) In: Amino Acids as Chemical Transmitters (F. Fonnum, ed.) pp. 467-486, Plenum Press, New York.
(5) Enna, S. J. (1978) In: Amino Acids as Chemical Transmitters (F. Fonnum, ed.) pp. 445-456, Plenum Press, New York.
(6) Leach, M. J. and Wilson, J. A. (1978) Eur. J. Pharmacol. 48, 329-330.
(7) Snodgrass, S. R. (1978) Nature (Lond.) 273, 392-394.
(8) Iversen, L. L. and Neal, M. J. (1968) J. Neurochem. 15, 1141-1149.
(9) Iversen, L. L. and Johnston, G. A. R. (1971) J. Neurochem. 18, 1939-1950.
(10) Enna, S. J. and Snyder, S. H. (1976) Brain Res. 115, 174-179.
(11) Wong, D. T. and Horng, J. S. (1977) Life Sci. 20, 445-452.
(12) Chan-Palay, V. (1978) Proc. Nat. Acad. Sci., U.S.A. 75, 1024-1028.
(13) Curtis, D. R., Duggan, A. W., Felix, D. and Johnston, G. A. R. (1971) Brain Res. 32, 69-96.
(14) Young, A. B., Enna, S. J., Zukin, S. R. and Snyder, S. H. (1976) In: GABA in Nervous System Function (E. Roberts, T. N. Chase and D. B. Tower, eds.) pp. 305-317, Raven Press, New York.
(15) DeFeudis, F. V., Madtes, P., Ojeda, A. and DeFeudis, P. A. (1976) Exp. Neurol. 52, 285-294.
(16) Greenlee, D. V., Van Ness, P. C. and Olsen, R. W. (1978) Life Sci. 22, 1653-1662.
(17) Lowry, O. H., Rosebrough, N. J., Farr, A. L. and Randall, R. J. (1951) J. Biol. Chem. 193, 265-275.

RECENT STUDIES ON THE MECHANISM WHEREBY BENZODIAZEPINES FACILITATE GABA-ERGIC TRANSMISSION

E. Costa and A. Guidotti

Laboratory of Preclinical Pharmacology, National Institute of Mental Health, Saint Elizabeths Hospital, Washington, D.C. 20032, U.S.A.

INTRODUCTION

Drugs can modify the activity of neurotransmitters by a number of mechanisms: they can mimic or block transmitter action at receptors; they can facilitate or block the spontaneous or the nerve impulse-mediated release of transmitters; they can decrease or increase transmitter catabolism or synthesis; they can block or facilitate the inactivation of transmitters at receptor sites. When the above-mentioned mechanisms were studied to elucidate how benzodiazepines facilitate GABA-ergic transmission, none could be shown to be operative (1-5). Since the pharmacological profile of benzodiazepines strongly suggests that they act by facilitating GABA-ergic transmission, we studied whether benzodiazepines change the affinity binding of [^{3}H]GABA to crude synaptic membrane preparations from rat brain (6).

Current studies describe neuronal membranes as a fluid mosaic where post-synaptic receptors are located (7). Thus, the cross section of the membrane is formed by the classic phospholipid bilayer whose continuity is interrupted by the presence of globular molecules of proteins, alone or associated with glycosaccharides or lipids, to form glycoproteins and lipoproteins, respectively. The polar end of the protein is usually in contact with the aqueous phase, and the extent to which the protein molecule is embedded in the membrane is under thermodynamic control. Among these proteins, certain ones have a high affinity for specific ions and can carry these ions across the membrane. These proteins, termed ionophores, can be appropriately modified by allosteric mechanisms which are regulated by membrane changes that follow the interactions of transmitters with specific receptors. Usually this phenomenon is referred to as an opening or closing of the ion gate. The existence of such a mechanism is substantiated by direct measurements of membrane impedance changes that occur synchronously with the changes in the ionophore function.

The response of post-synaptic neurones to GABA released at GABA-ergic synapses involves an opening of Cl^- channels. In practice, this means inhibition of the specific ionophore protein which constantly keeps an artificial Cl^- distribution across the membrane. When GABA becomes associated with its recognition sites, the Cl^- gates open and Cl^- equilibrates, causing hyperpolarization or depolarization according to the Cl^- content of the extracellular milieu (8). The tight coupling between GABA recognition sites and the mechanisms that regulate Cl^- channels strongly suggests that GABA-receptors, like many other post-synaptic receptors, are supramolecular units that function through a coordination of their molecular subunits which is regulated by the interactions of the transmitter with its specific receptor. Since the receptor response is a cooperative interaction of a number of membrane proteins, the efficiency of the interaction of GABA with its receptor may differ with respect to changes in the properties of the membrane and the abundance or quality of the proteins present. The present report will summarize our work (6) which has revealed that the affinity of recognition sites for GABA is regulated by a modulator protein present in the membrane. Benzodiazepines change the affinity of recognition sites for GABA by interfering with the function of this protein (9). Preliminary data suggest that phosphorylation of membrane protein may be involved in this regulatory mechanism.

ACTION OF BENZODIAZEPINES ON THE AFFINITY OF CRUDE SYNAPTIC MEMBRANE PREPARATIONS FOR [^{3}H]GABA

A basic feature of the fluid membrane model is the mobility of its membrane constituents; the most frequent motion is lateral diffusion, with diffusion constants of about 10^{-8} - 10^{-9} cm^2 sec^{-1} (10). However, within the limitations imposed by the fluid mosaic model it is possible to envision that the motility of a group of protein molecules is restricted when the structure changes. For instance, phosphorylation can bring about new interactions or prevent molecular activities that were operative in the dephosphorylated state. It is therefore possible to envision how frequency changes in phosphorylation-dephosphorylation reactions contribute to the formation of one of two populations of recognition sites for the transmitter within the membrane. For instance, one receptor population could be phosphorylated while the other would not be, and the recognition site of the population of phosphorylated receptors could have a higher affinity for GABA.

In brain, there are two types of GABA recognition site, both of which have a high affinity for GABA and one of which requires Na^+. The Na^+-dependent recognition site is associated with the pre-synaptic structures of synaptosomes; the one that is Na^+-independent is probably the recognition site associated with post-synaptic GABA-receptors (11-13).

The data in Table I show that the addition of various concentrations of diazepam to crude synaptic membrane preparations changes the kinetics of GABA binding (14). Normally, there exists only one population of receptors with a dissociation constant (K_D) of about 200 nM. However, in the presence of diazepam, at concentrations of 5 x 10^{-7} M, or higher, a second population of receptors is un-

Table I - Kinetics of GABA Binding to Freshly Prepared Crude Synaptic Membranes from Rat Cerebral Cortex

Drug addition	Concentration (M)	Affinity binding	
		K_{D1}	K_{D2}
None	-	*	210
Diazepam	5×10^{-9}	*	190
"	5×10^{-7}	61	200
"	5×10^{-6}	34	321
"	5×10^{-5}	25	285

* Only one affinity constant was detectable.

Table II - Affinity Changes for GABA Binding in Different Types of Membrane Preparations

Membrane preparation*	Supernatant added	µl	Affinity binding	
			K_{D1}	K_{D2}
A	None	0	**	210
B	None	0	20	180
B	A	20	**	200
A	B	20	**	215

* Type A membranes were prepared as described in the text; Type B membranes represent a further purification of Type A. Membranes were frozen and thawed and incubated with Triton-X 100 (0.1%) for 30 minutes. ** Only one K_D could be measured.

masked with a K_D of 26 - 60 nM. The binding of GABA to recognition sites of post-synaptic receptors is modified in a similar fashion by other benzodiazepines. For a small group of compounds, we tested the potency ranking of this facilitatory action on GABA binding to its post-synaptic receptors; we found that it parallels the potency ranking of their ability to prevent convulsions elicited by a blockade of GABA synthesis.

REGULATION OF GABA AFFINITY BINDING BY AN ENDOGENOUS MODULATOR PROTEIN

When crude synaptic membrane preparations from rat cerebral cortex were repeatedly washed after freezing and thawing, as suggested by Enna et al. (13), benzodiazepines failed to change the kinetics of GABA affinity binding. These procedures remove proteins incorporated into membranes. As shown in Table II, removal of these proteins elicited biphasic kinetics. Indeed, from the standpoint of GABA binding, these membranes resembled normal membranes treated with benzodiazepines. From the data of Table II, it became clear that crude synaptic membrane preparations contain a substance that inhibits high-affinity GABA binding. This substance, which was found to be thermostable, sensitive to destruction by trypsin, and free of phospholipids, was purified 1500-fold as an homogeneous protein weighing about 15,000 daltons (9). When the purified protein was added to Type B membranes (see Table II) it changed the kinetics of GABA binding from biphasic to monophasic. This protein regulates GABA binding by a non-competitive mechanism. The data available do not oppose the possibility that the protein modulates GABA binding by an allosteric mechanism. A property of this modulator protein that co-purified with its inhibitory action on GABA binding is its ability to inhibit protein kinase (15). This inhibition extends to the cyclic-AMP-dependent and cyclic-AMP-independent enzymes. When the purified modulator (0.1 - 0.5 μg) was added to crude synaptic membrane preparations (1 mg) the autophosphorylation of the membranes was completely blocked.

Table III - K_i of the Modulator to Reduce GABA Binding in the Presence of Various Benzodiazepines

Drug	K_i (μg)
Solvent	10
Clonazepam	120
Flunitrazepam	110
Diazepam	30
Chloridazepoxide	20

INTERACTION OF THE BENZODIAZEPINES WITH THE PROTEIN MODULATOR

Inhibition of the high-affinity binding of GABA to crude membrane preparations, that had been treated with detergents and repeatedly frozen and thawed, by this protein modulator was modified by benzodiazepines. This mechanism appeared to be competitive. In addition, the modulator protein interfered with the high-affinity binding of benzodiazepines to crude synaptic membrane preparations. This antagonism also appeared to be competitive. Interaction of various benzodiazepines with the inhibition of high-affinity GABA binding produced by the modulator was estimated by calculating the K_i of the modulator in the presence of 10^{-6} M concentrations of the benzodiazepines. Each benzodiazepine was pre-incubated with frozen Triton-treated membranes for 15 minutes at 0°C (Table III). The potency ranking of this antagonism was reminiscent of that concerning their abilities to antagonize convulsions elicited by blockade of GABA synthesis.

The role of the endogenous modulator in regulating diazepam binding to crude membrane preparations was evaluated by comparing diazepam binding to type A or B (see Table II) membrane preparations (see Table IV). It can be seen that the modulator content of the membrane determines the affinity of the ligand for the receptor.

Table IV - $[^3H]$Diazepam Binding in Fresh (A) and "Modulator"-Deprived (B) Crude Synaptic Membrane Preparations from Rat Cerebral Cortex[1]

	GABA binding		Diazepam binding		"Modulator"
	K_D (nM)	B_{max} (p-mole/mg protein)	K_D (nM)	B_{max} (p-mole/mg protein)	(Units/g)
A	220	1.8	12	0.25	110
B	20	5.3	6.8	0.25	N.D.

[1] From ref. 6, modified. N.D. = not detectable (<1 unit).

While the absence of the modulator increases the B_{max} for GABA, it fails to change that of diazepam. This is in accord with an allosteric modification of GABA binding and a competitive inhibition of diazepam binding. Although diazepam or the endogenous inhibitor modify GABA binding when they are bound to the membranes, they do not bind to the GABA binding sites. Hence, the regulatory site for GABA binding may be the high-affinity site for benzodiazepine binding. The endogenous modulator might even represent a natural ligand for this benzodiazepine binding. However, this interpretation cannot be taken as definitive because we are only inferring these mechanisms from Scatchard analyses. Unfortunately, our reaction times appear to be too long and, for this reason, these inferences are not thoroughly supported by the data. Recent data have indicated that benzodiazepines may decrease the inhibition of membrane phosphorylation caused by the protein modulator. If this finding were to be substantiated, the following working model may apply:

1) R_{GABA} → $+PO_4^{-3}$ (K_D 20 nM)

R_{GABA} → $-PO_4^{-3}$ (K_D 200 nM)

The GABA recognition site (R_{GABA}) can exist in two states; one is phosphorylated and the other is not. The modulator "masks" the appearance of the high-affinity state because it blocks phosphorylation. Since this blockade is antagonized by benzodiazepines, the site in which the modulator protein binds to protein kinase may be the high-affinity site for diazepam binding. Thus, in the presence of diazepam less modulator molecules can act and more R_{GABA} are phosphorylated. This working model is currently being tested.

REFERENCES

(1) Costa, E., Guidotti, A., Mao, C. C. and Suria, A. (1975) Life Sci. 17, 167-185.

(2) Costa, E., Guidotti, A. and Mao, C. C. (1976) In: GABA in Nervous System Function (E. Roberts, T. N. Chase and D. B. Tower, eds.) pp. 413-426, Raven Press, New York.

(3) Guidotti, A. (1978) In: Psychopharmacology - A Generation of Progress (M. A. Lipton, A. DiMascio and K. F. Killam, eds.) pp. 1349-1358, Raven Press, New York.

(4) Biggio, G., Brodie, B. B., Costa, E. and Guidotti, A. (1977) Proc. Nat. Acad. Sci., U.S.A. 74, 3592-3596.

(5) Haefely, W. E. (1978) In: Psychopharmacology - A Generation of Progress (M. A. Lipton, A. DiMascio and K. F. Killam, eds.) pp. 1359-1374, Raven Press, New York.

(6) Costa, E., Guidotti, A. and Toffano, G. (1978) Brit. J. Psychiat. 133, 239-248.

(7) Singer, S. J. and Nicolson, G. L. (1972) Science, 175, 720-731.

(8) Takeuchi, A. (1976) In: GABA in Nervous System Function (E. Roberts, T. N. Chase and D. B. Tower, eds.) pp. 255-267, Raven Press, New York.

(9) Toffano, G., Guidotti, A. and Costa, E. (1978) Proc. Nat. Acad. Sci., U.S.A. 75, 4024-4028.

(10) Edidin, M. (1974) Amer. Rev. Bioph. Bioengineer. 3, 179-201.

(11) Zukin, S. R., Young, A. B. and Snyder, S. H. (1974) Proc. Nat. Acad. Sci., U.S.A. 71, 4802-4807.

(12) Enna, S. J. and Snyder, S. H. (1975) Brain Res. 100, 81-97.

(13) Enna, S. J. and Snyder, S. H. (1977) Mol. Pharmacol. 13, 442-453.

(14) Guidotti, A., Toffano, G., Grandison, L. and Costa, E. (1978) In: Proceedings of the Symposium on Amino Acids as Chemical Transmitters - A NATO Advanced Study Institute (F. Fonnum, ed.) Plenum Publ. Co., New York, pp. 517-530.

(15) Szmigielski, A., Guidotti, A. and Costa, E. (1977) J. Biol. Chem. 252, 3848-3853.

GABA BINDING PROCESSES IN RAT BRAIN AND LIVER

F. V. DeFeudis, M. Maitre, L. Ossola, A. Elkouby,
G. Roussel and P. Mandel

Centre de Neurochimie du CNRS, et Institut de Chimie Biologique, Faculté de Médecine, 67085 Strasbourg Cedex, France

INTRODUCTION

γ-Aminobutyric acid (GABA) likely serves as a major inhibitory neurotransmitter in the vertebrate CNS (see, e.g., refs. 1-8). The *in vivo* inhibitory actions of GABA in many CNS regions are antagonized by bicuculline, an agent which might interact with synaptic GABA-receptors (3,7,9-12). Since bicuculline and its derivatives also affect GABA binding to CNS subcellular particles, these agents have been used to determine the *in vitro* properties of GABA-receptors (see, e.g., ref. 3). However, since blockade of synaptic inhibition cannot, in itself, offer an adequate explanation for all of the *in vivo* effects of bicuculline (13), the effects of this agent on GABA binding to CNS subcellular particles in both the presence and absence of a physiological concentration of Na^+ have been further studied and compared with results obtained with subcellular particles of liver. Some studies with frozen/thawed tissues are also described.

MATERIALS AND METHODS

Studies with Fresh Tissues

After decapitating male Wistar rats (150-275 g), their cerebral cortices and livers were excised, weighed, and homogenized in 10 vols ice-cold 0.32 M sucrose solution. All further operations for preparing "synaptosomal-mitochondrial" (P_2) fractions were performed at 0 – 4°C as previously described (14,15). Briefly, 4-ml portions of homogenates were centrifuged at 1000 g for 10 minutes, and 3.0-ml portions of resultant supernatant fractions were re-centrifuged at 17,000 g, 30 minutes. P_2 pellets were re-suspended in 3.0 ml of either Na^+-free or Na^+-containing medium (see below) and centrifuged at 17,000 g, 30 minutes. Washed P_2 fractions were weighed, re-suspended in 3.0 ml of medium, and allowed to stand for 15 minutes; 0.25-ml aliquots of these suspensions were pipetted into previously-weighed centrifuge tubes. Then, 0.2 ml of medium, either free of added substance, or containing enough bicuculline-methiodide (BMI; Pierce Chem. Corp.) or unlabelled GABA (Calbiochem Corp.) to provide final concentrations of 10^{-3} M, was added to each tube. (These high concentrations of BMI and GABA were used to estimate total "BMI-sensitive" and "GABA-sensitive" [^{3}H]GABA binding sites.) Samples were mixed for 4 – 5 seconds and allowed to stand for 15 minutes. Then, 0.5 ml of medium, providing final concentrations of 1.5×10^{-9} – 5.3×10^{-4} M of [^{3}H]GABA (2,3-^{3}H(N)-γ-aminobutyric acid; New England Nuclear Corp.; 34.5 _Ci_/m-mole, supplemented with unlabelled GABA when necessary) and 3.4×10^{-8} – 4.2×10^{-7} M of [^{14}C]sucrose (U-^{14}C-sucrose; Radiochemical Centre, Amersham, U.K.; 381 m_Ci_/m-mole), was added. Suspensions were mixed and allowed to stand for 15 minutes before final centrifugation at 17,000 g, 30 minutes. Since solutions of BMI were not very stable in the media employed, these were prepared 3 – 5 minutes before use.

Media for Studies with Fresh Tissues

Glucose-free, bicarbonate-buffered media (pH 7.4) were used. The osmolarities of the ions in the Na^+-containing medium were: Na^+, 147.3; K^+, 3.5; Ca^{2+}, 1.3; Mg^{2+}, 1.2; Cl^-, 128.5; HCO_3^-, 24.55; PO_4^{3-}, 0.45; and SO_4^{2-}, 1.2 m-osmol/l; total osmolarity ≃ 308 m-osmol/l (14). For some experiments this medium was made Na^+-free by substituting sucrose for 123 m-equiv/l of its contents of Na^+ and Cl^- and $KHCO_3$ for its $NaHCO_3$ content. Thus, Na^+-free medium contained unchanged concentrations of Ca^{2+}, Mg^{2+}, SO_4^{2-}, PO_4^{3-} and HCO_3^-, but had Cl^-, 5.7; K^+, 28.05; and sucrose, 245.4 m-osmol/l. Bicarbonate-buffered media were used since Tris (Tris(hydroxymethyl)aminomethane) and certain other synthetic buffers appear to alter the properties of excitable membranes (19-21). However, in some cases, Tris-HCl (50 mM), made isosmotic with sucrose, was used to control for the high K^+ and low Cl^- concentrations of the Na^+-free bicarbonate-buffered medium (data not shown).

Studies with Frozen/Thawed Tissues

The method of Bondy and Harrington (16) was used, with slight modification. Whole brains and livers of rats were excised, weighed, and frozen at -70°C for 17 hrs. Tissues were then homogenized in 15 vols 0.32 M sucrose solution, and 10-ml portions of homogenates were centrifuged at 40,000 g, 10 minutes. Pellets were re-suspended in about 20 vols cold water, and re-centrifuged at 40,000 g, 10 minutes. After discarding supernatants, pellets were frozen at -20°C for an hour, thawed, and re-suspended in 20 ml cold water. Another cycle of centrifuging and freezing was carried out, and then pellets were thawed and re-suspended in 1.2 ml of 20 mM Tris-HCl buffer (pH 7.1). Aliquots (0.25 ml) of suspensions were mixed with 0.2 ml Tris-HCl buffer which was either free of added substance or which contained enough GABA to provide a final concentration of 10^{-3} M, and then 0.5 ml of Tris-HCl containing (as final concentrations) 1.53×10^{-9} –

1.53×10^{-8} M of [^{3}H]GABA and $3.5 \times 10^{-8} - 3.5 \times 10^{-7}$ M of [^{14}C] sucrose was added. Samples were mixed and incubated at 30°C for 15 minutes, and then centrifuged at 40,000 g, 10 minutes. Aliquots of supernatants were prepared for scintillation counting. Pellets were washed by re-suspension in 4 ml Tris-HCl and re-centrifuged at 40,000 g, 10 minutes, supernatants were discarded, and pellets were weighed.

Determination of Radioactivity and Protein

Radioactivity due to ^{3}H and ^{14}C was determined in 0.1-ml aliquots of supernatants and in 0.5-ml aliquots of pellets after their re-suspension in 1.0 ml of water (17). Protein contents of re-suspended pellet fractions were determined by the method of Lowry et al. (18).

Presentation of the Data

The data are presented as pellet/supernatant distribution ratios, as mole GABA bound/mg protein, as mole GABA bound/g pellet (corrected for sucrose space), and as mole GABA bound/g original wet weight of tissue (17). Those amounts of [^{3}H]GABA remaining in pellets in the presence of 10^{-3} M unlabelled GABA or BMI were subtracted from values for the total [^{3}H]GABA present in pellets to estimate GABA-sensitive and BMI-sensitive components of [^{3}H]GABA binding sites. Linear regression analyses (method of least squares) were used to obtain values for dissociation constants (K_B), maximal binding capacities (B_{max}) and Hill coefficients, except for K_B and B_{max} values of the sigmoidal plots of Figures 1 and 2 which were estimated "by eye".

Electron Microscopy

After taking aliquots for binding assays, P_2 fractions were re-centrifuged at 17,000 g, 30 minutes and crude membrane fractions

Table I - Characteristics of "Synaptosomal-Mitochondrial" Fractions of Rat Cerebral Cortex and Liver Prepared from Fresh Tissues

Parameter	Cerebral cortex With Na^+	Cerebral cortex Na^+-free	Liver With Na^+
Weight of initial P_2 fraction (g)	0.190 ± 0.015 ((18)	0.205 ± 0.019 (60)	0.112 ± 0.012 (12)
Weight of final pellet (g)	0.014 ± 0.001 (162)	0.013 ± 0.001 (532)	0.007 ± 0.001 (105)
Protein in final pellet (mg/g)	85.5 ± 5.1 (162)	91.8 ± 5.3 (522)	65.5 ± 10.4 (99)
[^{14}C]Sucrose distribution ratio[a]	0.60 ± 0.05 (158)	0.60 ± 0.03 (530)	0.80 ± 0.05 (99)

Experiments were conducted using bicarbonate-buffered medium containing physiological concentrations of Na^+ and other inorganic ions or in Na^+-free bicarbonate-buffered medium (see "Methods"). Means ± standard deviations; numbers of samples in parentheses; all values for cerebral cortex were significantly different from corresponding values for liver ($p < 0.001$; Student's *t*-test; two-tailed).

[a]Distribution ratio = $\frac{\text{dis/min per g pellet}}{\text{dis/min per g supernatant}}$.

were re-centrifuged at 40,000 g, 10 minutes. The resultant pellets were fixed with glutaraldehyde (5%, in 0.1 M phosphate buffer, pH 7.4), post-fixed with osmic acid (1%, in the same buffer), dehydrated using graded dilutions of ethanol, and embedded in Araldite. Ultrathin sections were cut using a Reichert ultra-microtome (Model OMU-2), stained with uranyl acetate and lead citrate, and then examined and photographed in a Philips 300 electron microscope.

RESULTS

Studies with Fresh Tissues

Distribution Ratios. The weights and protein contents of pellet fractions and [^{14}C]sucrose distribution ratios of cerebral cortex were very similar for experiments conducted in Na^+-free and Na^+-containing media; however, the weights of initial and final pellet fractions and the protein contents of final pellets of cerebral cortex were greater than those of liver, and the [^{14}C]sucrose distribution ratio (a measure of supernatant fluid trapped by the pellet) was greater for liver than for cerebral cortex (Table I). Distribution ratios for [^{3}H]GABA differed markedly, values of only about unity in Na^+-free media and about 40 in Na^+-containing media for cerebral cortex and about 2.0 for liver in Na^+-containing medium being achieved at the lowest concentrations of [^{3}H]GABA used. In all cases, unlabelled GABA (10^{-3} M) decreased [^{3}H]GABA distribution ratios to values which approached those of respective [^{14}C]sucrose distribution ratios. BMI (10^{-3} M) also caused significant decreases in [^{3}H]GABA distribution ratios in all cases, these effects being most pronounced at the lowest concentrations of [^{3}H]GABA.

Experiments with Na^+-Free Medium. No saturable binding process existed in the range of 10^{-9} – 10^{-7} M of [^{3}H]GABA, whether data were plotted as total ^{3}H in the pellets, as GABA-sensitive sites, or as BMI-sensitive sites (data not shown). However, a high-affinity binding process ($K_B \simeq 7.4 \times 10^{-5}$ M; $B_{max} \simeq 58$ n-moles/g P_2 fraction) was detected using higher concentrations of [^{3}H]GABA and this process was inhibited by both unlabelled GABA and BMI (see Table II). As expected, inhibition by unlabelled GABA was competitive; however inhibition by BMI was non-competitive. Reciprocal plots of GABA-sensitive and BMI-sensitive components of this process revealed that BMI-sensitive sites accounted for about 65% of the GABA-sensitive [^{3}H]GABA binding sites (Table II).

Experiments with Na^+-Containing Medium. Using a physiological concentration of Na^+ and other inorganic ions, two high-affinity [^{3}H]GABA binding processes were detected in particles of cerebral cortex (Table II). Scatchard plots of these data revealed that the lower-affinity component had $K_B \simeq 4 \times 10^{-6}$ M in both the presence and absence of BMI, and that its B_{max} was reduced by about 0.5 n-moles/mg protein by BMI, suggesting non-competitive inhibition (Table II). Sigmoidal and Lineweaver-Burk plots of such data are shown in Figures 1 and 2. Using lower concentrations of [^{3}H]GABA, a higher-affinity component was detected which appeared to be competitively inhibited by BMI. BMI-sensitive and GABA-sensitive components of this process had $K_B \simeq 2 \times 10^{-8}$ M and 4×10^{-8} M, respectively, when data were plotted on a pellet weight basis, whereas the B_{max} of BMI-sensitive sites was only about 8% that of GABA-sensitive sites (Fig. 3; Table II). The lower affinity binding process was not detected in the liver P_2 fraction (Figs. 1 and 2) but the higher-affinity process was found (Fig. 4).

Hill Coefficients. Hill plots of [^{3}H]GABA binding to P_2 fractions of cerebral cortex and of the BMI- and GABA-sensitive components of this binding, in both the presence and absence of added Na^+, revealed that neither positive nor negative cooperativity appeared to be involved in these interactions (see also refs. 15, 22).

Studies with Frozen/Thawed Tissues. The weight of initial pellet fractions and the [^{14}C]sucrose distribution ratios of frozen/

Table II - K_B and B_{max} Values for Components of [^{3}H]GABA Binding to a Synaptosomal-Mitochondrial Fraction of Rat Cerebral Cortex

Component	K_B (M)		B_{max} (n-moles)		
	per mg protein	per g pellet[a]	per mg protein	per g pellet[a]	per g original tissue wet wt.[b]
Na$^+$-free medium					
GABA-sensitive sites	1.8×10^{-5}	1.1×10^{-5}	0.04	5.8	3.5
BMI-sensitive sites	$\simeq 10^{-5}$	1.7×10^{-5}	0.025	3.7	2.2
Na$^+$ medium					
Lower-affinity					
GABA-sensitive sites	$\simeq 4 \times 10^{-6}$	$\simeq 10^{-5}$	2.6	550	350
BMI-sensitive sites	$\simeq 4 \times 10^{-6}$	$\simeq 10^{-5}$	0.5	80	48
Higher-affinity					
GABA-sensitive sites	$\simeq 10^{-7}$	3.8×10^{-8}	0.03	2.7	1.7
BMI-sensitive sites	4×10^{-8}	2.1×10^{-8}	0.002	0.225	0.143

[a]These values were corrected for [^{3}H]GABA present in supernatant fluid that was trapped in pellets using [^{14}C]sucrose distribution ratios.

[b]These values were calculated from B_{max} values given in mole GABA/g pellet, using the weights of initial P_2 fractions given in Table I (see ref. 17).

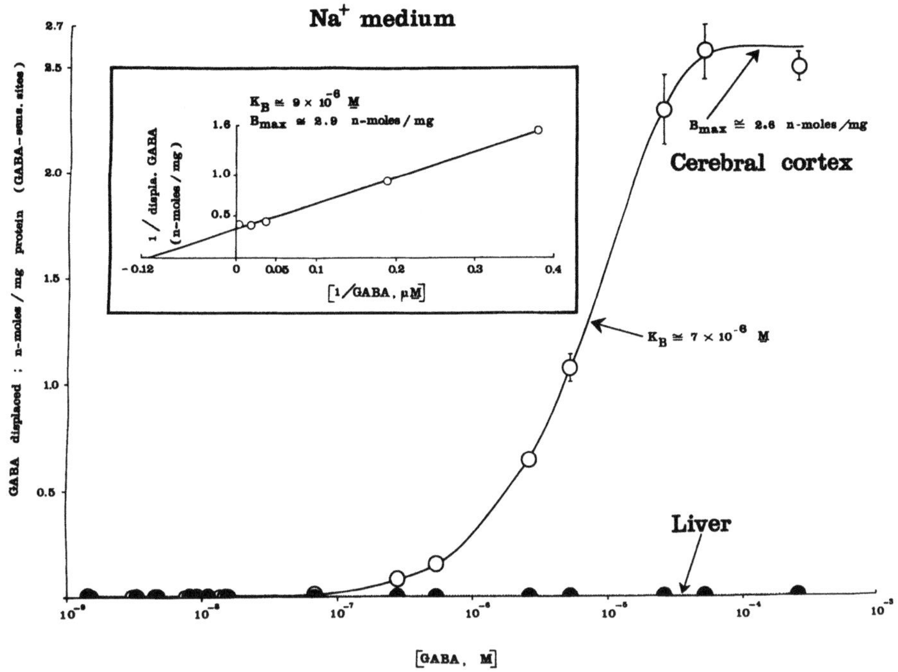

Fig. 1. Semilogarithmic plot and Lineweaver-Burk plot (inset) of the binding of [^{3}H]GABA to "synaptosomal-mitochondrial" fractions of rat cerebral cortex and liver, represented as GABA-sensitive sites (i.e., those amounts of [^{3}H] GABA in pellets that were sensitive to 10^{-3} M unlabelled GABA). Experiments were conducted in a balanced, bicarbonate-buffered medium. Means ± standard deviations; 3 samples for each point; standard deviations not indicated were within the symbols themselves. K_B and B_{max} are indicated. Note that no GABA-sensitive component, with an affinity in the range shown here, was detected with the liver preparation. Reproduced from ref. 33 with permission from Plenum Press.

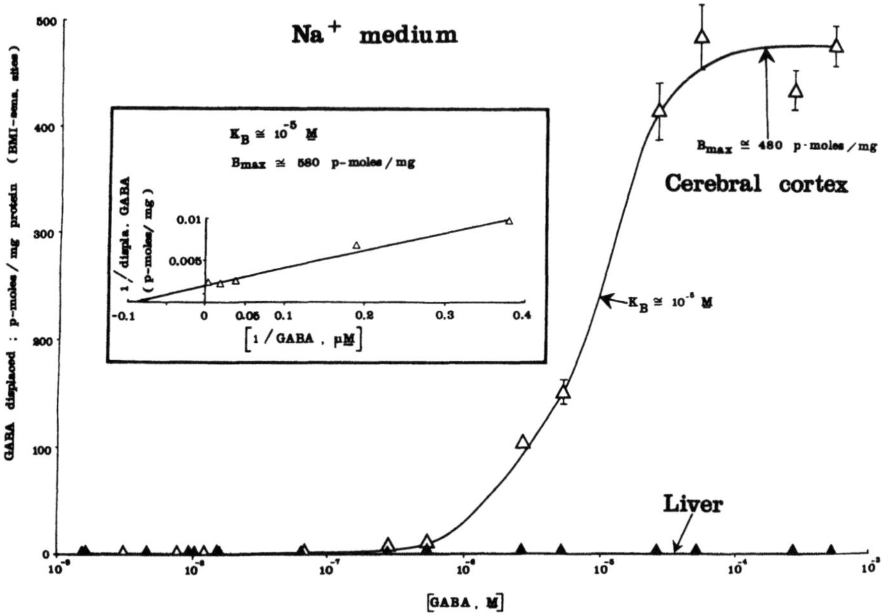

Fig. 2. Semilogarithmic plot and Lineweaver-Burk plot (inset) of the binding of [^{3}H]GABA to "synaptosomal-mitochondrial" fractions of rat cerebral cortex and liver, represented as BMI-sensitive sites (i.e., those amounts of [^{3}H] GABA in pellets that were sensitive to 10^{-3} M BMI). Experiments were conducted in a balanced, bicarbonate-buffered medium. Means ± standard deviations; 3 samples for each point; standard deviations not indicated were within the symbols themselves. K_B and B_{max} are indicated. Note that no BMI-sensitive component, with an affinity in the range shown here, was detected with the liver preparation. Reproduced from ref. 33 with permission from Plenum Press.

thawed brain preparations were greater than those of liver; the protein content of final liver pellets was greater than that of brain pellets, and final pellet weight did not differ between the two tissues (data not shown). [^{3}H]GABA binding was saturable in both brain

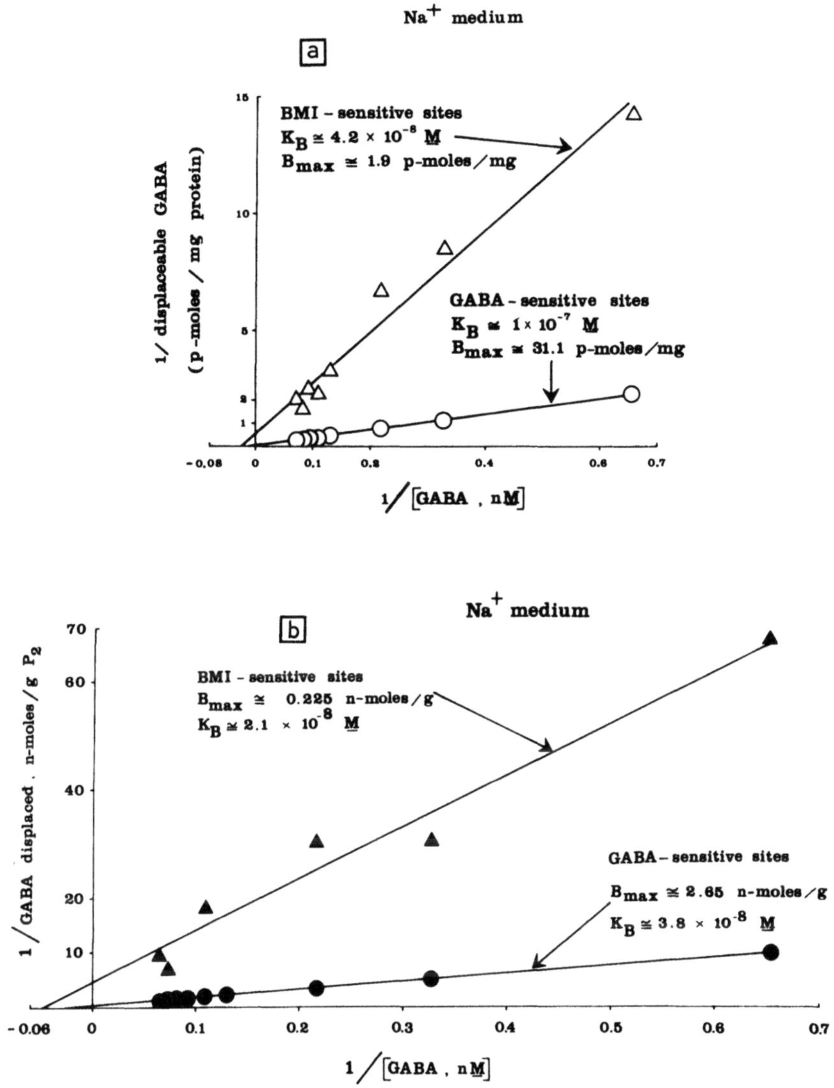

Fig. 3. Lineweaver-Burk plots of highest-affinity binding of [^{3}H]GABA to a synaptosome-enriched fraction of rat cerebral cortex in a balanced, bicarbonate-buffered medium expressed on protein (a) and weight (b) bases. GABA-sensitive and BMI-sensitive components of the binding are indicated. Each point represents the reciprocal of the difference

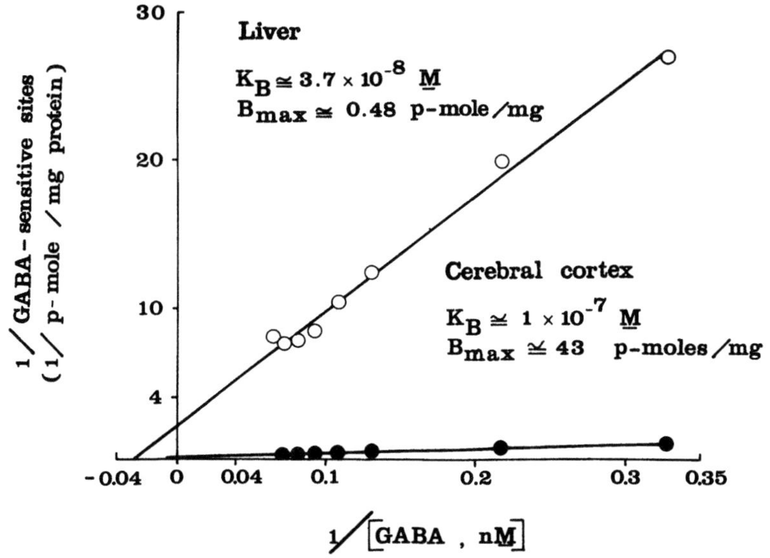

Fig. 4. Lineweaver-Burk plots of high-affinity binding of $[^3H]$GABA to "synaptosomal-mitochondrial" fractions of rat cerebral cortex and liver, represented as bound $[^3H]$GABA that was sensitive to 10^{-3} M unlabelled GABA. K_B and B_{max} values are indicated. Note that although K_B values were quite similar for both tissues, B_{max} for cerebral cortex was nearly 100 times greater than B_{max} for liver. Each point represents the reciprocal of the difference between 2 – 4 samples incubated in the absence or presence of 10^{-3} M unlabelled GABA. Reproduced from ref. 33 with permission from Plenum Press.

(Fig. 3, continued) between 3 – 4 values obtained in the absence of added substance and 3 – 4 values obtained in the presence of either unlabelled GABA (10^{-3} M) or BMI (10^{-3} M). K_B and B_{max} values are indicated. Reproduced from refs. 34 and 35 with permission from Pergamon Press.

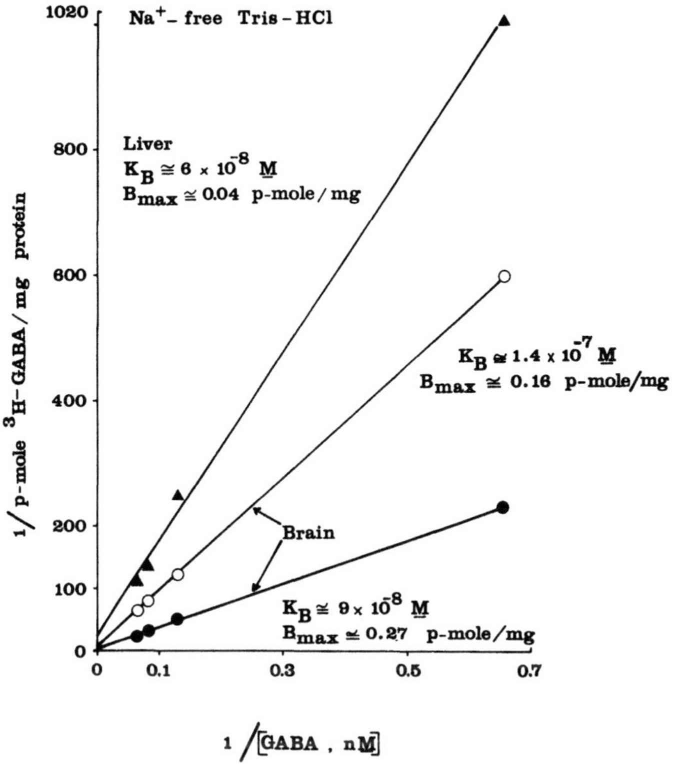

Fig. 5. Lineweaver-Burk plots of high-affinity binding of [^{3}H]GABA to crude membrane fractions of rat whole brain and liver, expressed on a protein basis. These experiments were conducted with a Na^+-free Tris-HCl medium. Each point represents the mean of 3 values; K_B and B_{max} are indicated. With this method of presentation of the data, high-affinity [^{3}H]GABA binding appeared to exist in both brain (●) and liver (▲). However, when displacement by 10^{-3} M unlabelled GABA was used to provide an estimate of "specific" binding, a high-affinity process was evident in brain (○), but not in liver. Reproduced from ref. 33 with permission from Plenum Press.

and liver preparations when data were plotted on a protein basis (Fig. 5). However, comparison of [^{14}C]sucrose and [^{3}H]GABA distribution ratios for these data revealed that [^{3}H]GABA was bound significantly to brain particles but not to those of liver. Also, [^{3}H]GABA binding was sensitive to 10^{-3} M unlabelled GABA in brain, but not in liver (Fig. 5). Furthermore, a saturable high-affinity binding process existed for [^{3}H]GABA in brain, when the data were plotted on a pellet weight basis (corrected for sucrose space), and a component of this process was sensitive to 10^{-3} M GABA; similar calculations for the liver preparation revealed no significant binding.

Electron Microscopy. Electron micrographs of initial P_2 pellet fractions revealed that subcellular structures (e.g., synaptosomes, mitochondria) were not as well preserved after exposure of the particles to Na^{+}-free medium. The greater amount of OsO_4 precipitate found in samples exposed to Na^{+}-free medium was perhaps due to a more pronounced reaction of OsO_4 with lipids in the membrane bilayer. Representative electron micrographs of frozen/thawed preparations of brain and liver are shown in Figure 6.

DISCUSSION

High-Affinity Binding of [^{3}H]GABA to a P_2 Fraction of Cerebral Cortex in the Absence of Added Na^{+}

Three bicuculline-sensitive, high-affinity binding processes for GABA were detected in a synaptosome-enriched fraction of rat cerebral cortex. The single high-affinity process detected with Na^{+}-free media had $K_B \simeq 10^{-5}$ M, which is similar to that for high-affinity GABA uptake by slices and homogenates of rat cerebral cortex (23,24). Although this process occurred in the absence of added Na^{+}, a substantial portion of it could be related to GABA transport since nipecotic acid (10^{-3} M) inhibited this process to a greater extent than BMI (data not shown). However, this process was markedly

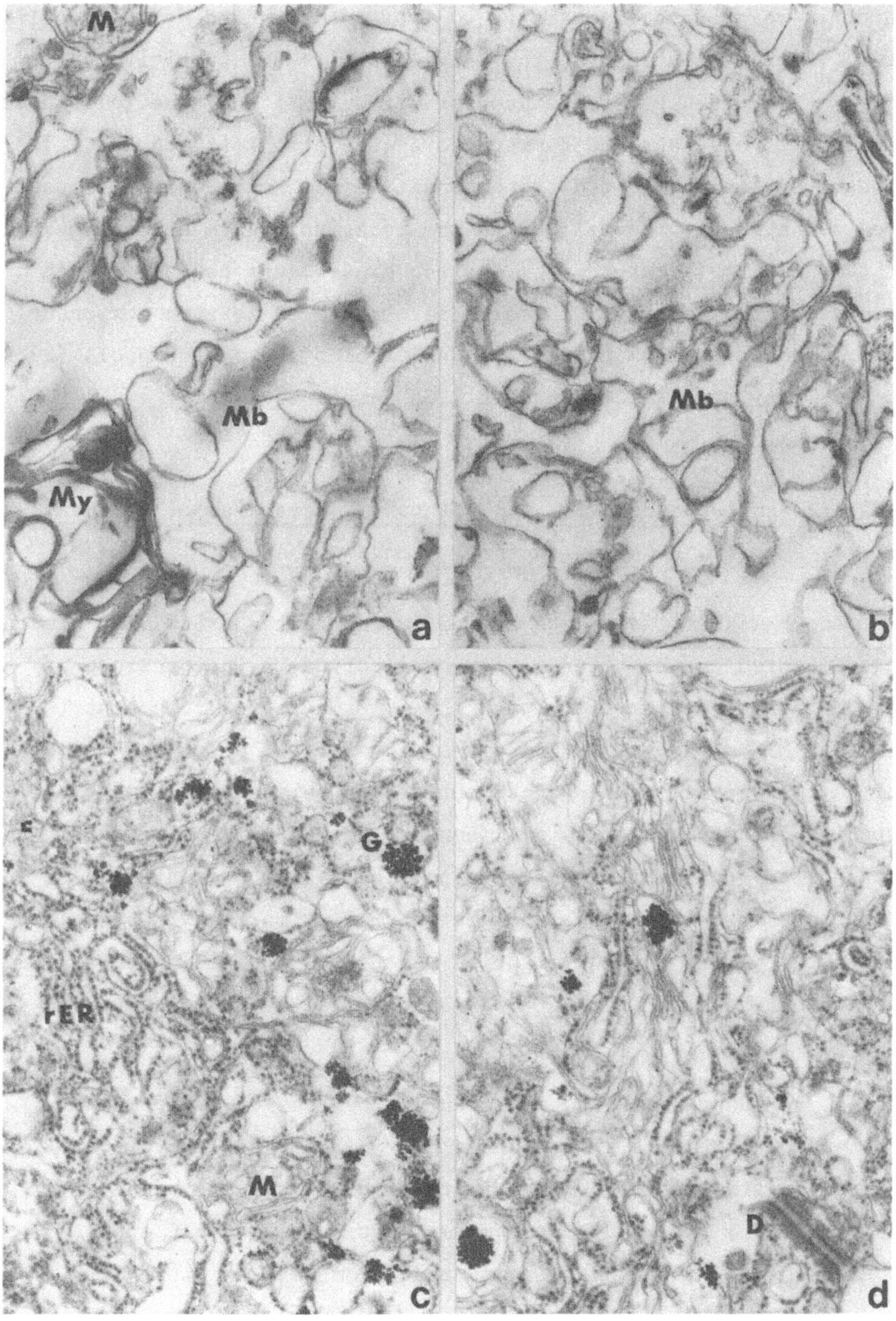

Fig. 6

inhibited by BMI (see Table II), unlike high-affinity GABA transport which is relatively insensitive to (or not affected by) bicuculline (e.g., 24,25). Therefore, this high-capacity BMI-sensitive system (B_{max} ≃ 2 n-moles/g cerebral cortex wet weight) could be related to BMI-sensitive synaptic GABA-receptors (see also refs. 26-28). The absence of a higher-affinity component in the Na^+-free medium used does not appear to be related to the relatively high K^+ concentration of this medium (about 28 mEq/l) since experiments conducted with Na^+-free isotonic Tris-HCl medium also revealed no such component. Also, high K^+ appears to release GABA from pre-synaptic pools (29, 30) but not from presumed synaptic GABA-receptors (31).

High-Affinity Binding of [^{3}H]GABA to a P_2 Fraction of Cerebral Cortex in the Presence of Na^+

Two BMI-sensitive, high-affinity components of [^{3}H]GABA binding were detected in particles of cerebral cortex in the presence of Na^+-containing medium. Although the lower-affinity component (K_B ≃ 4 x 10^{-6} M) had BMI-sensitive sites with B_{max} 48 n-moles/g cerebral cortex wet weight, this process could also be associated with a high-capacity GABA-receptor system. However, as GABA is bound mainly to uptake sites in the presence of a physiological concentration of Na^+, it appears more likely that BMI interacted primarily with these uptake sites. This finding might be useful for explaining the *in vivo* GABA-potentiating actions of bicuculline-like agents observed in

Fig. 6. Representative electron-micrographs of crude membrane preparations of rat whole brain (a,b) and liver (c,d). Magnification = 40,000 X; M, mitochondrion; Mb, membrane; My, myelin; rER, rough endoplasmic reticulum; G, glycogen granules; D, desmosome. Reproduced from ref. 33 with permission from Plenum Press.

iontophoretic studies (13,32). Highest-affinity $[^3H]$GABA binding detected in Na^+-containing medium had kinetic constants that are more generally considered to be related to synaptic GABA-receptors ($K_B \simeq 10^{-8}$ M; $B_{max} \simeq 140$ p-moles/g cerebral cortex wet weight for the BMI-sensitive component of this process).

Studies with a P_2 Fraction of Liver

A high-affinity binding process for $[^3H]$GABA existed in a freshly-prepared fraction of liver, namely in crude mitochondria. However, this process of liver can account for only about 1/100th the capacity of GABA-sensitive sites and for only about 1/20th that of BMI-sensitive sites of cerebral cortex. This binding of $[^3H]$GABA could represent sites involved in its transport into mitochondria for subsequent metabolism, or could be related to the unmyelinated sympathetic nerve fibers which accompany blood vessels and bile ducts in this organ.

Studies with Frozen/Thawed Membrane Fractions

The finding of Bondy and Harrington (16), that a specific high-affinity $[^3H]$GABA-binding process exists in liver, was not confirmed. $[^3H]$GABA binding to liver membranes, when expressed on a protein basis, was not displaced by 10^{-3} M unlabelled GABA, and when the data were corrected for $[^3H]$GABA present in trapped supernatant fluid of pellets, no GABA binding existed in the liver preparation in the absence of added Na^+.

CONCLUSION

Although these studies have provided some further insight into GABA binding processes associated with transport and receptor-activation, such attempts remain limited due to the lack of specificity of the ligands used to displace these components. Thus, BMI appeared to interact with both transport and synaptic receptor sites for GABA. Perhaps certain GABA-agonists (e.g., muscimol, isoguvacine) will be shown to interact more specifically with synaptic GABA-receptors? So-called "specific" binding of GABA and its analogues, or antagonists might not be the issue to be kept in focus in this type of experiment, especially since it cannot be separated into presynaptic, postsynaptic and non-synaptic components. The specificity of an interaction of ligand with membrane site may be determined by other requirements of the particular system concerned. In any case, it seems essential to examine the binding sites, themselves, in order to define more clearly which of these are involved in receptor-activation and in uptake. An approach involving the isolation and purification of GABA-receptors should provide evidence for specific binding of GABA to its postsynaptic receptors.

REFERENCES

(1) Krnjević, K. (1974) Physiol. Rev. 54, 418-540.
(2) Curtis, D. R. and Johnston, G. A. R. (1974) Ergebn. Physiol. 69, 97-188.
(3) DeFeudis, F. V. (1977) Progr. Neurobiol. 9, 123-145.
(4) DeFeudis, F. V. (1978) Neurochem. Res. 3, 263-280.
(5) Oja, S. S., Kontro, P. and Lähdesmäki, P. (1977) Progr. Pharmacol. 1, 1-119.
(6) Hösli, L. and Hösli, E. (1978) Rev. Physiol. Biochem. Pharmacol. 81, 135-188.
(7) Johnston, G. A. R. (1978) Ann. Rev. Pharmacol. Toxicol. 18, 269-289.
(8) Roberts, E., Chase, T. N. and Tower, D. B. (1976) (eds.) GABA in Nervous System Function. Raven Press, New York.
(9) Curtis, D. R., Duggan, A. W., Felix, D. and Johnston, G. A. R. (1970) Nature (Lond.) 226, 1222-1224.
(10) Curtis, D. R., Duggan, A. W., Felix, D. and Johnston, G. A. R. (1970) Nature (Lond.) 228, 676-677.
(11) Curtis, D. R., Duggan, A. W., Felix, D. and Johnston, G. A. R. (1971) Brain Res. 32, 69-96.
(12) Curtis, D. R., Duggan, A. W., Felix, D., Johnston, G. A. R. and McLennan, H. (1971) Brain Res. 33, 57-73.
(13) Krnjević, K., Puil, E. and Werman, R. (1977) Can. J. Physiol. Pharmacol. 55, 670-680.
(14) DeFeudis, F. V., Balfagón, G., Sagarra, M. R. de, Madtes, P., Somoza, E. and Gervas-Camacho, J. (1975) Exp. Neurol. 49, 497-505.

(15) DeFeudis, F. V. and Somoza, E. (1977) Gen. Pharmacol. 8, 181-187.
(16) Bondy, S. C. and Harrington, M. E. (1978) Biochem. Biophys. Res. Commun. 80, 161-168.
(17) DeFeudis, F. V. (1974) Can. J. Physiol. Pharmacol. 52, 138-147.
(18) Lowry, O. H., Rosebrough, N. J., Farr, A. L. and Randall, R. J. (1951) J. Biol. Chem. 193, 265-275.
(19) Ko, K.-C., Gimeno, A. L. and Berman, D. A. (1969) Amer. J. Physiol. 216, 853-859.
(20) Mandrino, M. (1969) J. Physiol. (Paris) 61, Suppl. 2, 346-347.
(21) Davidoff, R. A. and Sears, S. (1975) Amer. J. Physiol. 229, 831-837.
(22) Enna, S. J. and Snyder, S. H. (1975) Brain Res. 100, 81-97.
(23) Iversen, L. L. and Neal, M. J. (1968) J. Neurochem. 15, 1141-1149.
(24) Iversen, L. L. and Johnston, G. A. R. (1971) J. Neurochem. 18, 1939-1950.
(25) Olsen, R. W., Ban, M. and Miller, T. (1976) Brain Res. 102, 283-299.
(26) Peck, E. J., Schaeffer, J. M. and Clark, J. H. (1973) Biochem. Biophys. Res. Commun. 52, 394-400.
(27) Peck, E. J., Schaeffer, J. M. and Clark, J. H. (1976) In: GABA in Nervous System Function (E. Roberts, T. N. Chase and D. B. Tower, eds.) pp. 319-336, Raven Press, New York.
(28) Fiszer de Plazas, S. and DeRobertis, E. (1975) J. Neurochem. 25, 547-552.
(29) DeBelleroche, J. S. and Bradford, H. F. (1973) Progr. Neurobiol. 1, 275-298.
(30) Nelson-Krause, D. C. and Howard, B. D. (1978) Brain Res. 147, 91-105.
(31) Enna, S. J. and Snyder, S. H. (1977) Mol. Pharmacol. 13, 442-453.
(32) Hill, R. G., Simmonds, M. A. and Straughan, D. W. (1973) Brit. J. Pharmacol. 49, 37-51.
(33) DeFeudis, F. V., Ossola, L., Maitre, M., Elkouby, A., Roussel, G. and Mandel, P. (1979) Neurochem. Res. (in press).
(34) DeFeudis, F. V., Maitre, M., Ossola, L., Elkouby, A., Roussel, G. and Mandel, P. (1979) Neuroscience, 4, 897-912.
(35) DeFeudis, F. V., Maitre, M., Ossola, L., Elkouby, A. and Mandel, P. (1979) Gen. Pharmacol. 10, 193-194.

SODIUM-INDEPENDENT, HIGH-AFFINITY BINDING OF [^{3}H]γ-AMINOBUTYRIC ACID IN HUMAN NEUROLOGICAL DISORDERS

K. G. Lloyd[1], S. Dreksler[2], L. Shemen[2] and L. Davidson[2]

[1] Synthélabo-L.E.R.S., Department of Biology, Neuropharmacology Unit, 31 avenue Paul Vaillant Couturier, 92220 Bagneux, France

[2] Department of Psychopharmacology, Clarke Institute of Psychiatry, 250 College Street, Toronto, Ontario, Canada M5T 1R8

INTRODUCTION

The development of new drugs for the treatment of different neurological disorders will, in great probability, depend upon the identification of the neurotransmitter systems involved in the neuropathological bases of these disease states. In a manner similar to that used in the development of L-DOPA therapy for Parkinson's disease (cf. ref. 1), work is progressing in many laboratories to identify the neurochemical pathologies of various nervous diseases such as Huntington's chorea, Lesch-Nyhan syndrome, Gilles de la Tourette's disease, the various dystonias and the epilepsies. Of the numerous neurotransmitter systems which have been proposed to function in the brain, it is likely that those neurones utilizing γ-aminobutyric acid (GABA) are involved in some of these CNS disease states. One of the reasons for this belief is the widespread distribution of GABA-ergic neurones in the brain. Evidence suggests that GABA-ergic neurones serve as interneurones in cerebral and cerebellar cortices and in the hippocampus (2-5) and possibly in the striatum (6). Additionally, GABA acts as the neurotransmitter in the cerebellar efferent path (3, 7,8). GABA-ergic neurones are present in the striato- and/or pallido-nigral tracts, being integrated into a feedback loop controlling the activity of the nigrostriatal dopamine path (cf. ref. 9).

In addition to their widespread distribution in the brain, there are some specific findings which indicate that GABA-ergic neurones are altered in various neurological diseases. The best known of these is Huntington's chorea, in which GABA and its synthesizing enzyme (L-glutamate decarboxylase, GAD) are decreased in the substantia nigra and other basal ganglia regions (6,10-12). In Parkinson's disease GAD is deficient in the striatum and substantia nigra. However, this alteration in GAD activity does not appear to be a permanent deficit as it is apparently reversed by prolonged L-DOPA therapy (13). The fact that in both Parkinson's and Huntington's disease these biochemical alterations occur specifically within the regions of maximum neuropathology and are not generalized throughout the brain enhances the probability that the neurochemical changes are related to the disease processes.

The measurement of changes in GABA levels or GAD activity provides information which is related mainly to the presynaptic function of GABA-mediated synapses. However, for the proper understanding of the pathological process and (in our view) more importantly for the design of drugs which will be effective in these synapses, it is necessary also to have precise information on the functional state of GABA-receptors. For example, if in a particular disease state the GABA input is destroyed but the GABA-receptor remains intact, then a GABA-mimetic may be an efficacious form of therapy. However, if GABA-receptors are also destroyed, then GABA-mimetics may be, at best, of limited value.

Recently, work by Enna, Snyder and colleagues (14-16) has made possible the study of [^{3}H]GABA binding sites which, in the rat brain, have the pharmacological characteristics of a postsynaptic GABA-receptor. With this method it has been possible to examine [^{3}H]GABA binding in post-mortem material from patients who had various neurological diseases in an attempt to elucidate possible alterations in the functional state of GABA-receptors.

MATERIALS AND METHODS

Human brains were obtained post-mortem, frozen and dissected as previously described (17-19). [^{3}H]GABA binding was examined by the method of Enna and Snyder (20) with modifications, as previously reported (18). Briefly, frozen material was homogenized (Polytron) in bi-distilled water, left at room temperature for 30 minutes and then centrifuged at 48,000 g for 15 minutes. The resultant pellet was washed twice with bi-distilled water and re-suspended in phosphate buffer (0.1 M; pH, 7.2). The membranes were then "incubated" for 5 minutes at 0°C in the presence of 25 nM [^{3}H]GABA and any drug to be studied. The material was then centrifuged (48,000 g, 10 min) and the pellet surface and tube walls were rinsed with ice cold bi-distilled water. After solubilization of the pellet, its content of radioactivity was estimated in the presence of 10 ml Aquasol in a Nuclear-Chicago, Mark III scintillation counter. For the studies utilizing Triton X-100 (0.02 percent) or phospholipase C (0.001 units per incubation; from *Cl perfringens*, Serdary Research Labs), the preparative step of 30 minutes standing at room temperature was substituted by a 30-minute incubation at 37°C in the presence of these substances.

Protein concentrations were estimated by the method of Lowry et al. (21) and drug sources have been previously reported (22).

RESULTS AND DISCUSSION

Control Series of Patients

Previous studies have shown that normal physiological correlates or post-mortem factors have little influence on [^{3}H]GABA binding to membranes prepared from various brain regions. Thus, the age of the patient (6 months — 80 years), the time between death and freezing of the brain (4 — 33 hours), or the storage time before analysis (0 — 60 months) were not related to the levels of [^{3}H]GABA binding in the putamen, hippocampus or cerebellum. Also, [^{3}H]GABA binding did not appear to vary with the sex of patient or with respect to several common causes of death (e.g., pneumonia, carcinoma, or cardiac arrest; ref. 22).

The pharmacology of the [^{3}H]GABA binding site in membranes prepared from human cerebellar cortex is similar to that seen in rat brain and is consistent with that observed for the neurophysiologically identified GABA-receptor. Thus, the affinity of the binding site for [^{3}H]GABA in two different groups of control patients has been determined by three methods: the displacement of [^{3}H]GABA by unlabelled GABA (IC_{50}), the dissociation constant (K_D) using the Lineweaver-Burk plot, and the K_D using the Scatchard plot. All produced similar results (Table I) for both groups of patients. This binding affinity was similar to that seen in rat brain, for which the IC_{50} and K_D were about 2×10^{-7} M (15,18,20). Several compounds displaced [^{3}H]GABA binding with a relatively high affinity. These included muscimol, 3-aminopropanesulfonic acid, imidazoleacetic acid, δ-amino-n-valeric acid and 4-amino-3-hydroxybutyric acid. Compounds with lesser activity included bicuculline and homocarnosine. Inhibitors of GABA synthesis or uptake, other neurotransmitters, picrotoxin, baclofen and many other substances were inactive (19,22, 23). These results were very similar to those obtained in neurophysiological investigation of the GABA-receptor (24,25) and in similar binding studies on rat brain (14,15,20,26).

The distribution of [^{3}H]GABA binding in the normal human brain (Fig. 1) does not parallel that of either the GABA-synthesizing enzyme, GAD, or the GABA levels. This has also been observed by other workers for [^{3}H]GABA binding in the brains of both humans (27) and experimental animals (14) and also for other putative transmitter

Table I - Kinetics of $[^3H]$GABA Binding in Membranes Prepared from Human ("Control") Cerebellar Cortex

Group	Binding constants (10^{-7} M)			Amount bound (p-mole/mg tissue)		
	IC_{50}	K_D (LWB)	K_D (Scatchard)	$[^3H]$GABA (at 25 nM)	B_{max} (LWB)	B_{max} (Scatchard)
I	1.96 ±0.29 (5)	2.30 ±0.43 (5)	2.79 ±1.11 (5)	0.011 ±0.003 (5)	0.124 ±0.014 (5)	0.129 ±0.021 (5)
II	1.19 ±0.13 (4)	1.16 ±0.07 (4)	1.00 ±0.14 (4)	0.013 ±0.003 (4)	0.061 ±0.10 (4)	0.065 ±0.010 (4)

Data are expressed as means ± S.E.M.; numbers of brains examined in parentheses.

LWB = Lineweaver-Burk plot.

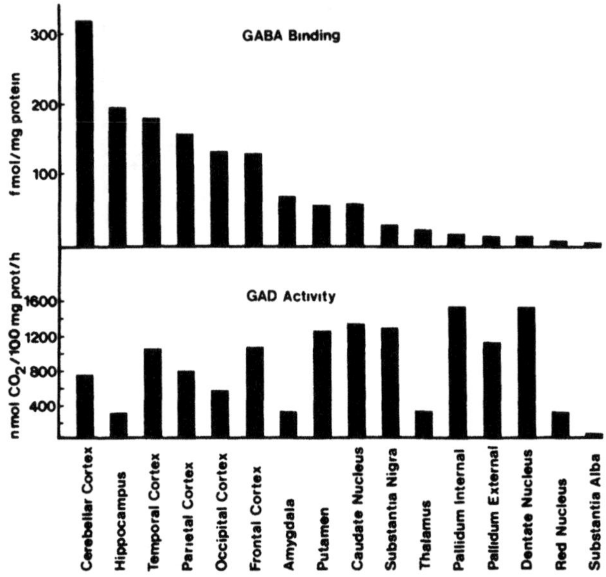

Fig. 1. Distribution of GABA binding (f-mole/mg protein) and GAD activity (n-mole CO_2 produced/mg protein/hr).

receptors in comparison with endogenous levels of the transmitter (28,29). However, the distribution of [^{3}H]GABA-binding does correlate with neurophysiologically-identified GABA-receptors. Thus, although the cerebellum has relatively low GAD activity, it has the highest [^{3}H]GABA binding capacity and at least four different putative GABA synapses (cf. 3,4). Similarly, the hippocampus and cerebral cortex have neurophysiologically-identified GABA-receptors (3, 30). Some areas possessing high levels of GABA and high GAD activity (e.g., striatum and substantia nigra) have disproportionately low levels of [^{3}H]GABA-binding. This might indicate that in these regions much of the GABA and GAD present is not related to the activity of GABA-ergic synaptic terminals or that there is a tremendous convergence of GABA-ergic terminals to relatively few GABA-receptors.

These data indicate that [^{3}H]GABA-binding in post-mortem human brain preparations likely reflects the state of the GABA-receptor, as defined in the rat brain by similar binding techniques as well as by neurophysiological investigations. Therefore, the use of this technique for studying pathological material may provide us with an insight for the usefulness of GABA-mimetics in these disease states, which the study of GABA levels and GAD activity does not provide.

Parkinson's Disease

The major neuropathological observation in Parkinson's disease is a loss of the melanin-containing cell bodies in the pars compacta of the substantia nigra. This is parallelled by the disappearance of dopamine, its synthesizing enzymes and its metabolites in the substantia nigra, caudate nucleus and putamen (cf. 31-33). Involvement of GABA pathways in Parkinson's disease is indicated by the decrease in GAD in extrapyramidal areas (cf. 13,34,35). As mentioned previously, this change is not likely related to cell loss, but to the reduced dopamine input to the striatum, as the GAD activity of patients chronically treated with L-DOPA is similar to that of control patients. Furthermore, a direct linear correlation exists between the striatal GAD activity and the duration of the L-DOPA treatment (13). This suggests (together with a multitude of evidence from animal experiments) that GABA-ergic neurones are involved in the feedback loop mediating the activity of nigrostriatal dopaminergic neurones (cf. ref. 9).

$[^3H]$GABA binding in post-mortem material from the brains of Parkinsonian patients (Table II) has been examined (18). The major loss (70%) of $[^3H]$GABA binding sites in the substantia nigra is the outstanding finding. From this it seems reasonable to conclude that the $[^3H]$GABA binding sites occur on the melanin-containing dopamine cell bodies and, furthermore, that normally there exists a direct synaptic GABA input to these dopamine cell bodies in the substantia nigra.

In view of the effect of L-DOPA on GAD activity, it was of interest to compare results on $[^3H]$GABA binding in the substantia

Table II - [^{3}H]GABA Binding to Membranes Prepared from the Brains of Control and Parkinsonian Patients[1]

Region	Control patients Mean ± S.E.M. (No. of cases)	Parkinson's disease Mean ± S.E.M. (No. of cases)	Parkinson's disease Percent control	Parkinson's disease Significance *vs* control[2]
Cerebellar cortex	328.1 ± 50.1 (10)	645.8 ± 242.6 (3)	199	NS
Frontal cortex	133.9 ± 28.1 (9)	236.7 ± 30.8 (3)	177	< 0.05
Temporal cortex	187.8 ± 27.6 (12)	240.8 ± 14.8 (5)	129	NS
Hippocampus	204.6 ± 18.6 (8)	121.5 ± 10.2 (6)	59	< 0.01
Caudate nucleus	66.3 ± 8.2 (9)	70.6 ± 9.4 (8)	107	NS
Putamen	67.8 ± 13.7 (17)	91.7 ± 19.6 (8)	135	NS
Substantia nigra	30.8 ± 5.0 (11)	9.7 ± 2.9 (6)	31	< 0.01

[1] Data are from Lloyd *et al*. (18) and are expressed in f-mole/mg protein.
[2] Two-tailed *t*-test.

Table III - Analysis of [^{3}H]GABA Binding in the Substantia Nigra of Parkinsonian Patients

Patient group	[^{3}H]GABA binding (f-mole/mg protein)	Percent control	p vs controls (two-tailed t-test)
Controls	30.8 ± 5.0 (11)	100	-
Parkinson's disease			
Non-L-DOPA			
Untreated	0	0	
Anticholinergic	8.3	26.9	
Amantadine	3.7	12.0	
All non-L-DOPA	4.0 ± 2.4	13.0	< 0.01
L-DOPA-treated	15.4 ± 1.8 (3)	50.0	< 0.05
All Parkinson's	9.7 ± 2.9 (6)	31.4	< 0.001

nigra with the different types of therapy used to treat these Parkinsonian patients (Table III). Although the L-DOPA-treated patients had somewhat greater [^{3}H]GABA binding than did the patients not receiving L-DOPA, too few patients were available to consider the possible significance of this finding. However, it can be said that, at least in the rat cerebellum, acute or prolonged (6-month) L-DOPA administration does not alter [^{3}H]GABA binding (36).

In Parkinsonian patients the striatum (caudate nucleus and putamen) is greatly depleted of dopamine, its metabolites and its synthesizing enzymes (33), reflecting the degeneration of nigrostriatal dopaminergic neurones. Interestingly, [^{3}H]GABA binding is not reduced in the caudate nucleus or putamen in these Parkinsonian patients (Table II). This might indicate that the number of [^{3}H]GABA binding sites on presynaptic dopamine terminals is relatively small in comparison with the total number of [^{3}H]GABA binding sites in the striatum. However, before this can be taken as fact detailed kinetic analyses of the binding should be performed. That such presynaptic GABA contacts on dopamine terminals do likely exist is evidenced by the alterations in striatal dopamine release produced in vivo after topically-applied GABA-related drugs (37-39).

Regarding the usage of GABA-related drugs in Parkinson's disease, it would seem that, in spite of the loss of [^{3}H]GABA binding sites, GABA-mimetic drugs are not appropriate as therapy. Thus, it has been proposed (40) that the decrease in GAD activity seen in non-DOPA-treated Parkinson's disease is a compensatory response aimed at reducing inhibition of dopamine cell bodies. Furthermore, as it is known that GABA-mimetics potentiate neuroleptic-induced catalepsy in animal models (41,42), it seems that it would be illogical and irresponsible to suggest these as therapeutic agents in Parkinson's disease. From this, it might appear that GABA-antagonists might have a place in the treatment of Parkinsonism. Although this may be a theoretical possibility, it is dangerous at the practical level since inhibition of GABA-ergic synaptic activity via a variety of actions (e.g., direct receptor blockade by bicuculline, blockade of Cl^- channels by picrotoxinin, blockade of synthesis by allylglycine) rapidly produces a convulsive state (cf. 43). If it is possible to produce a non-convulsant GABA-antagonist, then this may be of some use in Parkinson's disease.

Huntington's Chorea

Huntington's chorea (Huntington's disease) is a hereditary extrapyramidal disease which involves massive degeneration of the caudate nucleus and putamen and severe cortical atrophy (cf. 44). In some studies the substantia nigra has also been found to have severe degeneration, but these findings are inconsistent. The major neurochemical findings in Huntington's disease have been the decrease in GAD activity and GABA levels, especially in the striatum and substantia nigra (6,10-12). Other neurochemical systems, such as the striatal cholinergic neurones, also appear to be involved

(45). With regard to the massive cell loss in the striatum of Huntington's patients, it was somewhat surprising to see that [^{3}H]GABA binding was at most only slightly altered in this tissue (46). In a separate study (19), we reported a severe loss of [^{3}H]GABA binding in the caudate nucleus or putamen of Huntington's patients. These latter findings correlated well with the cell losses and gliosis seen in these structures. The discrepancy between the results of these two separate studies has not yet been explained. In this regard, Iversen et al. (47) have recently reported that two populations of Huntington's patients might exist, one with low striatal [^{3}H]GABA binding and one with apparently normal [^{3}H]GABA binding.

The decreases in GABA levels and GAD activities in striatum and substantia nigra of Huntington's patients has raised the possibility that GABA-mimetics would be a useful form of therapy in Huntington's disease. To date, however, reports of agents which either increase GABA concentrations (e.g., Depakine) or which are direct acting GABA-mimetics (imidazoleacetic acid, GABA, muscimol) indicate that these are inactive in Huntington's disease (48,49). This may be due to the loss of GABA-receptors, and if so, the future of GABA-mimetics in this disease appears bleak. However, there are reasons to be somewhat optimistic rather than pessimistic. Thus, muscimol and GABA enter the brain very poorly (50; cf. 49), and Depakine produces only a limited elevation of brain (51) and CSF (48) GABA levels. Furthermore, there have been some reports of positive actions of GABA-mimetics in Huntington's chorea. Thus, iproniazid (an inhibitor of GABA metabolism; refs. 52,53) and SL-76 002 (a GABA-mimetic and prodrug; see ref. 54) provide a rationale for continuing the use of GABA-mimetics in Huntington's disease.

Even when considering the loss of striatal [^{3}H]GABA binding sites in Huntington's chorea, there are reasons to believe that GABA-mimetics will be useful in this disease. If the remaining receptors have an increased affinity, then administration of GABA-mimetics might still produce an increase in receptor activity. For the brains

Table IV - [^{3}H]GABA Binding to Cerebellar Membranes from Two Independent Groups of Patients

	IC_{50} (10^{-7} M)		Binding (at 25 nM [^{3}H]GABA)			
			f-mole/mg protein		f-mole/mg tissue	
	Controls	Huntington's	Controls	Huntington's	Controls	Huntington's
Group I[1]	2.33	0.40^2	--	--	11.60	22.48^2
	± 0.22	± 0.13			± 1.59	± 3.42
	(17)	(12)			(15)	(12)
Group II	0.89	0.38^2	401.7	1222.8^2	17.22	32.73^2
	± 0.09	± 0.04	± 57.1	± 107.8	± 3.46	± 4.37
	(12)	(9)	(12)	(9)	(12)	(9)
All patients combined	1.71	0.39^2	--	--	14.10	26.87^2
	± 0.13	± 0.07			± 1.73	± 2.64
	(29)	(21)			(27)	(21)

All results are expressed as means ± S.E.M.; numbers of brains examined in parentheses.

[1] From Lloyd et al. (18).

[2] $p < 0.01$ vs appropriate controls.

Table V - Binding Constants for Displacement of $[^3H]$GABA by Unlabelled GABA from Cerebellar Membranes of Two Different Groups of Huntington's Patients

	IC_{50} (at 25 nM $[^3H]$GABA)	K_D (Lineweaver-Burk plot)	K_D (Scatchard plot)
Group I (n = 3)	18 ± 11 nM	29 ± 9 nM	43 ± 12 nM
Group II (n = 4)	46 ± 1 nM	46 ± 6 nM	41 ± 7 nM

Data are expressed as means ± S.E.M.

of Huntington's patients which have been studied there was insufficient striatal material to enable a detailed kinetic analysis. Therefore, cerebellar cortex was used as a source of material rich in GABA synapses. In the adult form of Huntington's chorea this region is less severely affected than it is in the juvenile disease. The initial report (19) indicated that the affinity for $[^3H]$GABA and also the amount of $[^3H]$GABA bound at 25 nM were greatly increased. These studies have been repeated on the cerebellar cortices of a second group of patients with essentially similar results (Table IV). Furthermore, for some of these patients in the second group the affinity of GABA for the $[^3H]$GABA binding site has been analyzed by three different methods — log dose-response curve (IC_{50}), Lineweaver—Burk plot (K_D) and Scatchard plot (K_D). The results, shown in Table V, substantiate our earlier findings. Thus, the binding affinity is similar by the three methods of analysis and in each case is significantly ($p < 0.01$ or less) greater than that for the respective controls (Table I).

These findings indicate that in Huntington's chorea the receptors might have a greater affinity for GABA than those which exist in control brains. Support for this hypothesis is found in the reports of studies on intra-striatal administration of kainic acid, which is currently used as an animal model for Huntington's disease. This procedure reduces GAD activity, GABA levels and choline acetylase activity in a manner similar to that observed in Huntington's patients (55,56). Furthermore, the affinity of striatal [^{3}H]GABA binding was increased (decreased K_D) shortly after kainic acid injection. Total [^{3}H]GABA binding in the striatum was decreased, in accord with our results with caudate nucleus or putamen of Huntington's patients, but this occurred only 9 months after kainic acid injection (57). For shorter time periods large decreases in [^{3}H]GABA binding were not observed, probably due to the long-lasting attachment of postsynaptic fragments (which possibly contain GABA receptor sites) to presynaptic terminal membranes (58).

The question of the nature of the membrane alteration in Huntington's chorea which produces this alteration in [^{3}H]GABA binding is naturally of interest. Phospholipoproteins are a major membrane component (59), and in this regard it is interesting that in the brains of both humans (22) and experimental animals (15) Triton-X-100 (a non-ionic detergent) increased the binding of [^{3}H]GABA and that in the rat brain this detergent increased the affinity of the binding site. We have repeated these studies using Triton-X-100 on membranes prepared from either control or Huntington's patients. As can be seen from Table VI, pre-incubation with 0.02 percent Triton-X-100 increased the affinity of [^{3}H]GABA binding in membranes from control brains whereas it did not alter that in Huntington's material. Whether this is an alteration of the affinity of one site for [^{3}H]GABA or if it represents the unmasking of a higher affinity site is not yet clarified. However, upon analyzing 15 control brains by Lineweaver-Burk and/or Scatchard plots, only two brains have yielded evidence for the existence of two different binding affinities. Even in this event, the high-affinity process ($K_D \simeq$ 90 nM) was of lower affinity than that observed after Triton X-100 treatment ($K_D \simeq$ 40 nM).

Table VI - Effect of Triton-X-100 or Phospholipase-C Treatment on [^{3}H]GABA Binding to Membranes Prepared from the Cerebellar Cortex of Control or Huntington's Disease Patients

Patient group	IC_{50} (nM)	K_D from LWB plot (nM)	K_D from Scatchard plot (nM)
<u>Controls (n = 4)</u>			
Untreated	112 ± 13	116 ± 7	100 ± 14
Triton-X-100 (0.02%)	42 ± 3*	32 ± 4*	34 ± 4*
Phospholipase-C (0.001 U)	39 ± 5*	36 ± 9*	36 ± 10*
<u>Huntington's chorea (n = 4)</u>			
Untreated	46 ± 1	46 ± 6	41 ± 7
Triton-X-100 (0.02%)	48 ± 6	39 ± 4	37 ± 3
Phospholipase-C (0.001 U)	35 ± 1	21 ± 3*	21 ± 7

Results are expressed as means ± S.E.M.; LWB = Lineweaver-Burk plot.

* $p < 0.01$ <u>vs</u> untreated membranes.

Thus, Triton-X-100 may be removing from the membranes a lipid substituent which is associated with the GABA-receptor and which may have a functional role in controlling either the accessibility of GABA to its receptor or the active configurations of the receptor. In this regard, it is known that alterations of membrane phospholipids alter GABA receptor binding in a highly specific manner. Phospholipase-A (which removes the fatty acid in position 1 or 2 from phosphoglycerides) decreases GABA binding in either rat whole brain (15) or cerebellar preparations (60). However, it has been reported that phospholipase-C (which hydrolyzes the bond between phosphoric acid and glycerol) increases GABA binding to rat cerebellar membranes (60) in a manner reminiscent of that seen with Triton-X-100.

Accordingly, membranes from control and Huntington's patients were incubated with 0.001 units of phospholipase-C (Table VI). In material from control brains this enzyme altered the affinity of [^{3}H]GABA binding in a manner very similar to that seen with Triton-X-100. In the material from Huntington's cerebellum the affinity of [^{3}H]GABA-binding was increased slightly, but to a much lesser extent than that seen in the control brains. From these findings it seems highly possible that the difference in [^{3}H]GABA binding that exists between control and Huntington's brain is related to an alteration in the phospholipid components of the membrane localized near or at the binding site. Furthermore, this alteration may involve the site which is sensitive to phospholipase-C activity; i.e., the bond between the phosphoric acid and glycerol components (59).

These data suggest that the affinity or accessibility of the [^{3}H]GABA binding site (and therefore the GABA-receptor?) is at least partially controlled by an endogenous phospholipid, which may act as an endogenous inhibitor. In this regard, Watkins (61) has proposed that phosphatidylethanolamine competes with GABA for binding. This substance and phosphatidylglycerol are very susceptible to phospholipase-C activity (62). Preliminary data have experimentally verified this using junctional complexes from rat cerebellum (60). As glycerophosphoethanolamine has been found to be increased in the brains of Huntington's patients (11) the effects of this compound and phosphatidylethanolamine are presently being studied on [^{3}H]GABA-binding in phospholipase-C-treated control and Huntington's cerebellar material.

From the meager literature available, it would appear that an augmentation of binding by phospholipase-C is not an effect generally applicable to neurotransmitter receptors. Thus, phospholipase-C decreases β-adrenergic receptor binding capacity without altering receptor affinity (63) and greatly decreases muscarinic receptor binding (64).

SUMMARY

With respect to [^{3}H]GABA-binding in material prepared from human post-mortem brain, the following observations have been made: (1) The [^{3}H]GABA binding site in the cerebellum has the pharmacological characteristics of the physiological GABA-receptor observed in other species. Together with the post-mortem stability exhibited for [^{3}H]GABA-binding, this provides an approach for determining the functional state of the GABA-receptor in various disease states; (2) In Parkinson's disease [^{3}H]GABA-binding in the substantia nigra is significantly decreased whereas that in the putamen and caudate nucleus is unaltered. The former finding likely indicates that GABA binding sites (receptors) occur on nigral dopaminergic cell bodies and/or dendrites. The latter finding may signify that relatively few of the striatal [^{3}H]GABA binding sites occur on dopaminergic nerve terminals in the human caudate or putamen; (3) In Huntington's disease [^{3}H]GABA binding was decreased in the caudate nucleus and putamen, in parallel with the massive cell loss and gliosis observed in this condition. Membranes prepared from cerebellar tissue of these patients possessed an increased affinity for [^{3}H]GABA-binding; (4) Pre-treatment of cerebellar membranes from control brains with Triton-X-100 (0.02%) or phospholipase-C (0.001 units) results in kinetic changes very similar to those observed in Huntington's brains. In contrast, such treatment was virtually without effect on the IC_{50} or K_D for [^{3}H]GABA on cerebellar membranes prepared from Huntington's brains; (5) These results imply that a phospholipid, possibly related to phosphoglycerolethanolamine, is altered in the membranes of Huntington's patients and that this phospholipid normally has a role in controlling accessibility to the GABA-receptor.

REFERENCES

(1) Hornykiewicz, O. (1973) Fed. Proc. 32, 183-190.
(2) Bloom, F. E. and Iversen, L. L. (1971) Nature (Lond.) 229, 628.
(3) Ito, M. (1976) In: GABA in Nervous System Function (E. Roberts,

T. N. Chase and D. B. Tower, eds.) pp. 427-448, Raven Press, New York.

(4) Obata, K. (1972) Int. Rev. Neurobiol. 15, 167-187.

(5) Storm-Mathisen, J. (1972) Brain Res. 40, 215-235.

(6) McGeer, P. L. and McGeer, E. G. (1976) In: GABA in Nervous System Function (E. Roberts, T. N. Chase and D. B. Tower, eds.) pp. 487-495, Raven Press, New York.

(7) Obata, K. (1976) In: GABA in Nervous System Function (E. Roberts, T. N. Chase and D. B. Tower, eds.) pp. 217-222, Raven Press, New York.

(8) Storm-Mathisen, J. (1976) In: GABA in Nervous System Function (E. Roberts, T. N. Chase and D. B. Tower, eds.) pp. 149-168, Raven Press, New York.

(9) Lloyd, K. G. (1978) In: Essays in Neurochemistry and Neuropharmacology (M. B. H. Youdim, W. Lovenberg, D. F. Sharman and J. R. Lagnado, eds.) Vol. 3, pp. 129-207, Wiley, New York.

(10) Bird, E. D. and Iversen, L. L. (1974) Brain, 97, 457-472.

(11) Perry, T. L., Hansen, S., Lesk, D. and Kloster, M. (1973) In: Advances in Neurology (A. Barbeau, T. N. Chase and G. W. Paulson, eds.) Vol. 1, pp. 609-618, Raven Press, New York.

(12) Stahl, W. L. and Swanson, P. D. (1974) Neurology, 24, 813-819.

(13) Lloyd, K. G., Möhler, H., Bartholini, G. and Hornykiewicz, O. (1976) In: Advances in Parkinsonism (W. Birkmayer and O. Hornykiewicz, eds.) pp. 186-192, Editiones Roche, Basel.

(14) Enna, S. J. and Snyder, S. H. (1975) Brain Res. 100, 81-97.

(15) Enna, S. J. and Snyder, S. H. (1977) Mol. Pharmacol. 13, 442-453.

(16) Young, A. B., Enna, S. J., Zukin, S. R. and Snyder, S. H. (1976) In: GABA in Nervous System Function (E. Roberts, T. N. Chase and D. B. Tower, eds.) pp. 305-317, Raven Press, New York.

(17) Lloyd, K. G. and Kornykiewicz, O. (1972) J. Neurochem. 19, 1549-1559.

(18) Lloyd, K. G., Shemen, L. and Hornykiewicz, O. (1977) Brain Res. 127, 269-278.

(19) Lloyd, K. G., Dreksler, S. and Bird, E. D. (1977) Life Sci. 21, 747-754.

(20) Enna, S. J. and Snyder, S. H. (1976) J. Neurochem. 26, 221-224.

(21) Lowry, O. H., Rosebrough, N. J., Farr, A. L. and Randall, R. J. (1951) J. Biol. Chem. 193, 265-275.

(22) Lloyd, K. G. and Dreksler, S. (1978) Brain Res. (in press).

(23) Iversen, L. L. (1978) In: Psychopharmacology: A Generation of Progress (M. A. Lipton, A. DiMascio and K. F. Killam, eds.) pp. 25-38, Raven Press, New York.

(24) Bowery, N. G. and Brown, D. A. (1974) Brit. J. Pharmacol. 50, 205-218.

(25) Johnston, G. A. R. (1976) In: GABA in Nervous System Function (E. Roberts, T. N. Chase and D. B. Tower, eds.) pp. 305-317, Raven Press, New York.

(26) Krogsgaard-Larsen, P. (1978) In: Amino Acids as Chemical Transmitters (F. Fonnum, ed.) pp. 305-321. Plenum Press, New York.

(27) Enna, S. J., Bennett, J. P., Bylund, D. B., Creese, I., Burt, D. R., Charness, M. E., Yamamura, H. T. and Snyder, S. H. (1977) J. Neurochem. 28, 233-236.
(28) Rall, T. W. and Gilman, A. G. (1976) Neurosci. Res. Bull. 8, 223-232.
(29) Yamamura, H. T., Kuhar, M. J., Greenberg, D. and Snyder, S. H. (1974) Brain Res. 66, 541-546.
(30) Krnjević, K. and Schwartz, S. (1968) Exp. Brain Res. 3, 320-336.
(31) Bernheimer, H., Birkmayer, O., Hornykiewicz, O., Jellinger, K. and Seitelberger, F. (1973) J. Neurol. Sci. 20, 415-455.
(32) Hornykiewicz, O. (1966) Pharmacol. Rev. 18, 925-962.
(33) Lloyd, K. G., Davidson, L. and Hornykiewicz, O. (1975) J. Pharmacol. Exp. Therapeut. 195, 453-464.
(34) McGeer, P. L., McGeer, E. G., Wada, J. A. and Jung, E. (1971) Brain Res. 32, 425-431.
(35) Rinne, U. K., Sonninen, V., Riekkinen, P. and Laaksonen, H. (1974) Med. Biol. 52, 208-217.
(36) Lloyd, K. G., Shibuya, M., Davidson, L. and Hornykiewicz, O. (1977) In: Advances in Biochemical Psychopharmacology (E. Costa and G. L. Gessa, eds.) Vol. 16, pp. 409-415, Raven Press, New York.
(37) Bartholini, G., Stadler, H., Gadea-Ciria, M. and Lloyd, K. G. (1976) Neuropharmacology, 15, 515-519.
(38) Cheramy, A., Nieoullon, A. and Glowinski, J. (1978) Eur. J. Pharmacol. 48, 281-295.
(39) Starr, M. (1978) Eur. J. Pharmacol. 48, 325-328.
(40) Lloyd, K. G. (1977) In: Parkinson's Disease: Concepts and Prospects (J. P. W. F. Lakke, J. Korf and H. Wesseling, eds.) pp. 61-72, Excerpta Medica, Amsterdam.
(41) Kaariainen, I. (1976) Acta Pharmacol. Toxicol. 40, 188-192.
(42) Worms, P., Willigens, M. T. and Lloyd, K. G. (1978) J. Pharmac. Pharmacol. (in press).
(43) Meldrum, B. S. (1975) Int. Rev. Neurobiol. 17, 1-36.
(44) Forno, L. S. and Jose, C. (1973) In: Advances in Neurology (A. Barbeau, T. N. Chase and G. W. Paulson, eds.) Vol. 1, pp. 453-470, Raven Press, New York.
(45) Aquilonius, S. M., Eckernas, S. A. and Sundvall, A. (1975) J. Neurol. Neurosurg. Psychiat. 38, 669-677.
(46) Enna, S. J., Bird, E. D., Bennett, J. A., Bylund, D. B., Yamamura, H. I., Iversen, L. L. and Snyder, S. H. (1976) New Engl. J. Med. 294, 1305-1309.
(47) Iversen, L. L., Spokes, E. and Bird, E. (1978) In: GABA-Neurotransmitters: Pharmacochemical, Biochemical and Pharmacological Aspects (H. Kofod, P. Krogsgaard-Larsen and J. Scheel-Krüger, eds.) Munksgaard, Copenhagen (in press).
(48) Chase, T. N. and Tamminga, C. A. (1978) In: GABA-Neurotransmitters: Pharmacochemical, Biochemical and Pharmacological Aspects (H. Kofod, P. Krogsgaard-Larsen and J. Scheel-Krüger, eds.) Munksgaard, Copenhagen (in press).

(49) Chase, T. N. and Walters, J. (1976) In: GABA in Nervous System Function (E. Roberts, T. N. Chase and D. B. Tower, eds.) pp. 497-513, Raven Press, New York.
(50) Enna, S. J., Ferkany, J. and Krogsgaard-Larsen, P. (1978) In: GABA-Neurotransmitters: Pharmacochemical, Biochemical and Pharmacological Aspects (H. Kofod, P. Krogsgaard-Larsen and J. Scheel-Krüger, eds.) Munksgaard, Copenhagen (in press).
(51) Godin, Y., Heiner, L., Mark, J. and Mandel, P. (1969) J. Neurochem. 16, 869-873.
(52) Wood, J. D. and Peekser, S. J. (1973) Can. J. Physiol. Pharmacol. 51, 959-965.
(53) Perry, T. L., MacLeod, P. M., and Hansen, S. (1977) New Engl. J. Med. 297, 840.
(54) Bartholini, G., Lloyd, K. G., Scatton, B. and Zivkovic, B. (1978) In: GABA-Neurotransmitters: Pharmacochemical, Biochemical and Pharmacological Aspects (H. Kofod, P. Krogsgaard-Larsen and J. Scheel-Krüger, eds.) Munksgaard, Copenhagen (in press).
(55) Coyle, J. T., Schwarcz, R., Bennett, J. P. and Campochiaro, P. (1977) Progr. Neuro-Psychopharm. 1, 13-30.
(56) McGeer, E. G., Innane, V. T. and McGeer, P. L. (1976) Brain Res. 118, 356-358.
(57) Zaczek, R., Schwarcz, R. and Coyle, J. T. (1978) Brain Res. 152, 626-632.
(58) Olney, J. W. and DeGubareff, T. (1978) Brain Res. 140, 340-343.
(59) Lehninger, A. L. (1976) Biochemistry. 2nd Ed. Worth, New York.
(60) Giambalvo, C. T. and Rosenberg, P. (1976) Biochim. Biophys. Acta, 436, 741-756.
(61) Watkins, J. C. (1965) J. Theoret. Biol. 9, 37-49.
(62) Mavis, R. D., Bell, R. M. and Vagelos, P. R. (1972) J. Biol. Chem. 247, 2835-2841.
(63) Limbird, L. E. and Lefkowitz, R. J. (1976) Mol. Pharmacol. 12, 559-567.
(64) Aromstam, R. S., Abood, L. G. and Baumgold, J. (1977) Biochem. Pharmacol. 26, 1689-1695.

GABA-RECEPTORS IN THE SUBSTANTIA NIGRA CONTROLLING CYCLIC-GMP IN THE CEREBELLAR CORTEX

G. Biggio, M. G. Corda, M. Casu, M. L. Mulas and G. L. Gessa

Institute of Pharmacology, University of Cagliari, Italy

INTRODUCTION

The cerebellum controls posture and modulates muscular movement by means of GABA released from Purkinje cell terminals onto neurons in cerebellar deep nuclei (12). In fact, the main function of Purkinje neurons is that of transducing the excitatory input from parallel and climbing fibers into an inhibitory signal to the neurons of cerebellar deep nuclei. While the nature of the neurons in the cerebellar cortex and the morphology of their synaptic contacts with climbing or mossy fibers are well established, the excitatory transmitters released from parallel and climbing fibers are still unknown (12). However, strong evidence indicates that cyclic 3',5'-guanosine monophosphate (cyclic-GMP) is the second messenger generated by excitation of the GABA-ergic neurons in the cerebellar cortex (4,8,9). Therefore, the content of this nucleotide is a good biochemical indicator both of the activity of such neurons and of the function of the parallel and climbing fibers.

Recent evidence indicates that dopamine (DA) receptors in the caudate nucleus have a major role in controlling cyclic-GMP content in the cerebellar cortex. Blockade of striatal DA-receptors, induced by the systemic or local injection of different neuroleptics, results in a fall in the cyclic-GMP content (8,10). On the contrary, the systemic or intra-striatal injection of apomorphine, a DA-receptor stimulant, increases cerebellar cyclic-GMP content (8,10). Finally, the destruction of striatal DA-receptors induced by local injection of kainic acid is associated with an irreversible, almost complete, disappearance of cyclic-GMP in the cerebellar cortex and prevents the increase in this nucleotide normally elicited by systemic administration of apomorphine (5). Cerebellar cyclic-GMP content seems to be a sensitive index of the state of activation or blockade of DA-receptors in the striatum. Other studies have revealed that the content of this nucleotide is decreased by doses of haloperidol (8) or sulpiride (3) much lower than those required to elicit catalepsy. Following chronic treatment with haloperidol, tolerance develops to its effect of decreasing cyclic-GMP content; such tolerance to neuroleptics is associated with supersensitivity to the enhancing effect of apomorphine on the nucleotide content (6).

On the basis of the above observations and of results obtained in different anatomical and electrophysiological studies (1,14), we postulated the existence of a multi-synaptic striato-cerebellar connection exerting a tonic inhibitory control on cerebellar cyclic-GMP formation and that the stimulation of striatal DA-receptors would result in "switching-off" the activity of such a connection (7). The present results are part of a study in progress in our laboratory that is being aimed at identifying the site and nature of the synapses along the chain of neurons of the striato-cerebellar pathway. The evidence presented here shows that the striatal output system to the cerebellum has a GABA-ergic synapse located in the substantia nigra.

STRIATO-NIGRAL GABA-ERGIC PATHWAY

Numerous GABA-ergic terminals are present in both the pars compacta and the pars reticulata of the substantia nigra (2,13). Studies based on the effects of CNS lesions have revealed that most of these terminals originate in the striatum and globus pallidus, and it has been suggested that they synapse mainly on nigral dopaminergic neurons and inhibit their activity (17). However, recent evidence from our laboratory and from other laboratories has shown that the striato-nigral GABA-ergic pathway is not only a feedback loop for controlling the activity of dopaminergic neurons, but that it is mainly an output pathway conveying the effect of the stimulation of striatal DA-receptors (11,15,18,20).

According to these studies, the stimulation of striatal DA-receptors would result in a release of GABA from the above-mentioned terminals onto non-dopaminergic neurons located in the pars reticulata of the substantia nigra. Thus, different behavioral changes (e.g., stereotypy) elicited by the stimulation of striatal DA-receptors would be mediated by the GABA-ergic inhibition of the above-mentioned non-dopaminergic nigral neurons. We have investigated whether the GABA-ergic striato-nigral pathway conveyed also the effect of dopaminergic stimulation on cerebellar cyclic-GMP content; i.e., if a GABA-ergic synapse of the above-mentioned nigral-cerebellar connection might reside in the substantia nigra. If the substantia nigra is a station of the striato-cerebellar pathway and if GABA is the neurotransmitter that mediates, via this synapse, the effect of stimulation of DA-receptors in the striatum, local injections of GABA-mimetics into this area should increase cyclic-GMP in the cerebellar cortex.

In agreement with this contention, the results shown in Table I revealed that bilateral injections of 100 ng of muscimol, a potent GABA-mimetic, into the rat substantia nigra produced a marked increase in the cyclic-GMP content of cerebellar cortex. This effect was completely prevented by bilateral intranigral injections of 50 ng of picrotoxin, a specific GABA-antagonist. We investigated further the importance of the striato-nigral GABA-ergic pathway in controlling the cyclic-GMP content of cerebellar cortex either by destroying, with intra-striatal kainic acid, the neurons originating in the caudate nucleus or by blocking striatal dopamine-receptors

Table I - Increase of Cerebellar Cyclic-GMP Content Elicited by Bilateral Intra-nigral Injection of Muscimol; Antagonism by Picrotoxin

	Cerebellar cyclic-GMP (p-moles/mg protein)	
	Saline	Picrotoxin 50 ng
Saline 0.5 µl	8.6 ± 0.04	5.5 ± 0.09*
Muscimol 100 ng/0.5 µl	16.5 ± 0.07*	6.4 ± 0.05*

Each point is the mean ± S.E.M. of 4 experiments; * $p < 0.01$ with respect to saline-treated rats. Rats were killed by a 3-second exposure to a focussed high intensity microwave radiation (16). Muscimol and picrotoxin were injected 30 and 45 minutes before sacrifice, respectively, into the substantia nigra of Sprague-Dawley rats (270 - 300 g); (coordinates A = 3; L = 2; DV = 7.8) (19).

with systemic injections of haloperidol. Both these treatments produced an almost complete disappearance of cyclic-GMP in the cerebellar cortex but failed to block the increase of cyclic-GMP elicited by intra-nigral muscimol, indicating that the nigral synapse is downstream with respect to the striatum (Fig. 1). Consistent with the above hypothesis, bilateral intra-nigral injections of picrotoxin not only significantly decreased cyclic-GMP but also prevented the effect of systemic injection of apomorphine (Fig. 2).

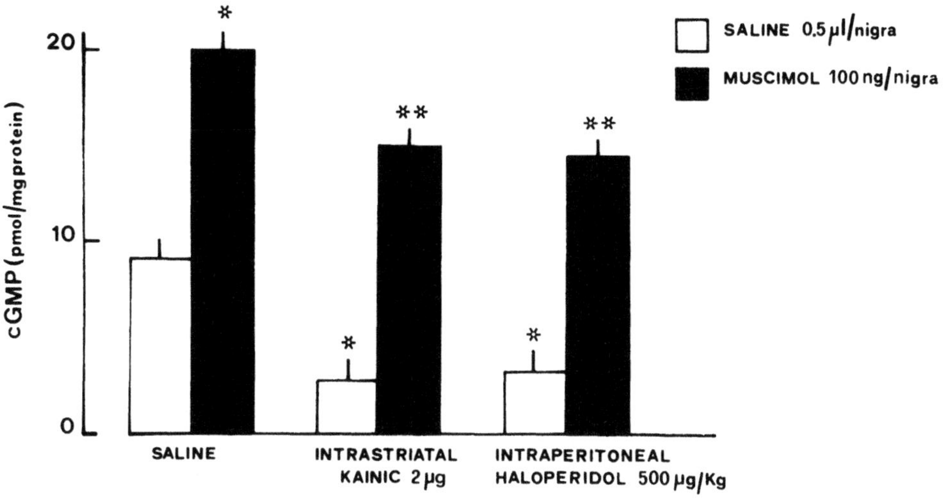

Fig. 1. Increase of cerebellar cyclic-GMP content by bilateral intra-nigral muscimol in rats with both striata lesioned with kainic acid or treated with haloperidol. Each point is the mean ± S.E.M. of 4 experiments; *p < 0.01 with respect to saline-treated rats; **p < 0.01 with respect to kainic-acid- or haloperidol-treated rats. Rats were killed 30 minutes, 45 minutes and 3 days after muscimol, haloperidol and kainic acid, respectively.

INTRA-NIGRAL MUSCIMOL IN RATS TREATED WITH 3-ACETYLPYRIDINE

It has been previously shown that climbing fibers are not involved in mediating the effect of dopamine agonists and antagonists on cerebellar cyclic-GMP content; in fact, the destruction with 3-acetylpyridine of the inferior olivary nucleus, the site of origin of climbing fibers, failed to prevent the effects of apomorphine and haloperidol on cerebellar cyclic-GMP content (4,8). If a GABA-ergic neurotransmitter mechanism in the substantia nigra mediates the changes of cyclic-GMP elicited by dopamine receptor stimulation, the intra-nigral injection of muscimol should increase cyclic-GMP also in animals treated with 3-acetylpyridine. As shown in Table II, bilateral intra-nigral injections of muscimol still increased

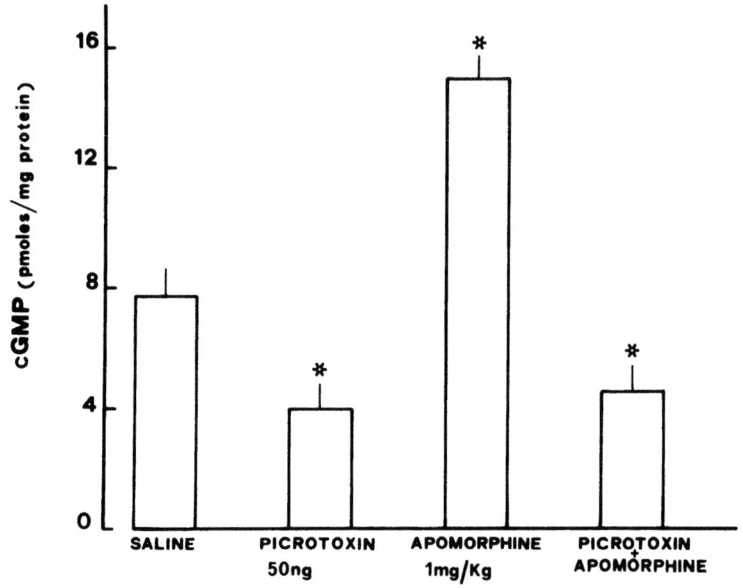

Fig. 2. Increase of cerebellar cyclic-GMP content elicited by apomorphine; antagonism by bilateral intra-nigral picrotoxin. Each point is the mean ± S.E.M. of 4 experiments; *$p < 0.01$ with respect to saline-treated rats. Rats were killed 12 and 45 minutes after apomorphine and picrotoxin, respectively.

cyclic-GMP in animals whose climbing fibers had been destroyed by 3-acetylpyridine. Thus, climbing fibers are not a part of the striato-nigral-cerebellar pathway.

CONCLUSION

The results presented herein indicate that the stimulation of GABA-receptors in the substantia nigra mimics the effect of striatal DA-receptor stimulation on cerebellar cyclic-GMP levels. Blockade of GABA-receptors causes not only the same changes in cyclic-GMP levels as DA-receptor blockade, but also prevents the effect of DA-receptor stimulation. These results suggest that the cerebellar

Table II - Increase of Cerebellar Cyclic-GMP Content Elicited by Bilateral Intra-nigral Injection of Muscimol in Rats Treated with 3-Acetylpyridine

	Cerebellar cyclic-GMP (p-moles/mg protein)	
	Saline	3-Acetylpyridine 80 mg/kg
Saline 0.5 µl	9.2 + 0.07	4.6 + 0.08*
Muscimol 100 ng/0.5 µl	19.2 + 0.09*	12.8 + 0.06*

Each point is the mean ± S.E.M. of 4 experiments; *$p < 0.01$ with respect to saline-treated rats. Rats were killed 30 minutes and 3 days, respectively, after muscimol and 3-acetylpyridine.

responses to striatal DA-receptor stimulation are mediated, as are other behavioral and neurological responses (11,15,18,20), by the release of GABA within the substantia nigra. Conversely, the effect of DA-receptor blockade would depend on a reduction of nigral GABA release with the consequent activation of nigral non-dopaminergic neurons, which are part of the striato-cerebellar connection. The findings that intra-nigral muscimol still increases cerebellar cyclic-GMP content after destruction of DA-receptors with kainic acid, and that intra-nigral picrotoxin prevents the effect of apomorphine, suggest that the GABA-ergic synapse is downstream with respect to the striatum along the striato-cerebellar connection.

Our results support the contention that the substantia nigra exerts not only a regulatory function on the activity of striatal neurons through the nigro-striatal system, but also an output function through its non-dopaminergic neurons. It is likely that the output from the substantia nigra to the cerebellum is conveyed via

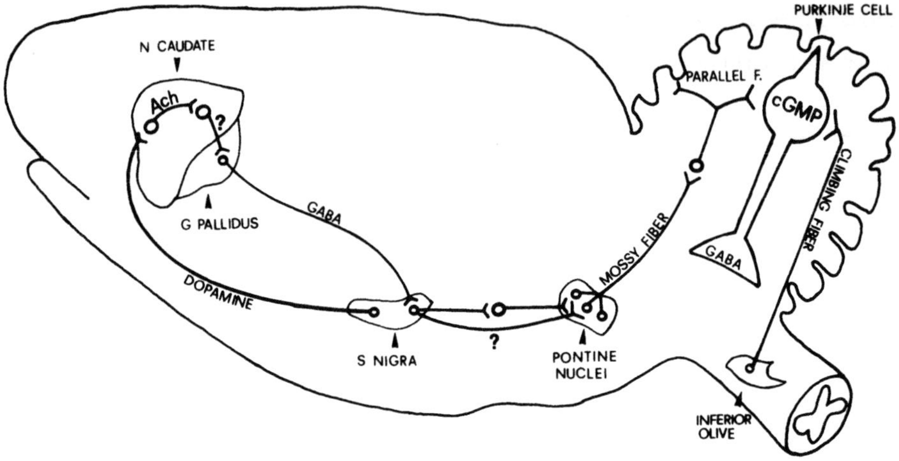

Fig. 3. Model of the hypothetical connection between striatum and cerebellum.

the nigro-pontine pathway to the mossy fiber system. On the other hand, the climbing fiber system does not participate in the cerebellar regulation by the striatum. Figure 3 shows our model representing the striato-cerebellar multi-synaptic pathway.

In conclusion, since the striatum, substantia nigra and cerebellum are involved in the regulation of motor function and in the coordination of movements, the functional link among these areas suggests that changes in cerebellar function may participate in the extrapyramidal disturbances produced by pathological and pharmacological modification of nigro-striatal activity.

REFERENCES

(1) Allen, G. I. and Tsukahara, N. T. (1974) Physiol. Rev. 14, 957-1006.

(2) Bak, I. J., Choi, W. B., Hassler, R., Usinoff, K. G. and Wagner, A. (1975) In: Advances in Neurology (A. Calne, T. N. Chase and A. Barbeau, eds.) Vol. 9, pp. 15-41, Raven Press, New York.

(3) Biggio, G. (1978) Unpublished results.

(4) Biggio, G., Brodie, B. B., Costa, E. and Guidotti, A. (1977) Proc. Nat. Acad. Sci., U.S.A. 74, 3592-3596.

(5) Biggio, G., Corda, M. G., Casu, M. and Gessa, G. L. (1978) Naunyn-Schmiedeberg's Arch. Pharmacol. 304, 5-7.

(6) Biggio, G., Corda, M. G., Casu, M. and Gessa, G. L. (1978) Life Sci. 23, 649-652.
(7) Biggio, G., Corda, M. G., Casu, M. and Gessa, G. L. (1978) Adv. Biochem. Psychopharm. 18, 227-244.
(8) Biggio, G., Costa, E. and Guidotti, A. (1977) J. Pharmacol. Exp. Therapeut. 200, 207-215.
(9) Biggio, G. and Guidotti, A. (1976) Brain Res. 107, 365-373.
(10) Biggio, G. and Guidotti, A. (1977) Nature (Lond.) 265, 240-242.
(11) Di Chiara, G., Porceddu, M. L., Morelli, M., Mulas, M. L. and Gessa, G. L. (1978) Naunyn-Schmiedeberg's Arch. Pharmacol. In press.
(12) Eccles, J. C., Ito, M. and Szentagothai, J. (1967) The Cerebellum as a Neuronal Machine. Springer-Verlag, New York.
(13) Fonnum, F., Grofova, I., Rinvik, E., Storm-Mathisen, J. and Walberg, F. (1974) Brain Res. 71, 77-92.
(14) Fox, M. and Williams, T. D. (1970) Brain Res. 20, 140-144.
(15) Garcia-Munoz, M., Nicolaou, N. M., Tulloch, A. K. and Arbuthnott, G. W. (1977) Nature (Lond.) 265, 363-365.
(16) Guidotti, A., Cheney, D. L., Trabucchi, M., Doteuchi, M., Wang, C. and Hawkins, R. C. (1974) Neuropharmacology, 13, 1115-1122.
(17) Hattori, T., McGeer, P. L., Fibiger, H. C. and McGeer, E. G. (1973) Brain Res. 54, 103-114.
(18) Olianas, M. C., De Montis, G., Mulas, G. and Tagliamonte, A. (1978) Eur. J. Pharmacol. 49, 233-241.
(19) Pellegrino, L. and Cushman, A. J. (1967) A Stereotaxic Atlas of the Rat Brain. Appleton-Century-Crofts, New York.
(20) Scheel-Kruger, J. J., Arnt, A. and Magelund, G. (1977) Neurosci. Lett. 4, 351-353.

SECTION IV

BEHAVIORAL CORRELATES

CORRELATION OF CHANGES IN THE GABA-ERGIC SYSTEM WITH THE DEVELOPMENT OF SPASTICITY IN PARAPLEGIC CATS

N. E. Naftchi*, W. Schlosser and W. D. Horst

*Laboratory of Biochemical Pharmacology, Institute of Rehabilitation Medicine, New York University Medical Center, 400 East 34th Street, New York 10016, U.S.A.

Hoffmann-La Roche Inc., Research Division, Nutley, New Jersey 07110, U.S.A.

INTRODUCTION

An increased excitability in segmental reflex pathways following spinal cord injury results from the loss of supraspinal control of spinal cord activity. The reason for this phenomenon remains obscure, although a number of explanations have been suggested. The reticulospinal tract, for example, is a descending pathway that exerts a marked influence on γ-motoneurone activity. It may either excite or inhibit γ-motoneurones through a postsynaptic mechanism (32,33). Indirectly, it controls α-motoneurones by its action on γ-motoneurones (22). This latter system, through the well-known γ-control of muscle spindles, sets the bias for activation of the Ia afferents which synapse with motoneurones (20). More recent studies have shown the significance of descending noradrenergic and 5-hydroxytryptaminergic pathways as inhibitors of spinal interneurones. These interneurones control transmission from the flexor reflex afferents (cutaneous, joint and height threshold muscle afferents) to motoneurones (1,2,18). Removal of these descending monoaminergic inhibitory structures produces an increased afferent input from the periphery to the motoneuronal pool, thus enhancing the excitability of the system.

Teasdall and Stavraky (45) have suggested that at the segmental level the increase in reflex activity may be due to denervation supersensitivity. Thus, a motoneurone, deprived of a significant portion of its presynaptic input, may respond in an exaggerated fashion when exposed to an excitatory neurotransmitter. Other workers have suggested that increased neuronal activity may result from the sprouting of new excitatory synapses from the dorsal root ganglion to the motoneural pool (34).

A monosynaptic reflex in the cat was found to be suppressed by vibratory stimulation of the intact hindlimb (19) and to be mediated by "presynaptic" inhibition since: 1) it correlated with the onset of the dorsal root potential (DRP) evoked by vibration; 2) it was reduced by picrotoxin; and, 3) it produced primary afferent depolarization (PAD), as demonstrated by direct stimulation of the afferent fibers. Clinical studies for examining presynaptic inhibitory processes in man, based on studies performed in the cat, were undertaken by Burke and Ashby (8), and by Delwaide (14). Their findings also suggested that a reduction of the "presynaptic" inhibitory mechanism might be a factor in the pathophysiology of spasticity, since inhibition of the H-reflex resulting from vibratory stimulation was reduced in subjects exhibiting a state of spastic hyper-reflexia.

Thus, a number of factors might be involved in the genesis of spasticity: 1) release of spinal cord function from tonic inhibitory and excitatory supraspinal control; 2) supersensitivity of denervated structures; 3) collateral sprouting; and, 4) direct changes in segmental spinal inhibitory pathways.

In the present investigation, the chronic spinal cat was used to examine spinal reflex activity and the neuropharmacological and neurochemical changes that occur during the development of spasticity.

MATERIALS AND METHODS

Preparation of Animals

Female cats (2.1 – 3.2 kg) were anesthetized with 30 mg/kg Ketamine-HCl (1 M), supplementary injections being made if necessary. Under sterile conditions, the spinal cord was exposed by a dorsal laminectomy at T12 – L1. The dura was cut and the exposed cord was treated with xylocaine (2 ml of a 1% solution). The local anesthetic properties of xylocaine served to reduce bleeding when the cord was transected. Complete cord transection was made between T12 – L1. Small pieces of gelfoam were then placed into position between the retracted cut ends of the cord which served to reduce bleeding. The gelfoam was left in place and each layer, muscle, fascia and skin, was separately sutured. Terramycin was injected for the next four days. Cats with total cord transection are unable to urinate, and consequently, their bladders had to be expressed manually two to three times a day.

Neurophysiological Methods

At various times (two days to three months) following transection, each animal was anesthetized with Metofane (Methoxyflurane), the trachea was cannulated, and the blood supply to the head was ligated. The spinal cord was then transected at C1 and the animal was placed on a respirator and allowed to recover from anesthesia. The spinal cord was then exposed from L5 – S2. The most caudal rootlet of the L6 dorsal root was dissected free to the level of its entrance into the cord and placed on recording electrodes for measurement of the DRP. Either L7 and S1 ventral root was also placed on recording electrodes. Three ipsilateral hindlimb nerves, gastrocnemius-soleus (GS), posterior biceps semitendinosus (PBST) and sural, were dissected free and placed on stimulating electrodes. The mineral oil pools surrounding exposed tissue were kept at 36 ± 1°C. The end respiratory P_{CO_2} was monitored continuously and maintained at 3.5% – 4%. Preparations with a mean arterial blood pressure lower than 90 mm Hg were not used. Sets of 16 evoked responses were analyzed with a computer of averaging transients and displayed later on an x-y recorder to determine the amplitudes of various potentials. At the end of the experiment, the spinal cord was frozen in liquid nitrogen and samples from sections above and below the transection were removed for biochemical analysis.

Neurochemical Methods

Sample Preparation. All assays were performed on tissue samples which had been homogenized in 5 volumes of 0.167 M potassium-phosphate buffer (pH 6.5).

Glutamate Decarboxylase; Kinetic Measurements. Determinations were made on 50-μl portions of homogenates according to the method of Bird and Iversen (7) which involves entrappment and measurement of evolved $^{14}CO_2$. Specific activities of the glutamic acid substrates were adjusted so that each sample contained 0.10 μCi of DL [1-^{14}C]glutamic acid. Samples containing different molarities of the non-radioactive carrier L-glutamic acid (final concentrations = 3.35×10^{-3} M, 1.03×10^{-3} M, 3.48×10^{-4} M, and 1.39×10^{-4} M) were incubated for an hour. Protein was determined according to Lowry *et al*. (31).

Calculations

The amount of product formed per mg protein per hour was analyzed for K_m and V_{max} according to the method of Cleland (11) as modified by Ottaway (39).

RESULTS

The first signs of the onset of spasticity occurred within two to three weeks following transection. The onset was typically seen as a fanning of the toes of the hindlimbs with concomitant twitching of the feet when the cats were raised from the ground. As time progressed, the legs remained in an extended state and spontaneous "scissor-like" movements and rapid movements of the tail occurred when the preparation was held by the scruff of the neck. During the spastic state, the preparations exhibited an exaggerated flexor reflex. Furthermore, the placement reflex was missing, as evidenced by inability of the leg to respond when the dorsal part of the foot came in contact with an object.

Figure 1 illustrates the effect of complete cord transection on prolonged spinal inhibition ("presynaptic inhibition"). A volley of five pulses at twice threshold strength was delivered through the PBST nerve to condition a test monosynaptic response evoked by stimulation of the GS nerve. The ordinate shows the amplitude (% of control) of the monosynaptic discharge conditioned by inhibitory impulses at the various time intervals indicated on the abscissa. The inhibitory curve, marked "2 Weeks", represents the mean value from eight preparations which were tested at two days to two weeks following transection. This curve is similar to that observed in acute spinal preparations. The first sign of a decrease in the prolonged spinal inhibitory mechanism was observed at six weeks, when a reduction in inhibition occurred at the beginning and at the end of the curve. The remaining part of the curve did not differ from that of the two-week cats. The data for the six-week preparations represent mean values from four cats. Two preparations were studied at eight weeks. Their curve (not shown) falls between that of the six-week and 12-week preparations. At 12 weeks there occurred a marked attenuation of prolonged spinal inhibition. Note that inhibition endured for less than 50 mseconds, whereas, in the two-week cats it lasted more than 100 mseconds.

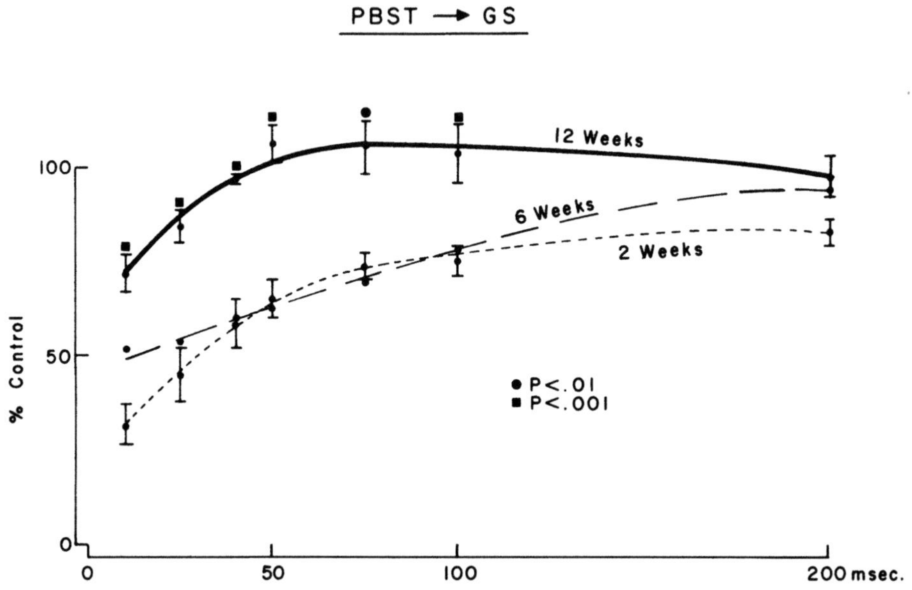

Fig. 1. Effect of complete spinal cord transection on prolonged ("presynaptic") inhibition of the monosynaptic reflex; the inhibitory curves relate the percent of control of the monosynaptic potential (ordinate) to the interval between conditioning and test volley (abscissa). The monosynaptic reflex was evoked by two volleys (1.5 msec intervals) in the nerve of the gastrocnemius-soleus (GS). The first volley produced no reflex but facilitated that produced by the second volley. The monosynaptic reflex was recorded from either L7 or S1 ventral root. The GS reflex was inhibited by five volleys (320/sec) in the posterior biceps semitendinosus (PBST) nerve (twice threshold). Each point represents the mean value; vertical bars indicate standard errors. The curve marked "2 Weeks" represents values from eight preparations studied two days to two weeks after transection,

One of the major effects of diazepam in the acute spinal cat is the prolongation and enhancement of presynaptic inhibition (41, 42,44). The effect of this drug on presynaptic inhibition in chronic spinal cats is illustrated in Figure 2. Mean values from four preparations at two weeks and from four other preparations at 10 – 12 weeks after transection are shown (Fig. 2A). Diazepam (1 mg/kg, i.v.) increased inhibition in each case (Fig. 2B). The degree of inhibition, however, was greatly reduced in the 12-week animals, as compared with two-week animals.

Since presynaptic inhibition is typically associated with a depolarization of primary afferent fibers (15), and as this depolarization is most easily observed as a DRP, we recorded the DRPs from most preparations. Typically, a reduction of prolonged spinal inhibition is associated with concomitant attenuation of the DRP. This, however, was not the case in chronic spinal cats (see Fig. 3). The differences in DRP recorded from the two- and 12-week cats (Fig. 3) were negligible.

Effect of Spinal Transection on γ-Aminobutyric Acid (GABA) Activity

The putative neurotransmitter involved in prolonged spinal inhibition is GABA. This type of inhibition is associated with axo-axonic synapses on primary afferent fibers and with axo-dendritic synapses on motoneurones (12). Our physiological data indicated that this inhibitory mechanism is decreased as time progresses after transection. Smith *et al*. (43) observed an initial decrease in GABA levels of the dorsal gray matter of the lumbar cord in dogs following

Fig. 1, continued: and that marked "6 Weeks" is from four animals, five to six weeks following transection. The curve labelled "12 Weeks" is from six preparations, ten weeks to 91 days after transection. The P-values compare the 2- and 12-week data.

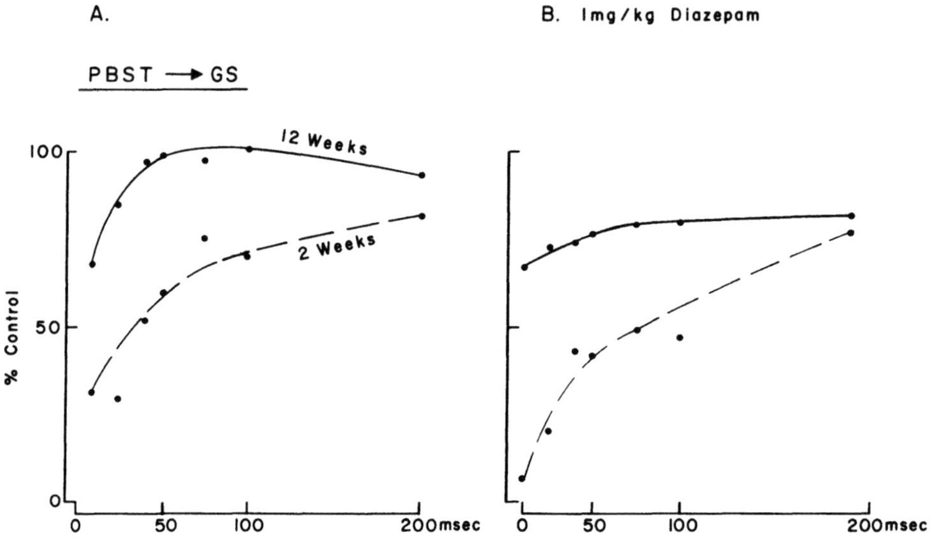

Fig. 2. Effect of diazepam on prolonged spinal inhibition in the chronic spinal cat. The test monosynaptic reflex was evoked by two pulses via the GS nerve, one subthreshold and the other supra-maximal for the monosynaptic spike. Inhibition was produced by five volleys (320/sec; twice threshold) via the PBST nerve. Each point represents the mean value from four preparations studied approximately 2 or 12 weeks following transection. A, Control inhibitory curves before diazepam; B, inhibitory curves determined 5 minutes after diazepam (1 mg/kg, i.v.).

mid-thoracic spinal cord transection which was correlated with the progressive development of spinal spasticity. Twelve weeks after transection, the GABA level was increased to 68% above control. They suggested that this increase in GABA might reflect a diminished release of the transmitter from the interneurones mediating presynaptic inhibition. Our data on L-glutamate decarboxylase (GAD) activity (K_m and V_{max}) correlate well with the changes in GABA concentration

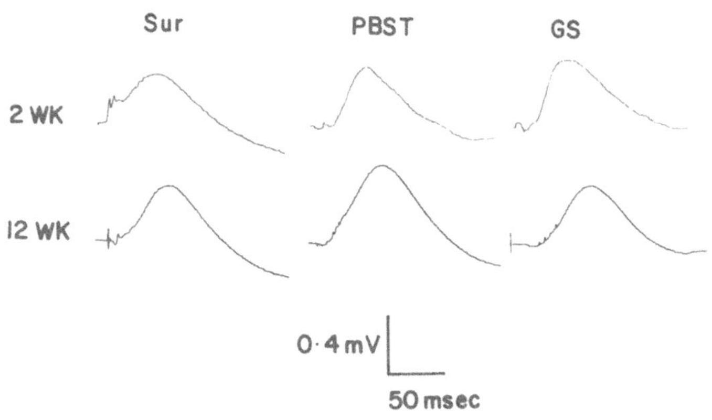

Fig. 3. Effects of complete spinal cord transection on dorsal root potentials and reflexes. Potentials were recorded from the caudal portion of the L6 dorsal root in response to impulses in the sural (Sur), PBST, and GS nerves. Note the lack of change in dorsal root potential.

observed by Smith et al. (43). In the acute spinal cat (Fig. 4), in regions below the transection, the V_{max} for GAD was approximately twice that found above the transection. On the other hand, the K_m was essentially the same. Following transection, specimens from both above and below the lesion showed a gradual decrease in K_m and V_{max}. Within 84 – 91 days after transection values for K_m and V_{max} of samples taken distal to the lesion became similar to values of acute spinal cats. In specimens taken from above the level of transection there appeared to be a tendency for a return to the control (acute) levels although this was delayed.

Alterations of Transmission Through the Flexor Reflex Afferents

Carlsson et al. (9) demonstrated noradrenergic nerve terminals in the spinal cord belonging to descending spinal pathways. Further work by Andén et al. (1,2) suggested that such descending noradrener-

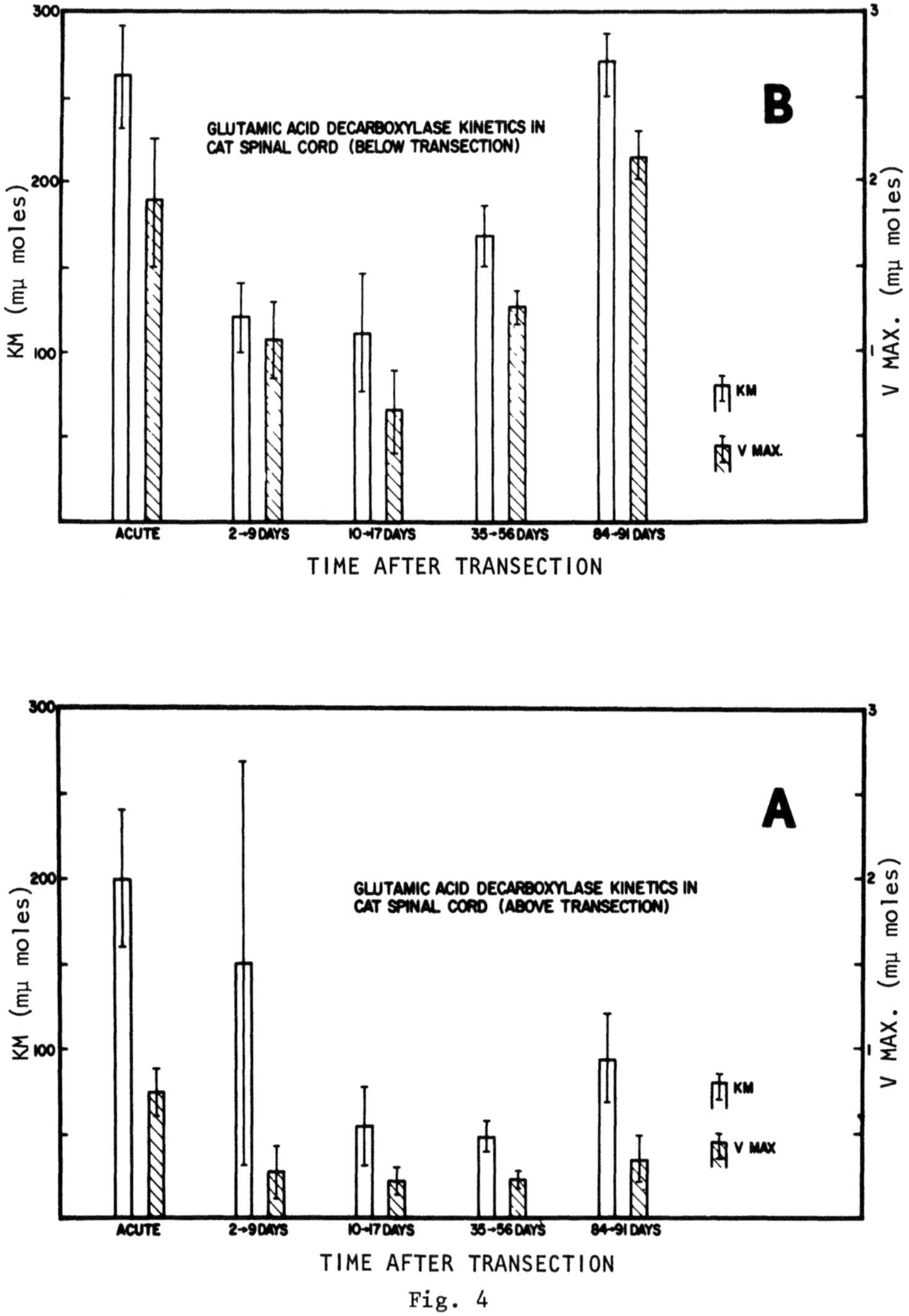
B
GLUTAMIC ACID DECARBOXYLASE KINETICS IN
CAT SPINAL CORD (BELOW TRANSECTION)
KM (mμ moles)
V MAX. (mμ moles)
300
200
100
3
2
1
KM
V MAX.
ACUTE
2-9 DAYS
10-17 DAYS
35-56 DAYS
84-91 DAYS
TIME AFTER TRANSECTION
A
GLUTAMIC ACID DECARBOXYLASE KINETICS IN
CAT SPINAL CORD (ABOVE TRANSECTION)
KM (mμ moles)
V MAX. (mμ moles)
300
200
100
3
2
1
KM
V MAX
ACUTE
2-9 DAYS
10-17 DAYS
35-56 DAYS
84-91 DAYS
TIME AFTER TRANSECTION

Fig. 4

Fig. 4. Effect of complete spinal cord transection on glutamate decarboxylase kinetics. A, Data from samples taken from above the level of transection (lower thoracic segments); B, data from samples below transection (lumbosacral region). Vertical lines indicate standard errors.

gic pathways function by inhibiting transmission of some fibers from flexor reflex afferents (FRA). (The FRA system includes high threshold muscle afferents and cutaneous afferents.) In the acute spinal cat, the test monosynaptic spike evoked by GS stimulation is blocked for a brief time if conditioned by a preceding impulse through the sural nerve, a cutaneous nerve (Fig. 5). When the time interval between conditioning and test stimuli approaches 25 mseconds, the test response is approximately at control level. A prolonged period (up to 100 msec) then follows, during which there occurs a slight rise in the test response. Of the six preparations which were examined 2 — 5 days following cord transection, four showed a reduction in the inhibition produced by sural stimulation. In contrast, the fifth cat exhibited an inhibitory curve similar to that of the acute spinal preparations, and the last cat's inhibitory response was similar to that of the long-term (1 — 12 weeks) spinal animals. The inhibitory curve for the long-term spinal preparations (Fig. 5) showed a very brief interval of complete inhibition of the monosynaptic spike, followed by a prolonged period of enhancement of this potential. Norepinephrine and 5-hydroxytryptamine increased above, and disappeared below the transection in the rat (13,37) and in the cat (this study).

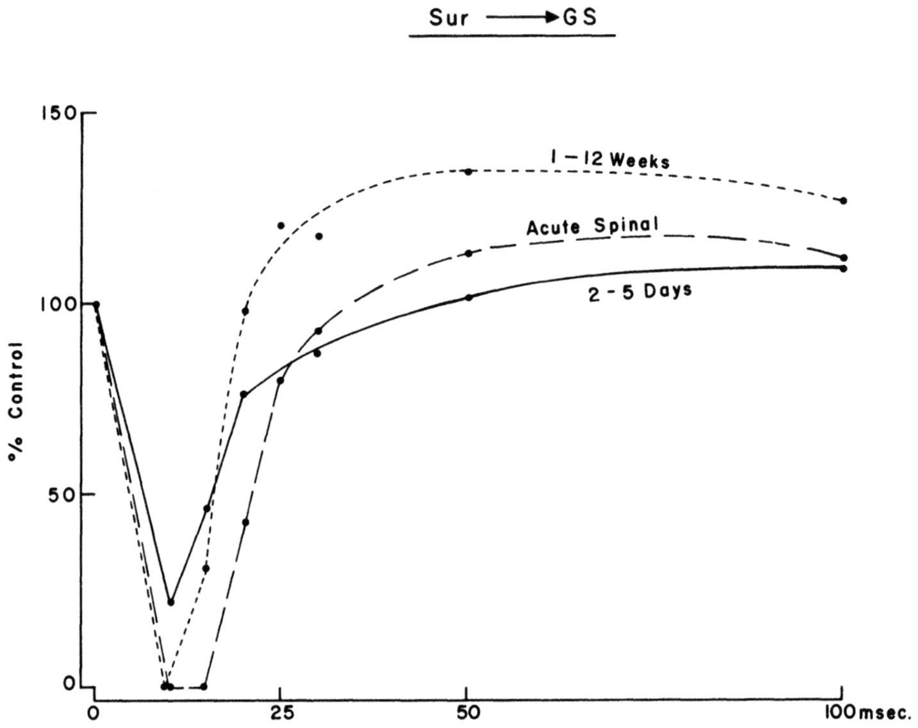

Fig. 5. Inhibition of extensor monosynaptic potentials by flexor-reflex afferents in acute and chronic spinal preparations. The monosynaptic spike was evoked by subthreshold and supra-maximal stimuli to the GS nerve. This was conditioned by a volley 4X threshold to the sural nerve. The inhibitory curves relate percent inhibition (ordinate) to the interval between the inhibitory and testing volley (abscissa). Each response represents the mean value from acute (n = 9), 2-5 days (4 of the 7 preparations showed this response) and 1-12 weeks (n = 14) preparations.

DISCUSSION

Effect of Chronic Spinal Cord Transection on GABA-Mediated Inhibition

The major inhibitory pathway in the spinal cord, in which GABA is the putative inhibitory transmitter, is that of prolonged, remote or "presynaptic" inhibition. Curtis *et al.* (12) have suggested that this inhibitory process might involve a number of components. First, there is the component suggested by Eccles (15), involving the synapses of GABA-ergic inhibitory interneurones impinging on the axons of incoming primary afferent fibers. At these axo-axonic synapses, release of the inhibitory transmitter evokes a depolarization of the primary afferent terminals with a resultant attenuation of the amount of excitatory transmitter released by an afferent volley. The second component consists of axo-dendritic synapses formed between the axons of the inhibitory GABA-ergic interneurones and the dendrites of motoneurones. At this junction the inhibitory transmitter produces a hyperpolarization of the dendrites, thus reducing the response of the motoneurone to the excitatory transmitter (21,23). Recent anatomical studies have revealed that GABA-ergic neurones form both axo-axonic and axo-dendritic synapses in the spinal cord (3,35). The third component is the accumulation of extracellular K^+ around the terminals of primary afferent fibers. This K^+ accumulation might modify the excitatory state of afferent terminals (29,48).

Our study in the chronic spinal cat demonstrated that there occurs a gradual decrease in this complex inhibitory system. These data support previous clinical findings in man (8,14) and other observations in cats (25). In conjunction with the decrease in prolonged spinal (presynaptic) inhibition, there occurred alterations in the K_m and V_{max} for GAD, the enzyme which converts glutamate to GABA. Below the level of transection (lumbar region) there occurred gradual reductions of the K_m and V_{max} followed by an eventual return to values which approximated those found in acute preparations. Similar reductions in K_m and V_{max} occurred in the thoracic segments above the transection but these persisted for 84 — 91 days after spinal cord lesion. In samples removed from cats 84 — 91 days after transection there was an increase in K_m but not in V_{max}. These results suggested that both ascending and descending neurones might function in the enzymatic control of GABA synthesis. Ascending fibers are known to impinge onto interneurones which inhibit, presynaptically, afferent fibers in the dorsal spino-cerebellar tract in Clarke's column (16). These interneurones are believed to be GABA-ergic. Twelve weeks after transection of the spinal cord, when presynaptic inhibition was reduced in both degree and duration, the K_m and V_{max} levels below the transection were similar to those found in acute preparations. These findings tend to suggest that the function of the GABA-ergic inhibitory system at twelve weeks after spinal lesion should be similar to that in the acute preparations. Smith _et al_. (43), however, have found that the GABA concentration at this time is markedly enhanced. Our findings showed that K_m and V_{max} above the lesion were markedly lower than those of the acute spinal cats. Therefore, a reduction in the release of the inhibitory transmitter would possibly explain the biochemical and physiological data.

Paradoxically, however, spinal transection had no effect on the DRP. These dorsal root antidromic potentials, which result from primary afferent depolarization, are usually associated, physiologically and pharmacologically, with presynaptic inhibition. There are, however, data which suggest that transmission through GABA-ergic interneurones involved in presynaptic inhibition are events that may, in part, be independent of DRP. For example, Vyklický _et al_. (48) demonstrated that DRPs evoked in frog spinal cord are not always completely blocked by the presence of a high concentration of Mg^{2+}. This observation suggests that PAD is not completely dependent upon transmitter-mediated events. Polc _et al_. (40) reported an augmentation of prolonged spinal inhibition without an accompanying enhancement of the DRP in cats treated with aminooxyacetic acid (AOAA). Further,

Bell and Anderson (6) found that AOAA doubled GABA levels of the spinal cord, while exerting no effect on the dorsal root reflex or DRP. In mammalian brain, AOAA specifically inhibits GABA-transaminase *in vivo* and thereby augments GABA levels (4). Further work is necessary to elucidate the reason(s) for the attenuation of presynaptic inhibition which occurred in chronic spinal cats without a concomitant reduction of the DRPs.

Verrier *et al*. (47) reported that diazepam failed to potentiate presynaptic inhibition in one patient with long-standing complete spinal lesions. This contrasts with the enhancement of presynaptic inhibition observed in acute spinal cats (41,42,44). Our data from chronic spinal cats shows that in this species there also occurs a reduction of diazepam action on presynaptic inhibition. The diazepam response, indeed, appears to be dependent on the normal functioning of inhibitory mechanisms; the gradual reduction in presynaptic inhibition after spinal cord transection is accompanied by a decrease in diazepam's effectiveness.

Effect of Chronic Spinal Cord Transection on Descending Monoaminergic Pathways and Flexor Reflex Afferents (FRA)

Dahlström and Fuxe (13), using the histochemical fluorescence technique, demonstrated in the rat that soon after lesion of the spinal cord, norepinephrine and 5-hydroxytryptamine accumulated above the level of transection and disappeared caudal to the lesion. They showed further that the monoaminergic terminals belong to bulbospinal pathways whose cell bodies are located in the medulla oblongata and lower pons.

The physiological significance of these descending monoaminergic pathways was demonstrated by Carlsson *et al*. (10), who observed that L-DOPA increased the flexor reflex evoked by pinching the skin in the acute spinal cat. Later investigation (2) revealed that the major effects of L-DOPA are to depress transmission in the short latency paths from FRA and to release the long latency paths from the FRA. A working hypothesis was proposed as follows: L-DOPA acts to inhibit transmission from the FRA by producing increased synthesis and release of transmitter from the terminals of descending noradrenergic pathways. It is noteworthy that many of the effects of L-DOPA resemble the tonic inhibition of transmission through the FRA system seen in the decerebrate preparation (17).

In our study, transmission through the FRA pathways was examined by the inhibitory effect produced via a conditioning volley through cutaneous afferents (sural) on a monosynaptic discharge of extensor motoneurones (GS). In the acute spinal animal, complete inhibition of the monosynaptic reflex from extensor motoneurones was seen for at least 5 mseconds, and this was followed by a rapid return of the test potential approximately to control values (Fig. 5; acute). In contrast, preparations examined one to 12 weeks after transection showed an inhibitory effect, in which the duration of complete inhibition of the monosynaptic spike was markedly reduced, followed by a period of enhanced reflex activity. Similar observations were made by Hancock _et al_. (25). In addition, a phase of transition occurred two to five days following spinal cord transection at which time a volley from the FRA did not completely inhibit the monosynaptic reflexes from extensor motoneurones. This incomplete inhibitory response (Fig. 5; 2 — 5 days) mimics that seen in decerebrate preparations where tonic inhibition of transmission from the FRA was reported (17), and is also similar to that seen in acute spinal preparations treated with L-DOPA (1). These data, therefore, suggest an action of catecholamines during this transitional phase, and an interaction of the neurotransmitter (norepinephrine) with interneurones involved in inhibition of transmission from FRA to extensor motoneurones.

Most probably, spinal spasticity results from dysfunction of a number of neurotransmitters. Descending monoaminergic pathways containing norepinephrine and 5-hydroxytryptamine are disrupted, thus removing the inhibitory function of these transmitters (1,2,10,37). Ascending neuronal pathways containing substance P are blocked, which produces an increase of this putative neurotransmitter (modulator) below the level of transection (36). A predominance of excitatory effects has been reported after iontophoretic application of substance P on brain and spinal cord neurones (26,28,38). Inhibitory effects of substance P were rarely observed, and when they were they consistently inhibited the excitatory action of acetylcholine on Renshaw cells, thus preventing the inhibitory action of these interneurones (5,28).

The concentrations of glycine (inhibitory) and aspartate (excitatory) rapidly declined in the canine spinal cord post-transection. Within eight weeks following transection, the levels of these possible transmitters were about 50% below control values (24). The decrease in glycine correlates with the decrement in postsynaptic and recurrent inhibition observed by Herman et al. (27) and Veale et al. (46) in humans with various neurological lesions. Therefore, in addition to changes observed in GABA-ergic transmission, spinal cord lesions also affect other interneuronal neurotransmitters.

SUMMARY

Following spinal cord transection there occurred decreases in K_m and V_{max} of glutamate decarboxylase (GAD) both above and below the lesion, and an initial decrease in the concentration of GABA. Concomitantly, there was a gradual decrease in presynaptic inhibition. Eight to 12 weeks after spinal cord transection, K_m and V_{max} for GAD returned to control values, but the GABA content of the spinal cord below the lesion increased significantly and presynaptic inhibition became maximally depressed. These results suggested that during the chronic phase of spinal cord injury there is a decrease in release of GABA, the interneuronal inhibitory neurotransmitter which mediates presynaptic inhibition. Diazepam, a GABA enhancer, increased presynaptic inhibition in acute and chronic spinal cats, this being accompanied by a reduction in somatic muscular spasticity. The degree of this enhancement by diazepam, however, is attenuated with gradual loss of presynaptic inhibition. In the acute spinal cat, a conditioning volley applied to cutaneous afferents blocked the inhibition of the monosynaptic response to extensor motoneurones. In contrast, in chronic spinal cats (eight to 12 weeks), the duration of complete blockade was markedly reduced and was followed by a prolonged period during which cutaneous nerve stimulation potentiated the monosynaptic discharge. Similar to GABA, there also occurred an increase of substance P below the level of the lesion. Other neurotransmitters (e.g., norepinephrine, serotonin) accumulated above and disappeared below the transection level. Although somatic muscular spasticity appears to be, to some extent, due to GABA dysfunction in the spinal cord, alterations in "normal" functioning of other neurotransmitters and the loss of supraspinal control also contribute to this state.

ACKNOWLEDGEMENTS

These studies were supported by the Edmund A. Guggenheim Clinical Research Endowment and in part by DHEW, OHD, RSA Grants Nos. 13-P57975/2 and 13-P55867/2.

REFERENCES

(1) Andén, N.-E., Jukes, M. G. M., Lundberg, A. and Vyklický, L. (1966) Acta Physiol. Scand. 67, 373-386.
(2) Andén, N.-E., Jukes, M. G. M. and Lundberg, A. (1966) Acta Physiol. Scand. 67, 387-397.
(3) Barber, R. P., Vaughn, J. E., Saito, K., McLaughlin, B. and Roberts, E. (1978) Brain Res. 141, 35-55.
(4) Baxter, C. F. and Roberts, E. (1961) J. Biol. Chem. 236, 3287-3294.
(5) Belcher, G. and Ryall, R. W. (1977) J. Physiol. (Lond.) 272, 105-119.
(6) Bell, J. A. and Anderson, E. G. (1974) Neuropharmacology, 13, 885-894.
(7) Bird, E. D. and Iversen, L. L. (1974) Brain, 97, 457-472.
(8) Burke, D. and Ashby, P. (1972) J. Neurol. Sci. 15, 321-326.
(9) Carlsson, A., Falck, B., Fuxe, K. and Hillarp, N.-Å. (1964) Acta Physiol. Scand. 60, 112-119.
(10) Carlsson, A., Magnusson, T. and Rosengren, E. (1963) Experientia,

19, 359-360.
(11) Cleland, W. W. (1967) Adv. Enzymol. 29, 1-32.
(12) Curtis, D. R., Duggan, A. W., Felix, D. and Johnston, G. A. R. (1971) Brain Res. 32, 69-96.
(13) Dahlström, A. and Fuxe, K. (1965) Acta Physiol. Scand. 64 (Suppl.), 247.
(14) Delwaide, P. J. (1973) In: New Developments in Electromyography and Clinical Neurophysiology (J. E. Desmedt, ed.) Vol. 3, pp. 508-522, Karger, Basel.
(15) Eccles, J. C. (1964) Progr. Brain Res. 12, 65-91.
(16) Eccles, J. C., Oscarsson, O. and Willis, W. D. (1961) J. Physiol. (Lond.) 158, 515-543.
(17) Eccles, R. M. and Lundberg, A. (1959) J. Physiol. (Lond.) 147, 565-584.
(18) Engberg, I. and Ryall, R. W. (1966) J. Physiol. (Lond.) 185, 298-322.
(19) Gillies, J. D., Lance, J. W., Neilson, P. D. and Tassinari, C. A. (1969) J. Physiol. (Lond.) 205, 329-339.
(20) Granit, R. (1955) Receptors and Sensory Perception. Yale University Press, New Haven.
(21) Granit, R. (1968) In: Structure and Function of Inhibitory Neuronal Mechanisms (C. von Euler, S. Skoglund and U. Söderberg, eds.) Vol. 10, pp. 183-195, Pergamon Press, Oxford.
(22) Granit, R., Kaada, B. R. (1952) Acta Physiol. Scand. 27, 130-160.
(23) Green, D. G. and Kellerth, J. O. (1966) Science, 152, 1097-1099.
(24) Hall, P. V., Smith, J. E., Campbell, R. L., Felten, D. L. and Aprison, M. H. (1976) Life Sci. 18, 1467-1472.
(25) Hancock, J., Knowles, L. and Gillies, J. D. (1973) Proc. Australian Assoc. Neurologists, 10, 139-143.
(26) Henry, J. L. (1976) Brain Res. 114, 439-452.
(27) Herman, R., Freedman, W. and Meeks, S. (1973) In: New Developments in Electromyography and Clinical Neurophysiology (J. E. Desmedt, ed.) Vol. 3, pp. 579-588, Karger, Basel.
(28) Krnjević, K. and Lekić, D. (1977) Can. J. Physiol. Pharmacol. 55, 958-961.
(29) Krnjević, K. and Morris, M. E. (1972) Can. J. Physiol. Pharmacol. 53, 923-934.
(30) Krnjević, K. and Morris, M. E. (1974) Can. J. Physiol. Pharmacol. 52, 736-744.
(31) Lowry, O. H., Rosebrough, N. J., Farr, A. L. and Randall, R. J. (1951) J. Biol. Chem. 193, 265-275.
(32) Magoun, H. W. (1950) Physiol. Rev. 30, 459-474.
(33) Magoun, H. W. and Rhines, R. (1946) J. Neurophysiol. 9, 165-171.
(34) McCouch, G. P., Austin, G. M., Liu, C. M. and Liu, C. Y. (1958) J. Neurophysiol. 21, 205-216.
(35) McLaughlin, B. J., Barber, R. P., Saito, K., Roberts, E. and Wu, J.-Y. (1975) J. Comp. Neurol. 164, 305-322.
(36) Naftchi, N. E., Abrahams, S. J., St. Paul, N. H., Lowman, E. W. and Schlosser, W. (1978) Brain Res. (in press).

(37) Naftchi, N. E., Wooten, G. F., Lowman, E. W. and Axelrod, J. (1974) Circulation Res. 35, 850-861.

(38) Otsuka, M. and Konishi, S. (1976) Cold Spring Harbor Symp. Quant. Biol. 40, 135-143.

(39) Ottaway, J. H. (1973) Biochem. J. (Lond.) 134, 729-736.

(40) Polc, P., Möhler, H. and Haefely, W. (1974) Naunyn-Schmiedeberg's Arch. Pharmacol. 284, 319-337.

(41) Schlosser, W. (1971) Arch. Int. Pharmacodyn. 194, 93-102.

(42) Schmidt, R. F., Vogel, M. E. and Zimmermann, M. (1967) Naunyn-Schmiedeberg's Arch. Pharmacol. 258, 69-82.

(43) Smith, J. E., Hall, P. V., Campbell, R. L., Jones, A. R. and Aprison, M. H. (1976) Life Sci. 19, 1525-1530.

(44) Stratten, W. P. and Barnes, C. D. (1971) Neuropharmacology, 10, 685-696.

(45) Teasdall, R. D. and Stavraky, G. (1953) J. Neurophysiol. 16, 367-375.

(46) Veale, J. L., Rees, S. and Mark, R. F. (1973) In: New Developments in Electromyography and Clinical Neurophysiology (J. E. Desmedt, ed.) Vol. 3, pp. 523-537, Karger, Basel.

(47) Verrier, M., MacLeod, S. and Ashby, P. (1975) Can. J. Neurol. Sci. 2, 179-184.

(48) Vyklický, L., Syková, E., and Mellerová, B. (1976) Brain Res. 117, 153-156.

RELEVANCE OF GABA IN THE NIGROSTRIATAL SYSTEM

G. Racagni[1], F. Cattabeni[1] and A. Groppetti[2]

[1]Institute of Pharmacology and Pharmacognosy,
[2]Institute of Pharmacology, University of Milan,
Via Andrea del Sarto 21, 20129 Milan, Italy

INTRODUCTION

Current data suggest that the basal ganglia contain pre-programmed neuronal circuits for controlling postural and motor functions. A variety of intra- and inter-system neurons whose transmitters are γ-aminobutyric acid (GABA), dopamine (DA), acetylcholine (ACh), serotonin and substance P (Sub P) have been proposed to be functionally present in these circuits (1,2). The neostriatum seems to be the computer center of the basal ganglia. The information that the striatum has filtered is conveyed to neurons in the globus pallidus and substantia nigra and from there to other structures. Connections between globus pallidus, subthalamic nucleus, centromediam nucleus of thalamus and the midbrain tegmentum have been found, and others exist between substantia nigra and thalamus (3,4,5; see Fig. 1).

Short- and long-axoned inhibitory GABA-ergic neurons appear to play decisive roles in these circuits (1,2,6,7,8). It has been postulated that the strionigral GABA-ergic pathways are involved in the neuronal feedback control of striatal DA terminals (9) and in the regulation of the nigrothalamic pathway (10). The dopaminergic system is an important inhibitory afferent to the striatum. When

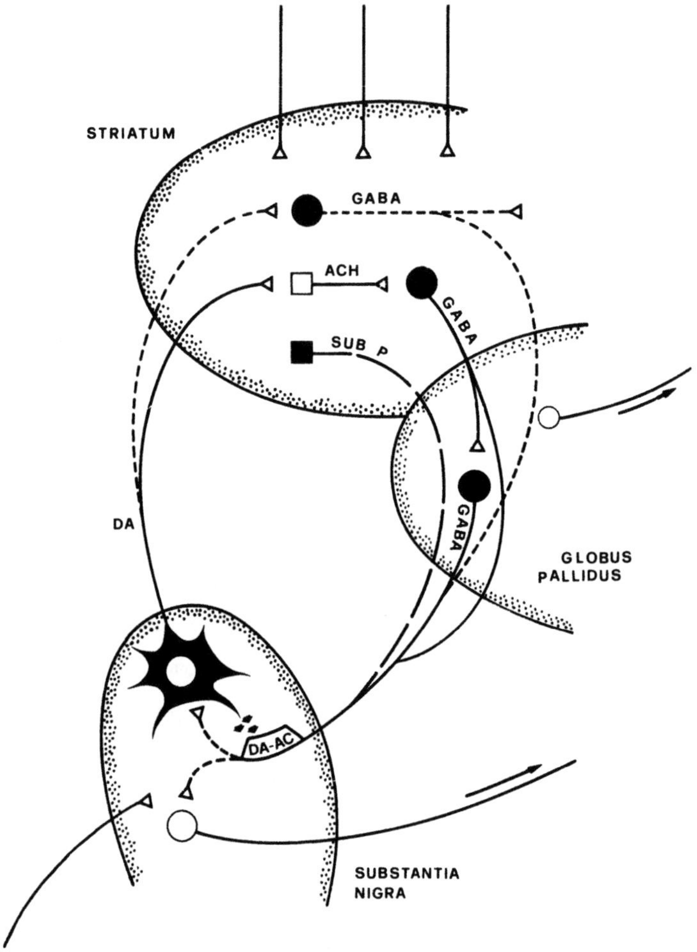

Fig. 1. Scheme of the current understanding of the possible interconnections between GABA-ergic and other neurons in the nigrostriatal system.

striatal DA-receptors are stimulated, motility is increased and stereotyped behavior may follow; when these receptors are blocked, motor activity is reduced and muscle tone increases. Therefore, the modulation of this system may affect striatal functions resulting in an alteration of the filtering activity operated by the striatum.

The link between the nigral zona reticulata and the thalamus suggests that this may well be the output pathway by which the striatum acts on motor function.

The purpose of this paper is to present a brief overview of our recent findings aimed at determining the functional roles of GABA in relation to other known neuronal systems present either in the striatum or in the substantia nigra.

DOPAMINERGIC INFLUENCE ON STRIATAL GABA-ERGIC NEURONS

The neostriatum contains ACh, GABA and Sub P. Therefore, any input to the caudate nucleus from the substantia nigra could ultimately be able to exert some effect on these neurons. Several lines of evidence have indicated that activation of the nigrostriatal dopaminergic pathway results in a depression of cholinergic interneurons (11,12,13,14). Moreover, Fuxe _et al._ (15) have shown that after 6-hydroxydopamine (6-OH-DA) the accumulation rate of GABA, after inhibition of GABA-transaminase (GABA-T) with L-γ-glutamyl-hydrazide (GAH), is inversely related to the degree of DA denervation in the caudate nucleus. Similarly, Vincent _et al._ (16) have found that 6-OH-DA lesions in substantia nigra increase glutamate decarboxylase (GAD) activity in the striatum. In agreement with these findings, we have reported (17) that interruption of the ascending dopaminergic fibers by hemi-transection increases GABA accumulation rate, as measured after inhibiting its catabolism by ethanolamine-O-sulphate (EOS) (18). In addition, we have found that an increase of striatal DA release after crus cerebri lesion (19) is followed by a decrease of GABA accumulation rate in this nucleus (Table I). These observations have indicated that the activity of striatal cholinergic and GABA-ergic neurons is mediated, at least in part, by the nigrostriatal DA pathway.

Pharmacological studies on the effects of anti-psychotics on the turnover of GABA in the striatum have, however, made evident the complexity of these DA-GABA interactions. Several studies have

Table I - Effects of Lesions and Different Pharmacological Treatments on Striatal Dopamine (DA) and GABA Neuronal Activities

Condition	Striatal DA release	Striatal GABA activity
Hemi-transection[a]	↓	↑
Crus cerebri[b]	↑	↓
Oxotremorine[a]	↓	no change
6-OH-DA[c]	↓	↑
6-OH-DA[d]	↓	↑
Haloperidol[e]	↑	↓
Clozapine[e]	no change	↑

[a] From Cattabeni et al. (25); [b] from Racagni et al. (19); [c] from Fuxe et al. (15); [d] from Vincent et al. (16); [e] from Maggi et al. (23).

shown that GABA-ergic neurones in striatum are differentially affected by cataleptogenic and non-cataleptogenic neuroleptics. In fact, clozapine, which is a potent anti-psychotic with muscarinic receptor blocking activity (20,21) and which is devoid of extrapyramidal side effects (22), increases activity of striatal GABA-ergic neurons, whereas haloperidol, which elicits extrapyramidal side effects, does not affect, or slightly decreases, the dynamics of the striatal GABA-ergic system (23,24). Moreover, recent studies from our laboratory seem to indicate that cholinergic neurons alone do not influence the activity of striatal GABA-ergic interneurons. In fact, cholinomimetics (e.g., oxotremorine; physostigmine) or cholinolytics (e.g., trihexyphenidyl) do not alter GABA accumulation rate in striatum (25). Therefore, besides a possible direct link between DA and GABA in the striatum, it appears possible that polyneuronal circuits or polysynaptic contacts may be interposed (Table I).

Since striatal DA-containing neurons seem to make synaptic contacts with a population of interneurons which are excitatory (possibly cholinergic) (26) or inhibitory (possibly GABA-ergic) (27), it is not surprising that it has remained controversial whether DA mediates an inhibitory (27,28,29) or excitatory (30,31,32) response in the nigrostriatal system. In fact, excitation of a target cell after the iontophoretic application of DA may be followed immediately by an inhibition induced by the collateral activation of an inhibitory interneuronal population. Moreover, the excitatory response may be the result of a disinhibition between two inhibitory neurons, as it might be in the case of a DA-GABA link. It is important, therefore, to understand how the neurons employing different transmitters relate to each other.

GABA-ERGIC INFLUENCE ON NIGRAL DA NEURONS

Studies utilizing retrograde axonal transport of horseradish peroxidase (33) have revealed the existence of a strionigral pathway from the head of the caudate nucleus to the zona reticulata of substantia nigra, as well as a direct connection between the tail of the caudate nucleus and the zona compacta of the substantia nigra (34-36). In addition, immunocytochemical studies on the localization of GAD have revealed that it is localized both in zona reticulata and compacta (7). These findings have suggested that a direct synaptic contact might exist between GABA-ergic terminals and the DA cell bodies in the zona compacta of substantia nigra. A further possible function of the descending GABA-ergic fibers to the reticulata is that of modulating the nigrothalamic pathway (5) or other nigral efferents (37).

Electrophysiological studies have shown that iontophoretically-applied GABA depresses the activity of cells in the substantia nigra in both zona compacta (38) and reticulata (39). Stimulation of the striatum leads to an inhibition of the firing of DA cells in the nigra which is sensitive to blockade of GABA-receptors (40).

Besides the electrophysiological data, pharmacological studies also support the idea that the strionigral GABA-ergic pathway may be involved in the negative feedback loop which controls DA metabolism (41). However, more recent data have questioned either the inhibitory effect of GABA terminals on DA cell bodies or the evidence for the importance of strionigral pathways in the feedback regulation of striatal DA neurons (42). Cheramy _et al_. (42) have reported recently that direct injections of GABA or GABA-agonists into the substantia nigra induced an increase of DA release from cat caudate nucleus. Furthermore, intra-nigral administration of muscimol, a potent GABA-agonist, has been found to elicit circling behavior contralateral to the side of injection (43). However, we have recently observed (19) that DA release in the striatum is significantly increased after a lesion at the level of the crus cerebri, which produces a similar decrease in the GABA content of the substantia nigra as extensive brain hemi-transection. Furthermore, when GABA levels are elevated in the substantia nigra (e.g., as they are 3 hours after intraventricular or intra-nigral administration of EOS) striatal DA release is consistently decreased (19) (Table II). However, Tanner _et al_. (48) have reported an increase of striatal homovanillic acid (HVA) content 24 hours after EOS treatment.

These data, in agreement with electrophysiological studies, seem therefore to suggest that GABA exerts an inhibitory influence on DA release in the striatum. However, it cannot be excluded that a more complex situation might be also operative, considering the findings reported by Cheramy _et al_. (42). These authors explained the excitatory effect elicited by GABA-ergic terminals on DA cell bodies by involving inhibitory nigral interneurons (possibly glycinergic). Our data indicate that, if this is the case, the summation of events is inhibitory when GABA variations in substantia nigra are drastically changed. Also the evidence for an excitatory role of nigral GABA-ergic terminals derived from the studies on circling behavior are in apparent contradiction. In fact, the effect elicited by muscimol is not antagonized by haloperidol (43), suggesting that neuronal pathways other than dopaminergic ones may be involved. This

Table II - Effect of Nigral GABA-ergic Activity on Striatal Dopamine (DA) Release

Condition	Striatal DA release	Nigral GABA activity
Crus cerebri lesion[a]	↑	↓
Kainate lesion[b]	↑	↓
Ethanolamine-O-sulphate (3 hours after treatment)[d]	↓	↑
Ethanolamine-O-sulphate (24 hours after treatment)[d]	↑	↑
GABA in substantia nigra[c]	↑	↑

[a] From Racagni et al. (19); [b] from Groppetti et al. (47); [c] from Cheramy et al. (42); [d] from Tanner (48).

inference is supported also by the findings that strionigral lesions produce behavioral effects similar to those of striatectomy (36).

Recently, the importance of the strionigral loop in regulating striatal DA activity has also been questioned. It has been reported that reduction of nigral GABA levels after lesioning the strionigral pathways does not counteract DA turnover and the increases of striatal 3,4-dihydroxyphenylacetic acid (DOPAC) and HVA concentrations (44).

In line with these findings we have shown (45) that the mechanisms which regulate DOPAC formation are intrinsic to the striatum. Therefore, the integrity of the strionigral fibers is not important for controlling this DA metabolite. In this context, Seeman and Lee (46) have also found that haloperidol changes DA metabolism in a preparation of striatal synaptosomes.

Furthermore, striatal DOPAC concentrations are not altered either by an increase of GABA levels in substantia nigra following EOS administration, or by the decrease of nigral GABA which follows interruption of strionigral fibers by a crus cerebri lesion (45,47). However, these experimental conditions can modify striatal 3-methoxytyramine (3-MT) content, taken as an index of DA release (19). Moreover, lesion of the strionigral pathways antagonized the increase of striatal 3-MT content elicited by haloperidol or other DA blocking agents (47). From these results we must admit that feedback pathways do play a functional role, since it seems that a dissociation exists between the mechanisms which modulate synthesis and release.

CONCLUDING REMARKS

In conclusion, even though the mutual integration of intrastriatal circuits remains an open question, the electrophysiological and neurochemical approaches seem to provide valid support toward understanding the complexity of these systems. Recent findings seem to indicate that, besides a dopaminergic and cholinergic link, interactions between DA and GABA neurons may be operative within the striatum. An understanding of these functional interplays may facilitate future attempts aimed at clarifying the electrophysiological discrepancies concerning the inhibitory or excitatory action of DA on the target cells in this brain structure.

Studies on the functional roles of pathways projecting from striatum to substantia nigra have indicated that these may be implicated in the regulation of nigral efferents to other brain nuclei and that they do control striatal DA activity through feedback mechanisms.

REFERENCES

(1) Hong, J. S., Yang, H. Y. I., Racagni, G. and Costa, E. (1977) Brain Res. 122, 541-544.
(2) Hökfelt, T., Kellerth, J. O., Nilsson, G. and Pernow, R. (1975) Science, 190, 889-890.
(3) Kemp, J. M. and Powell, T. P. S. (1971) Philos. Trans. Roy. Soc. Lond. 262, 441-462.
(4) Kemp, J. M. and Powell, T. P. (1971) Philos. Trans. Roy. Soc. Lond. 262, 383-401.
(5) Rinvik, E. (1976) Brain Res. 90, 313-318.
(6) Fonnum, F., Grofova, J., Rinvik, E., Storm-Mathisen, J. and Walberg, F. (1974) Brain Res. 71, 77-82.

(7) Ribak, C. E., Vaughn, J. E., Saito, K., Barber, R. and Roberts, E. (1976) Brain Res. 116, 287-292.
(8) Hattori, T., McGeer, P. L., Fibiger, H. C. and McGeer, E. G. (1973) Brain Res. 54, 103-114.
(9) Carlsson, A. and Lindqvist, M. (1963) Acta Pharmacol. Toxicol. 20, 140-144.
(10) Scheel-Krüger, J., Arnt, J. and Magelund, G. (1977) Neuroscience Lett. 4, 351-356.
(11) Racagni, G., Cheney, D. L., Trabucchi, M. and Costa, E. (1976) J. Pharmacol. Exp. Therapeut. 196, 323-332.
(12) Stadler, H., Lloyd, K. G. and Bartholini, G. (1974) Naunyn Schmiedeberg's Arch. Pharmacol. 283, 129-133.
(13) Consolo, S., Ladinsky, M. and Bianchi, S. (1975) Eur. J. Pharmacol. 33, 345-351.
(14) Guyenet, P. G., Agid, Y., Javoy, F., Beaujovan, J. C., Rossier, J. and Glowinski, J. (1975) Brain Res. 84, 227-244.
(15) Fuxe, K., Perez de la Mora, M., Hökfelt, T., Agnati, L., Ljungdahl, A. and Johansson, O. (1977) In: Psychopathology and Brain Dysfunction (C. Shagass, S. Gershon and A. J. Friedhoff, eds.) pp. 97-111, Raven Press, New York.
(16) Vincent, R. S., Nagy, J. I. and Fibiger, H. C. (1978) Brain Res. 143, 168-173.
(17) Racagni, G., Bruno, F., Cattabeni, F., Maggi, A., Di Giulio, A. M. and Groppetti, A. (1978) In: Interactions Between Putative Neurotransmitters in the Brain (S. Garattini, J. F. Pujol and R. Samanin, eds.) pp. 61-72, Raven Press, New York.
(18) Fowler, L. J. and Jhon, R. A. (1972) Biochem. J. (Lond.) 130, 569-573.
(19) Racagni, G., Bruno, F., Cattabeni, F., Maggi, A., Di Giulio, A. M., Parenti, M. and Groppetti, A. (1977) Brain Res. 134, 353-358.
(20) Miller, R. J. and Hiley, C. R. (1974) Nature (Lond.) 248, 596-598.
(21) Snyder, S. H., Banerjee, S. P., Yamamura, H. I. and Greenberg, D. (1974) Science, 184, 1243-1245.
(22) Shader, R. I. and DiMascio, A. (1970) In: Psychotropic Drug Side Effects (R. I. Shader and A. DiMascio, eds.) p. 92, Williams and Wilkins, Baltimore.
(23) Maggi, A., Cattabeni, F., Bruno, F. and Racagni, G. (1977) Brain Res. 133, 382-385.
(24) Marco, E., Mao, C. C., Cheney, D. L., Revuelta, A. and Costa, E. (1976) Nature (Lond.) 264, 363-365.
(25) Cattabeni, F., Bugatti, A., Groppetti, A., Maggi, A., Parenti, M. and Racagni, G. (1978) In: GABA-Neurotransmitters, Munksgaard, Copenhagen (in press).
(26) Bloom, F. E., Costa, E. and Salmoiraghi, G. C. (1965) J. Pharmacol. Exp. Therapeut. 150, 244-252.
(27) Aghajanian, G. K. and Bunney, B. S. (1973) In: Frontiers in Catecholamine Research (E. Usdin, ed.) pp. 643-648, Pergamon Press, New York.

(28) Connor, J. D. (1970) J. Physiol. (Lond.) 208, 691-698.
(29) Krnjević, K. (1975) In: Advances in Neurology; Dopaminergic Mechanisms (D. B. Calne, T. N. Chase and A. Barbeau, eds.) Vol. 9, pp. 13-24, Raven Press, New York.
(30) Feltz, P. and Albe-Fessard, D. (1972) Electroencephalogr. Clin. Neurophysiol. 33, 179-193.
(31) Frigyesi, T. L. and Purpura, D. P. (1967) Brain Res. 6, 440-456.
(32) Kitai, S. T., Sugimori, M. and Kocsis, J. D. (1976) Exp. Brain Res. 24, 351-363.
(33) LaVail, J. H. and LaVail, M. M. (1972) Science, 176, 1417-1418.
(34) Grofova, I. (1975) Brain Res. 91, 286-291.
(35) Bunney, B. S. and Aghajanian, G. K. (1976) In: The Basal Ganglia (M. D. Yahr, ed.) pp. 249-267, Raven Press, New York.
(36) Wright, A. K., Arbuthnott, G. W., Tulloch, I. F., Garcia-Munoz, M. and Nicalaou, N. M. (1977) In: Psychobiology of the Striatum (A. R. Cools, A. H. M. Lohman and J. H. L. van den Berchen, eds.) pp. 31-50, Elsevier/North Holland Biomedical Press, Amsterdam.
(37) Carpenter, M. B. (1976) In: The Basal Ganglia (M. D. Yahr, ed.) pp. 1-35, Raven Press, New York.
(38) Aghajanian, G. K. and Bunney, B. S. (1974) In: Proceedings 9th CINP, Excerpta Medica Foundation (J. R. Boissier, H. Hippius and P. Pichot, eds.) Excerpta Medica, Amsterdam.
(39) Crossman, A. R., Walker, R. J. and Woodruff; G. N. (1973) Brit. J. Pharmacol. 49, 696-700.
(40) Dray, A., Gonye, T. J. and Oakley, N. (1976) J. Physiol. (Lond.) 259, 825-849.
(41) Andén, N. E. and Stock, G. (1973) Naunyn-Schmiedeberg's Arch. Pharmacol. 279, 89-95.
(42) Cheramy, A., Nieoullion, A. and Glowinski, J. (1978) In: Interactions Between Putative Neurotransmitters in the Brain (S. Garattini, J. F. Pujol and B. Samanin, eds.) pp. 175-190, Raven Press, New York.
(43) Oberlander, C., Dumont, C. and Boissier, J. R. (1977) Eur. J. Pharmacol. 43, 389-390.
(44) Garcia-Munoz, M., Nicolaou, N. M., Tulloch, I. F., Wright, A. K. and Arbuthnott, G. W. (1977) Nature (Lond.) 265, 363-365.
(45) Racagni, G., Groppetti, A., Parenti, M., Bugatti, A., Bruno, F., Maggi, A. and Cattabeni, F. (1978) Life Sci. 23, 1757-1762.
(46) Seeman, P. and Lee, T. (1975) Science, 188, 1217-1219.
(47) Groppetti, A., Parenti, M., Galli, C. L., Bugatti, A., Cattabeni, F., Di Giulio, A. M. and Racagni, G. (1978) Life Sci. 23, 1763-1768.
(48) Tanner, T. (1978) J. Pharm. Pharmacol. 30, 158-161.

GABA IN HUNTINGTON'S CHOREA, PARKINSONISM AND SCHIZOPHRENIA

E. G. S. Spokes

Department of Neurological Surgery and Neurology,
Addenbrookes Hospital, Hills Road, Cambridge CB2 2QQ,
England

INTRODUCTION

γ-Aminobutyric acid (GABA) is believed to be a major inhibitory neurotransmitter in the mammalian central nervous system (CNS), possibly used by as many as one-third of all central synapses. Although GABA itself is unstable in brain tissue post-mortem, the biosynthetic enzyme glutamate decarboxylase (GAD) remains remarkably stable in human and animal brain after death, and serves as a useful marker for GABA-ergic neurones. This paper will describe recent findings on the marker enzyme GAD in Huntington's chorea, parkinsonism and schizophrenia, and will highlight the pitfalls that may be encountered in interpreting such data.

POST-MORTEM STABILITY OF GAD

In a normal heterogeneous population, wide variations are to be expected in any parameter under scrutiny. This means that large samples are needed to be adequately representative of the group as a whole. In order to obtain a sufficiently large sample for neurochemical studies on human post-mortem brain tissue, it is necessary to have flexible arbitrary criteria for acceptance of material. Since the interval between death and autopsy is a variable which is

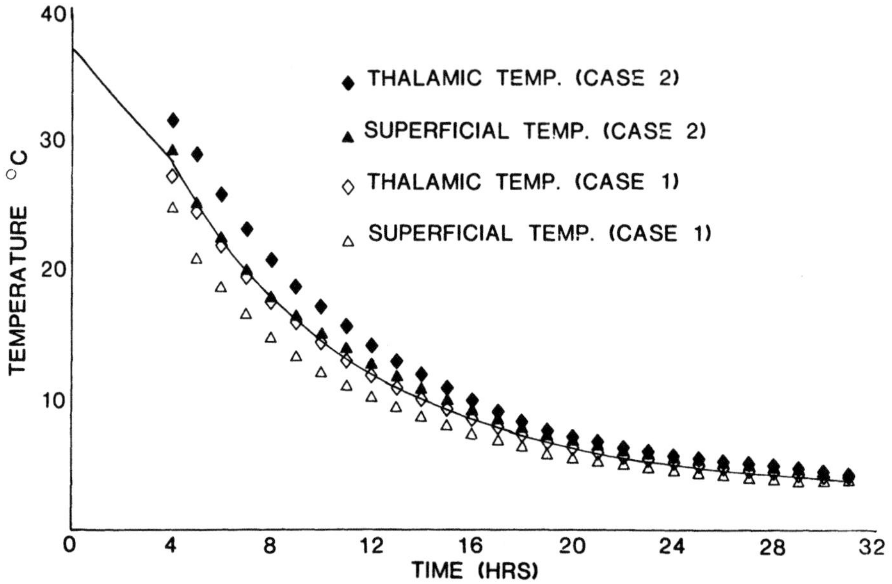

Fig. 1. Calculated mean cooling curve of human brain post-mortem. Case 1 entered the refrigerator 3 hours after death; case 2, 4 hours after death. Brain temperature at death was assumed to be 37°C.

not easily standardized, knowledge of the post-mortem stability of GAD is essential for the interpretation of measurements of enzyme activity.

We have examined this question by use of a simple animal model (1). The cooling of human brain in the course of the routine handling of cadavers has been monitored by the implantation of thermocouples, under X-ray control, into the thalamus and parietal cortex to measure core and superficial brain temperatures, respectively. The mean exponential curve thus obtained (Fig. 1) was reproduced in mice by maintaining them at controlled temperatures after death within a cooled incubator. Under these conditions there was a rapid initial decline in GAD activity within the first few hours to about 80% of initial values. Thereafter, GAD remained at a stable plateau

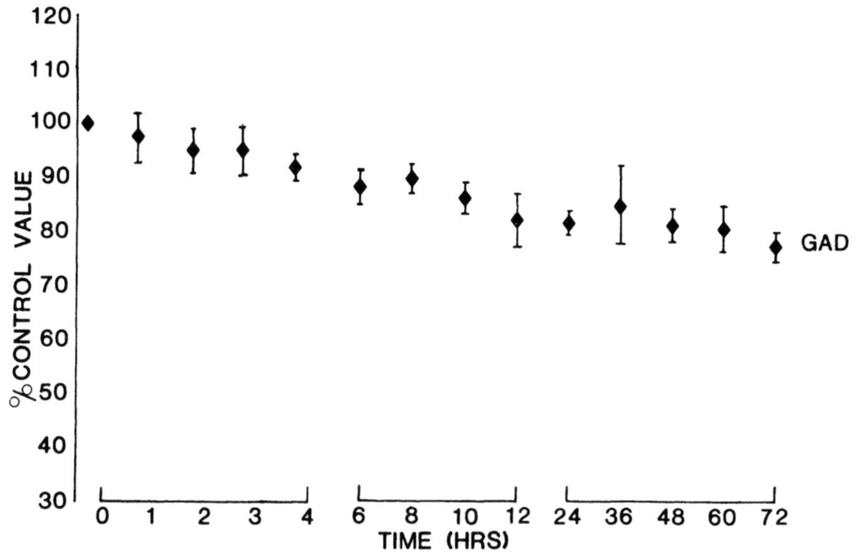

Fig. 2. Levels of GAD activity in mouse brain at various intervals after death when animals were stored under simulated temperature conditions. Each point is the mean ± S.E.M. of at least 4 cerebral hemispheres. Absolute value for GAD at time zero was 191 ± 5.5 μ-moles/hr/g protein.

(Fig. 2). These results confirmed earlier suggestions that GAD declines rapidly during the initial post-mortem period but is then remarkably stable (2-4). This conclusion is supported by comparisons of GAD values obtained from patients in which the interval between death and autopsy was more or less than 48 hours (Table I). It seems unlikely, therefore, that variable post-mortem delays play an important role in determining differences in GAD activity between groups.

We have also analyzed our human data as a function of the interval from death to refrigeration. As might be expected, over a range of 1 – 4 hours, this did not affect GAD levels since small variations in this interval do not greatly affect the rate of cooling of the brain.

Table I - Influence of the Interval Between Death and Necropsy on Measured GAD Values in Control Brain

Region	< 48 h[1]	> 48 h[2]
Caudate	47.4 ± 3.9 (31)	45.2 ± 6.2 (14)
Putamen	58.9 ± 5.3 (33)	56.9 ± 7.0 (16)
Nucleus accumbens	82.8 ± 9.1 (22)	80.0 ± 11.4 (10)

All values are expressed as means ± S.E.M.; numbers of samples in parentheses.

[1]Range 12 – 46 h; [2]range 48 – 94 h.

Table II - Influence of Agonal Status on GAD Levels in Post-Mortem Human Brain Tissues

Brain region	GAD (μ-moles/hr/g protein)		% Reduction
	Sudden death	Protracted illness	
Pars compacta	109.1 ± 10.0 (18)	25.7 ± 7.5 (11)	76
Pars reticulata	84.2 ± 9.2 (18)	28.8 ± 7.2 (11)	66
Medial pallidum	97.2 ± 8.1 (22)	32.9 ± 7.1 (14)	66
Lateral pallidum	92.9 ± 8.5 (22)	35.0 ± 8.2 (14)	62
Nucleus accumbens	84.4 ± 6.3 (36)	40.9 ± 6.3 (15)	52
Putamen	57.9 ± 3.9 (51)	34.9 ± 4.5 (24)	40
Caudate	48.0 ± 3.1 (49)	25.9 ± 3.8 (23)	46
Dentate	48.6 ± 3.0 (24)	23.3 ± 3.3 (11)	52
Cerebellar cortex	32.7 ± 3.2 (20)	15.2 ± 3.4 (12)	54
Red nucleus	21.4 ± 2.0 (26)	10.2 ± 2.4 (13)	52
Motor cortex	53.4 ± 3.9 (15)	22.7 ± 5.5 (9)	57
Hippocampus	26.2 ± 2.3 (37)	12.9 ± 1.9 (12)	51

All values are expressed as means ± S.E.M.; numbers of samples in parentheses.

Table III - Influence of Agonal Status and Cause of Death on Levels of GAD in the Caudate Nucleus

Agonal status	Cause of Death	GAD[1]
Protracted illness	Bronchopneumonia	27.7 ± 6.4 (10)
	Others	24.5 ± 3.9 (13)
	Total	25.9 ± 3.8 (23)
Sudden death	Various	48.0 ± 3.1 (49)

All values are expressed as means ± S.E.M.; numbers of samples in parentheses.

[1] μ-moles/hr/g protein.

ANTE-MORTEM FACTORS

Although post-mortem factors do not pose major interpretational problems with regard to GAD values, certain uncontrollable ante-mortem variables are of crucial importance in affecting post-mortem GAD activity. By far the most important of these appears to be the agonal status of the patient. Coma (4,5), pre-terminal hypoxia and bronchopneumonia (6) have been associated with reduced post-mortem GAD activity. Such ante-mortem events are often inter-related and thus it is difficult to implicate a particular agonal factor as being responsible for this observed decline. Moreover, prolonged hospitalization itself may be a contributory factor (4). Table II illustrates the effect of a protracted terminal illness on post-mortem GAD levels in several brain regions. Bronchopneumonia does not appear to be especially notorious in reducing GAD since losses of similar magnitude occurred with different immediate causes of death within this group (Table III), a detailed analysis of which has been reported elsewhere (18).

When the data from the sudden death cases were analyzed as a function of sex, no differences were found between male and female levels of GAD in any brain region. However, when GAD activity was

Table IV - GAD in Human Post-Mortem Brain: Statistically-Significant Correlations with Age

Brain region	Correlation coefficient	% Reduction (25 – 85 yr)
Amygdala (9)	-0.92***	69
Pars reticulata (18)	-0.466*	54
Red nucleus (26)	-0.562**	57
Cerebellar cortex (20)	-0.745***	58
Hippocampus (33)	-0.538***	60

All data were taken from sudden death cases only.

*, ** and *** indicate, respectively, $p < 0.05$, $p < 0.01$ and $p < 0.001$.

analyzed as a function of age, several statistically significant correlations appeared (Table IV). Although there was a tendency for GAD activity to decline with advancing age in the subdivisions of the corpus striatum, pars compacta, dentate nucleus and olive, the correlation coefficient did not reach statistical significance. Hence, it seems likely that certain brain areas are more susceptible to the ageing process than others with respect to loss of GAD activity.

GAD IN HUNTINGTON'S CHOREA

Bird and Iversen (2) reported a marked decrease in GAD activity in the corpus striatum and substantia nigra of patients dying with Huntington's chorea. This observation was consistent with the finding of Perry *et al.* (7) of a GABA deficit in this disease. GABA is believed to serve as an inhibitory neurotransmitter for small striatal interneurones as well as for striatal efferents to the globus pallidus and substantia nigra (8-10). Since loss of striatal neurones is a prominent pathological feature in Huntington's chorea, it seemed likely that the decreases in GAD activity and in GABA were reflections of this degenerative process. However, the value of post-mortem brain measurements of GAD has been brought into question by the finding that it is depleted in Parkinson's disease (5,11), senile dementia (4,6,12) and depression (4). We also reported a significant reduction in GAD activity in various brain regions from schizophrenic patients (13). It now appears that some of these findings are spurious, and are related to ante-mortem factors other than the disease under study. In this regard, the importance of the pre-terminal condition of the patient has already been discussed.

In view of these findings, the post-mortem GAD data for various brain regions of choreic patients have been re-examined. When values from control and choreic cases who had died suddenly were compared, there was a significant decrease in GAD activity in all regions from choreic brain (Tables II and V), called here the "primary depletion" in GAD. However, when values from choreic cases who had died suddenly were compared with those from choreic cases who had suffered a prolonged terminal illness (which in all cases was bronchopneumonia), bronchopneumonia was associated with a further significant loss of GAD activity, or "secondary depletion", in the caudate and putamen only (Table V). The failure of bronchopneumonia to produce a secondary depletion of GAD in the other areas may be due to several factors. It is possible that GAD in these regions had fallen to a basal level which was unaffected by bronchopneumonia. Alternatively, the striatum may be particularly susceptible to the bronchopneumonic process.

The primary reduction in GAD in the choreic cases may reflect non-specific factors such as prolonged hospitalization or drug treatment, rather than being related to the disease itself. If this were the case, a useful alternative control group would be provided by psychotic patients who have been chronically hospitalized and have been receiving neuroleptic medication, in common with the great

Table V - Influence of Agonal Status on GAD Levels in Post-Mortem Choreic Brain Tissues

Brain region	GAD (μ-moles/hr/g protein)		% Reduction
	Sudden death	Bronchopneumonia	
Pars compacta	22.8 ± 4.8 (8)	29.0 ± 4.9 (20)	-
Pars reticulata	18.8 ± 4.0 (8)	25.6 ± 4.3 (20)	-
Medial pallidum	30.4 ± 6.6 (7)	28.4 ± 4.6 (20)	-
Lateral pallidum	20.8 ± 2.7 (7)	18.5 ± 2.7 (20)	-
Nucleus accumbens	56.6 ± 17.8 (10)	30.3 ± 3.7 (20)	-
Putamen	21.6 ± 3.5 (15)	13.8 ± 2.0 (41)	36*
Caudate	18.6 ± 3.8 (13)	11.1 ± 1.7 (39)	40*
Dentate	20.9 ± 2.5 (6)	28.1 ± 4.7 (23)	-
Cerebellar cortex	16.9 ± 3.0 (7)	16.1 ± 3.1 (15)	-
Red nucleus	11.3 ± 2.6 (10)	15.6 ± 1.9 (23)	-
Olive	8.6 ± 1.8 (7)	9.1 ± 1.9 (21)	-
Motor cortex	27.9 ± 3.1 (10)	21.6 ± 3.1 (17)	-

All values are expressed as means ± S.E.M.; numbers of samples in parentheses; * indicates $p \simeq 0.05$.

majority of choreic patients. When data from the caudate nucleus were analyzed, no difference was found in GAD levels between hospitalized psychotic patients and non-hospitalized controls, indicating that neither hospitalization nor drug treatment can account for the primary reduction in GAD in the choreic caudate nucleus (Table VI).

However, regardless of the immediate cause of death, all choreic patients are suffering the ravages of a chronic, progressive, physically-debilitating disease. Such patients are usually emaciated and frail and their physical activity is considerably restricted. In this light, the distinction between sudden death and protracted terminal illness becomes blurred and, perhaps, Huntington's chorea,

Table VI - Influence of Chronic Hospitalization on GAD Levels in the Caudate Nucleus

Agonal status	Hospitalized		Non-hospitalized controls
	Choreic	Psychotic	
Sudden death	18.6 ± 3.8(13)	42.7 ± 6.9(13)	48.0 ± 3.1(49)
Protracted illness	11.1 ± 1.7(39)	27.0 ± 2.8(33)	30.5 ± 7.0(10)

All values are expressed as means ± S.E.M.; numbers of samples in parentheses.

per se, should be considered as a protracted terminal illness. In contrast, psychotic patients who die suddenly in hospital have generally been physically robust and active prior to death. The GAD values in the total choreic group (sudden death and protracted illness) have, therefore, been compared with those from control and psychotic cases suffering a protracted terminal illness (Table VII). When the data were treated in this fashion, significant decreases in GAD activity in choreic brain were confined to the caudate, putamen and lateral pallidal segment. If the premises for analyzing the data in this way are correct, only in these regions can the reduction in GAD activity be regarded as valid, which is in accord with the major pathological findings in Huntington's chorea. Moreover, it appears that the loss of GAD from the striatum in this disease may not be as severe as originally estimated.

GAD IN PARKINSONISM

We have studied a limited series of cases of parkinsonism which fall into two main diagnostic categories: idiopathic Parkinson's disease and striatonigral degeneration. A detailed report of the clinical, pathological and neurochemical findings in the cases of striatonigral degeneration has been published (14). In

Table VII - Comparison of GAD Values in Brain Tissues from Control and Psychotic Cases Dying from a Protracted Terminal Illness and from Cases of Huntington's Chorea

Brain region	GAD (μ-moles/hr/g protein)			% Reduction
	Control	Psychotic	Choreic	
Pars compacta	25.7 ± 7.5 (11)	28.3 ± 3.6 (12)	27.3 ± 3.6 (28)	-
Pars reticulata	28.8 ± 7.2 (11)		23.7 ± 3.1 (28)	-
Medial pallidum	32.9 ± 7.1 (14)		28.9 ± 4.5 (27)	-
Lateral pallidum	35.0 ± 8.2 (14)	33.8 ± 3.3 (12)	19.1 ± 2.1 (27)	45*
Putamen	34.9 ± 4.5 (24)	34.4 ± 4.1 (28)	15.9 ± 1.9 (56)	54**
Caudate	25.9 ± 3.8 (23)	27.0 ± 2.8 (33)	13.0 ± 1.6 (52)	50**
Nucleus accumbens	40.9 ± 6.3 (15)	48.8 ± 4.5 (36)	39.1 ± 6.6 (30)	-
Cerebellar cortex	15.2 ± 3.4 (12)	20.4 ± 6.5 (6)	16.3 ± 2.3 (22)	-
Dentate	23.3 ± 3.3 (11)	20.0 ± 4.6 (14)	26.5 ± 3.6 (29)	-
Red nucleus	10.2 ± 2.4 (13)		14.3 ± 1.7 (33)	-
Olive	6.6 ± 2.7 (10)	5.5 ± 2.1 (12)	9.0 ± 1.5 (28)	-
Motor cortex	22.7 ± 5.5 (9)	22.1 ± 5.5 (9)	23.9 ± 2.3 (27)	-

All values are expressed as means ± S.E.M.; numbers of samples in parentheses; * $p < 0.025$; ** $p < 0.001$.

Table VIII - GAD Values in Various Brain Regions in Parkinsonism

Brain region	Illness	GAD (μ-moles/hr/g protein	
		Sudden death	Protracted illness
Pars compacta	PD	71.0 ± 23.5 (3)	19.1 ± 8.2 (3)
	SND	22.4 (1)	6.9 ± 1.7 (3)
	C	109.1 ± 10.0 (18)	25.7 ± 7.5 (11)
Lateral pallidum	PD	61.9 ± 20.8 (4)	13.8 ± 5.4 (5)
	SND	48.3 (1)	11.8 ± 3.7 (3)
	C	92.9 ± 8.5 (22)	35.0 ± 8.2 (14)
Putamen	PD	39.9 ± 19.8 (4)	6.1 ± 1.7 (4)
	SND	52.1 (1)	8.6 ± 2.8 (3)
	C	57.9 ± 3.9 (51)	34.9 ± 4.5 (24)
Cerebellar cortex	PD	26.0 ± 8.3 (3)	8.6 ± 3.6 (3)
	SND	20.6 (1)	5.3 ± 3.2 (3)
	C	32.7 ± 3.2 (20)	15.2 ± 3.4 (12)
Dentate	PD	35.4 ± 9.2 (3)	13.3 ± 5.1 (3)
	SND	36.2 (1)	3.8 ± 1.7 (3)
	C	48.6 ± 3.0 (24)	23.3 ± 3.3 (11)

All values are expressed as means ± S.E.M.; numbers of samples in parentheses.

PD, Parkinson's disease; SND, striatonigral degeneration; C, controls.

common with other workers, we have observed a depletion of GAD in many brain regions in parkinsonism (Table VIII). However, as before, this may reflect mainly the patient's agonal status, although it must be stated that the numbers are too small to draw any firm conclusions.

GAD IN SCHIZOPHRENIA

Roberts (15) has suggested that abnormalities in GABA-ergic mechanisms may be associated with schizophrenic symptoms. In our preliminary communication on neurochemical abnormalities in psychosis, we reported a significant loss of post-mortem GAD activity in several brain regions (13). However, we realized subsequently that broncho-pneumonia was over-represented in the psychotic cases as a cause of death and, taking this ante-mortem factor into account, differences between the control and psychotic groups were no longer evident (16). In a more recent, and much more comprehensive study, we have been unable to observe any differences in GAD activity in a wide variety of regions from psychotic brain (17).

CONCLUSION

The results of biochemical studies on human post-mortem brain must be interpreted with caution because of the many uncontrollable factors which may influence such findings. Interpretation of GAD values, in particular, is especially fraught with problems because of the considerable individual variation in levels throughout the normal population and the effects of extraneous factors such as the agonal status of the patient. We can only speculate on the mechanisms underlying the effect of a protracted terminal illness, but such inter-related features as immobility, reduced calorie intake, hypoxia and lowered level of consciousness may all contribute to the observed depletion of cerebral GAD.

In Huntington's chorea, there is a widespread loss of GAD activity affecting many brain regions. However, when stringent ante-mortem criteria are applied for comparison with control material, this loss is restricted to the striatum and lateral pallidal segment. Loss of GAD activity is seen also in parkinsonism resulting from Parkinson's disease or striatonigral degeneration. However, this is most likely due to ante-mortem factors other than the disease processes. Schizophrenia is not associated with any deficit in cerebral GAD.

REFERENCES

(1) Spokes, E. G. and Koch, D. J. (1978) J. Neurochem. 31, 381-383.
(2) Bird, E. D. and Iversen, L. L. (1974) Brain, 97, 457-472.
(3) McGeer, P. L., McGeer, E. G. and Fibiger, F. C. (1973) Neurology (Minneap.) 23, 912-917.
(4) Perry, E. K., Gibson, P. H., Blessed, G., Perry, R. H. and Tomlinson, B. E. (1977) J. Neurol. Sci. 34, 247-265.
(5) McGeer, P. L. and McGeer, E. G. (1976) J. Neurochem. 26, 65-76.
(6) Bowen, D. M., Smith, C. B., White, P. and Davison, A. N. (1976) Brain, 99, 459-496.
(7) Perry, T. L., Hansen, S. and Kloster, M. (1973) New Engl. J. Med. 288, 337-342.
(8) Kim, J. S., Bak, I. J., Hassler, R. and Okada, Y. (1971) Exp. Brain Res. 14, 95-104.
(9) McGeer, P. L. and McGeer, E. G. (1975) Brain Res. 91, 331-335.
(10) Precht, W. and Yoshida, M. (1971) Brain Res. 32, 229-233.
(11) Lloyd, K. G., Möhler, H., Heitz, P. and Bartholini, G. (1975) J. Neurochem. 25, 789-795.
(12) Bowen, D. M., White, P., Flack, R. H. A., Smith, C. B. and Davison, A. N. (1974) Lancet, 1, 1247-1249.
(13) Bird, E. D., Spokes, E. G., Barnes, J., Mackay, A. V. P., Iversen, L. L. and Shepherd, M. (1977) Lancet, 2, 1157-1159.
(14) Spokes, E. G., Bannister, R., Oppenheimer, D. R. (1979) J. Neurol. Sci. (in press).
(15) Roberts, E. (1972) Neurosci. Res. Program Bull. 10, 468-481.
(16) Bird, E. D., Spokes, E. G., Barnes, J., Mackay, A. V. P., Iversen, L. L. and Shepherd, M. (1978) Lancet, 1, 156.
(17) Bird, E. D., Spokes, E. G. and Iversen, L. L. (1979) Brain (in press).
(18) Spokes, E. G. (1979) Brain (in press).

INVOLVEMENT OF CENTRAL GABA-ERGIC SYSTEMS IN CONVULSIONS AND AGGRESSIVE BEHAVIOR

P. Mandel, L. Ciesielski, M. Maitre, S. Simler, G. Mack and E. Kempf

Centre de Neurochimie du CNRS, and Institut de Chimie Biologique, Faculté de Médecine, 67085 Strasbourg Cedex, France

INTRODUCTION

Epilepsy occurs more frequently than any other neurological syndrome. It is also well known that impulsive and aggressive behaviors occur frequently in man suffering from chronic seizure. In some cases of epilepsy a very pronounced aggressive behavior was observed and was taken as an indication for psychosurgery.

The question that might be asked is: Can our knowledge of the molecular mechanisms which underlie seizure and aggressive behaviors be used to elaborate a justified physiological therapy? Unfortunately, epilepsy is a syndrome which appears in several different forms and which might result from multiple causes. Thus, in order to elaborate a therapy based on physiological mechanisms, we thought that it might be wise to follow a general biological rule; i.e., that the main biological mechanisms are controlled by an inhibitory-disinhibitory mechanism. This is the case for transcription of the genetic code, for the control of enzymatic activity through feedback inhibition and repression, for the control of hormonal secretions of adrenal and gonadal glands by the hypothalamus and hypophysis, and for some other processes (1).

For the vertebrate central nervous system (CNS), three inhibitory neurotransmitters have been postulated or demonstrated: γ-aminobutyric acid (GABA; see refs. 2 and 3 for review), taurine (see ref. 4 for review), and glycine (see ref. 5 for review). In the brain mainly GABA and taurine appear to serve such a role. A variety of studies (biochemical, physiological and pharmacological) have established that GABA is the major inhibitory neurotransmitter in the mammalian brain (2,3). Our own investigations have provided strong support for the hypothesis that taurine might act as a neurotransmitter in the visual pathway (4). Moreover, it has been shown that convulsions can be produced either by inhibition of the enzymatic activity of glutamate decarboxylase (GAD), which is responsible for GABA synthesis, by a variety of compounds (e.g., hydrazides, pyridoxine-antagonists, allylglycine) or by genetically determined or nutritionally produced vitamin B_6 deficiency. Compounds which block the inhibitory synaptic action of GABA, such as picrotoxin and bicuculline, are also potent convulsants (6). Also, in some cases in which the blood-brain barrier was probably altered, GABA exerted an anticonvulsant action (7). A good deal of evidence also suggests that the level of taurine might be altered in epileptic foci and that taurine might act as an antiepileptic agent (8,9). Finally, let us remember that a structural analogy between GABA and taurine is evident and that some highly purified enzyme preparations might act on the metabolism of both GABA and taurine (10). This report will deal only with mechanisms involved in GABA metabolism; those concerning taurine will be discussed elsewhere.

POTENTIATION OF GABA-ERGIC NEUROTRANSMISSION AS A TOOL FOR ANTICONVULSIVE THERAPY

Considering that a deficiency in GABA-ergic neurotransmission might produce seizures, it seemed logical to search for a method to potentiate GABA-ergic neurotransmission in order to inhibit convulsive states. Theoretically, there are several possible ways to favorize GABA-mediated inhibition: (a) by increasing the availability of its precursor glutamate; unfortunately, glutamate does not cross the blood-brain barrier; (b) by stimulating the activity of GAD; however, no compound is available that will ensure such activation; (c) by inhibiting the re-uptake of GABA; unfortunately, substances that have been shown to inhibit GABA re-uptake (e.g., nipecotic acid, L-2,4-diaminobutyric acid) do not cross the blood-brain barrier; (d) by using a GABA-agonist; however, the best GABA-agonists which have a high affinity for synaptic GABA-receptors are expected to bind to all receptors and therefore might produce many secondary effects; moreover, these agents might cause desensitization of GABA-receptors; (e) by inhibiting the degradation of GABA; this seems to us to be the most valid approach in view of the efficiency of inhibitors of monoamine oxidase in aminergic neurotransmission. Moreover, if convulsive behavior is controlled by the GABAergic system via liberation of GABA, then inhibition of GABA-α-ketoglutarate transaminate (GABA-T) should prolong and potentiate this inhibitory control.

Di-n-Propylacetate; An Inhibitor of GABA-T

By constructing Dreiding models for different compounds suspected to resemble GABA, we found that di-*n*-propylacetate (DPA), which was synthesized in 1882 by Burton (11), might be an interesting candidate as an inhibitor of GABA-T (Fig. 1). At about the same time we learned that a group of pharmacologists in Grenoble had succeeded in blocking pentylenetetrazol-induced convulsions with DPA (12). Later, Cohen-Addab et al. (13) found that a folded conformation of GABA resembled closely the structure of DPA.

By injecting DPA into mice that were exhibiting audiogenic seizures, we found that these seizures were blocked by DPA (14). Also, we observed that DPA increased brain GABA level in parallel to its inhibition of GABA-T (15). Furthermore, brain aspartate levels were decreased after DPA injection (15), but since this change

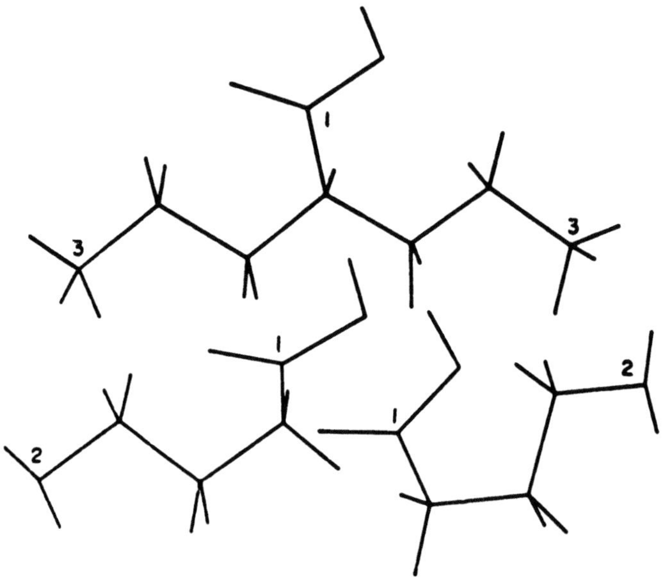

Fig. 1. Photographs of GABA and DPA. Top: one molecule of DPA; bottom: two molecules of GABA. (1) Carboxyl group; (2) amino group; (3) methyl group.

appeared to reflect the disturbance of the GABA shunt mechanism we did not consider it further; this finding has recently been confirmed by Schechter et al. (16). There is not any valid reason for involvement of aspartate in convulsive states, at least until now.

In order to establish clearly the mechanism of action of DPA and other branched-chain fatty acids which seem to possess similar properties (17,18), the following problems had to be investigated.

1. Effect of DPA on enzymes of the GABA shunt, and eventually on enzymes related to compounds of this shunt.
2. Effect of DPA on GABA uptake.
3. Effect of DPA on GABA binding to its receptors.

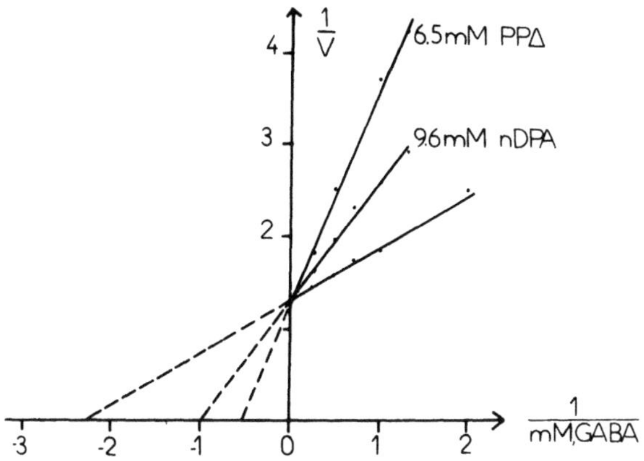

Fig. 2. Mouse GABA-T; DPA-GABA, PPΔ-GABA interactions; Lineweaver-Burk plot of 1/v = f(1/GABA); competitive inhibition.

Effect of DPA on Enzymes of the GABA Shunt

It was necessary to examine not only the effect of DPA on GABA-T, but also its effects on other enzymes of the GABA shunt, since some GABA-T inhibitors also inhibit GAD and since the action of DPA on GABA-T might also affect enzymes associated with GABA metabolites such as succinic semialdehyde dehydrogenase (SSADH). Also, the possible action of DPA on succinic semialdehyde reductase (SSAR) had to be studied.

Thus, we purified to homogeneity GAD from rat and human brain (19,20), GABA-T from human (21), rat (22), mouse (23) and calf brain (unpublished data), SSADH from human (24) and rat brain (25), and SSAR from human brain (26). As shown in Figure 2, DPA is a competitive inhibitor of GABA-T. The K_i for this action varies among different species, being lower for human brain GABA-T than for GABA-T of other species (27). DPA does not affect brain GAD, but does serve as a non-competitive inhibitor of rat brain SSADH (28) and as an uncompetitive inhibitor of human brain SSADH (26). However, as

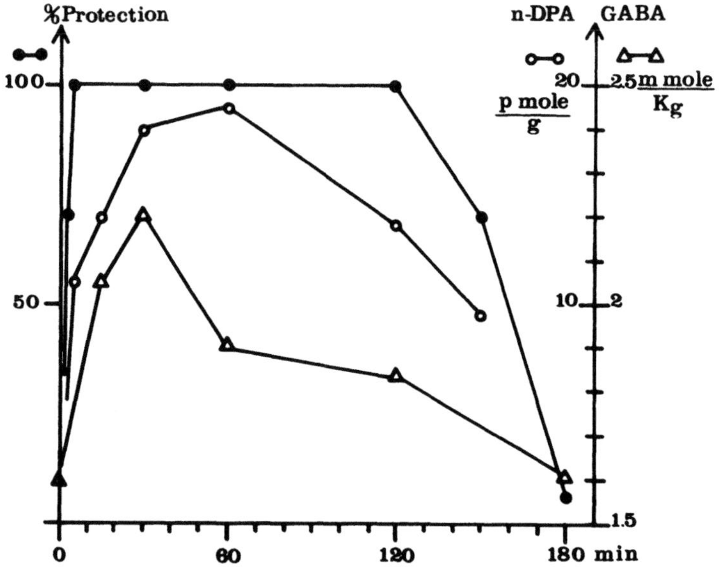

Fig. 3. Curves 1 and 2 show the time course of protection against audiogenic seizure and the change in cerebral GABA level after injection of 400 mg/kg DPA into audiogenetically-sensitive Swiss albino mice (RB strain). Curve 3 shows the cerebral fixation of [^{14}C]DPA (400 mg/kg + 50 μCi/kg, i.p.) in the same mouse strain.

will be shown below, this effect on SSADH cannot account for the in vivo effect of DPA. Finally, we found that DPA inhibits partially a non-specific SSAR of human brain, but not the specific enzyme (29).

Effect of DPA on Brain GABA Content

In 1969, we showed that injections of DPA produced an increase in brain GABA content in the rat (15). This finding was confirmed in mouse brain by our group (30) and in rat brain by other workers (31). Interestingly, the extent of this increase in GABA content

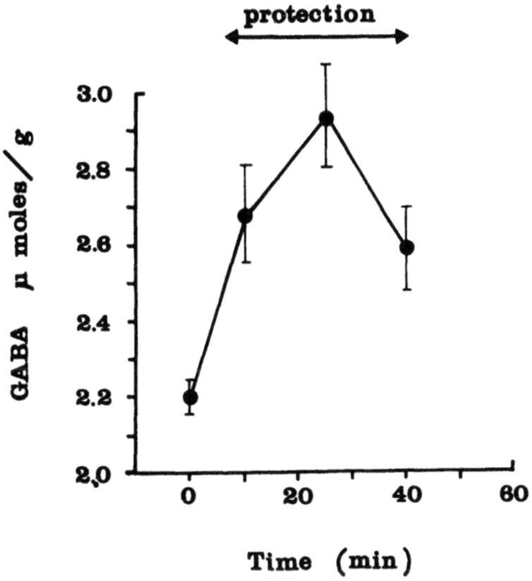

Fig. 4. Effect of DPA (400 mg/kg) on the cerebral GABA level and its protective effect on 20-day-old Quaking mice.

differed among various brain regions (32). In some regions there occurred only a slight increase whereas in others the increase was very pronounced, which explains the efficiency of the increased GABA level although the overall increase in brain GABA is not of great magnitude. Moreover, we found a good correlation between the increase in brain GABA and protection against audiogenic seizures (Fig. 3) or seizures produced in mutant mice with myelin deficiency (Fig. 4; ref. 17). The correlation between the DPA-induced increase in brain GABA and protection against audiogenic seizures has been confirmed by Schechter et al. (16).

Table I - Level of Incorporation of Tritium into GABA with [^{3}H]SSA as Precursor in the Presence of Various Branched-Chain Fatty Acids

	Specific activity of GABA / Specific activity of SSA
Control	0.20 ± 0.04 (4)
Dipropylacetate (10 mM)	0.17 ± 0.06 (4); n.s.
5,5-Dipropylpentanoate (10 mM)	0.20 ± 0.08 (4); n.s.
2-Propyl-2-pentenoate (10 mM)	0.081 ± 0.023 (4); $p < 0.01$

n.s., not significant

Effects of Other Branched-Chain Fatty Acids on GABA-T

Several branched-chain fatty acids which are structurally related to GABA and DPA, such as dimethylvalerionic acid, methylethylcaproic acid, tetramethylvalerionic acid and propylpentenoic acid, are competitive inhibitors of GABA-T and non-competitive or uncompetitive inhibitors of SSADH. All of these compounds are also anticonvulsants (33).

Mechanism of the Anticonvulsant Action of DPA

After we had postulated that DPA acts on convulsions by increasing brain GABA content, a hypothesis was advanced that DPA might act primarily by inhibiting SSADH and that the increase in GABA might be of a secondary nature. However, as shown by Baxter and Roberts (34) and recently by us (35), SSA is not a direct precursor of GABA (Table I). Moreover, comparison of different branched-chain fatty acids as inhibitors of SSADH and GABA-T revealed clearly that their relative efficiency for protection against seizures paralleled their inhibitory effect on GABA-T, but not on SSADH (Table II; see ref. 12).

It also appears clear that if the K_i of DPA for SSADH is lower than that of DPA for GABA-T the K_m of SSA for SSADH is much lower than that of GABA for GABA-T. Actually, the ratio of K_m GABA/K_i DPA, GABA

Table II - Correlation Between Anti-Epileptic Effectiveness (Threshold Dosage), Inhibition of GABA-T (K_i), and Inhibition of SSA Dehydrogenase (K_i) by Branched-Chain Fatty Acids

Anti-epileptic effectiveness	PPΔ ≃ MEC > DPP_n > TPP_r > DPA
Inhibition of GABA-T	PPΔ ≃ MEC > DPP_n > TPP_r > DPA
Inhibition of SSA dehydrogenase	DPP_n > TPP_r > MEC > DPA > PPΔ

PPΔ, 2-propyl-2 pentenoic; TPP_r, 2-3-3 tripropyl propionic; MEC, 2-methyl-2 ethyl caproic; DPA, dipropyl acetic; DPP_n, 5-5 dipropyl pentanoic.

is about 0.5 for GABA-T, whereas the ratio of K_m SSA/K_i DPA, SSA is about 160 for SSADH. Moreover, due to this very high activity of SSADH, SSA exists only in trace amount in brain. Thus, a mechanism of action of DPA involving SSADH inhibition can be excluded. It has also been suggested that DPA acts by increasing γ-hydroxybutyrate (GHBA). In this regard, it should be remembered that GHBA produces a typical seizure EEG pattern which resembles the clinical syndrome of petit mal epilepsy (12). Moreover, DPA effectively prevents seizures induced by GHBA (12). Therefore, a mechanism involving GHBA cannot be readily used to explain protection against seizures.

Attention has been drawn to the observation that the anticonvulsant effect of DPA outlasts the increase that it produces in brain GABA or the presence of DPA in the body. This seems to be a very weak objection since it is well known that long-lasting effects of substances or stimulation can occur in the nervous system under certain conditions. What is often difficult to understand and remains obscure is how a substance acting on a neurotransmitter pathway might produce such a long-lasting effect, rather than whether or not the effect is produced by the substance.

POTENTIATION OF GABA-ERGIC NEUROTRANSMISSION AS A TOOL FOR ANTI-AGGRESSIVE THERAPY

In order to approach the molecular concomitants and eventually the molecular mechanisms underlying aggressive behavior, we selected two reproducible models for this behavior: (a) muricidal (mouse-killing) behavior, described by Karli in 1956 (36) and then studied by several other workers (see ref. 37 for review); (b) aggressiveness produced by prolonged socio-environmental deprivation.

Muricidal behavior is a stereotyped event elicited without any previous experience characterized by: (a) pursuit of the mouse; (b) capture; (c) rapidly biting the back of the neck, sectioning the spinal cord and causing death. Aggressive behavior induced by prolonged isolation of adult laboratory animals is a well known model of aggressiveness.

Genetic Factors

Genetic control signifies an involvement of biochemical processes. In this regard, it is of interest to emphasize the genetic differences that exist in the two types of aggressive behavior that we studied. About 70% of wild Norway rats and only about 20% of male Wistar rats exhibited killing behavior. Moreover, about 58% of Long-Evans rats (which are derived from a Norway-Wistar cross) killed mice, whereas only about 5% of August rats exhibited this behavior (37). A striking difference also exists among various mouse strains with respect to isolation-induced agressive behavior. After 8 weeks of isolation the number of attacks made by the DBA strain was significantly greater than that of the C-57 Bl strain (unpublished results from our laboratory).

Neurotransmitter Alterations

Cerebral neurotransmitter systems were studied in killer rats and in mice that had been subjected to isolation. The main altera-

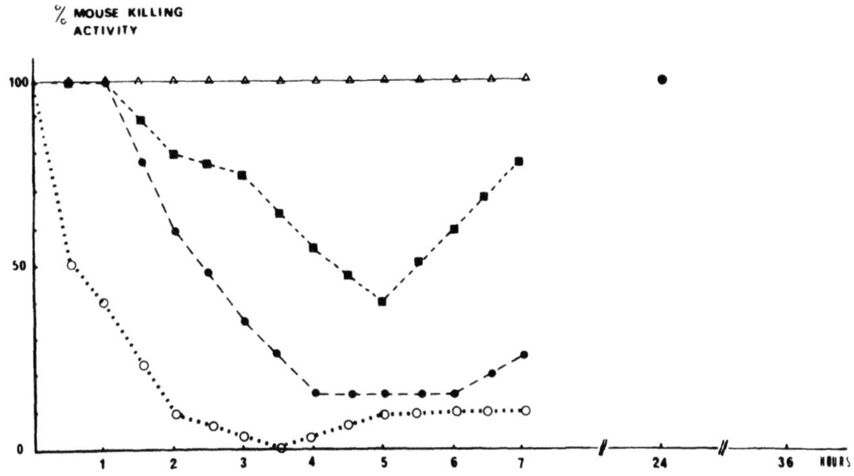

Fig. 5. Evolution of muricidal behavior in spontaneous killer rats after intraperitoneal injections of increasing doses of DPA; 20 rats per series. DPA: ■, 100 mg/kg; ●, 200 mg/kg; ○, 400 mg/kg. Δ, Saline control.

tions found in killer rats, as compared with non-killers, can be summarized as follows: (a) a decrease in GABA content of the olfactory bulbs; (b) an increase in dopamine turnover in olfactory tubercle; (c) a decrease in 5-hydroxytryptamine (5-HT) turnover rate; (d) an increase in choline acetyltransferase activity of the amygdala (37).

Compensation for the decrease in GABA content of the olfactory bulb, produced by direct injection of either GABA or DPA into the olfactory bulbs, suppressed killing behavior. Intraperitoneally-injected DPA also suppressed killing behavior in rats in a dose-dependent fashion. It is noteworthy that there exists a linear correlation between the inhibition of aggressive behavior, the K_i (mM), and the dose (mM) of the branched-chain fatty acids which block muricidal behavior. After ablation of the olfactory bulbs of killer rats, intraperitoneally-injected DPA was no longer effective (Fig. 5), in-

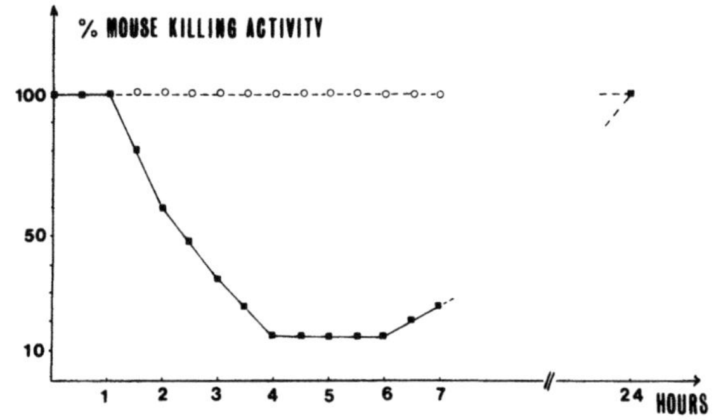

Fig. 6. Effect of DPA (200 mg/kg, i.p.) on spontaneous mouse-killing behavior (_filled squares_); n = 20 rats. _Open circles_: saline administration; n = 20 rats. Ordinate: percentage of rats showing killing behavior; abscissa, time after injection of DPA.

dicating that the presence of the olfactory bulbs is necessary for suppression of muricidal behavior by DPA (Fig. 6). Furthermore, intraperitoneal injections of DPA, which increase the GABA contents of olfactory bulb and other brain areas, also reversed the alterations in choline acetyltransferase and 5-HT in killer rats; values obtained after DPA were similar to those of non-killers (37).

Direct injections into the olfactory bulb of muscimol (a potent GABA receptor agonist) (Fig. 7) and L-2,4-diaminobutyric acid (an inhibitor of GABA uptake) (Fig. 8) and intraperitoneal injection of a derivative of nipecotic acid (an inhibitor of GABA uptake) also inhibited killing behavior. In contrast, direct introduction into the olfactory bulb of picrotoxin or bicuculline (agents which antagonize the synaptic action of GABA) or allylglycine (a GAD-inhibitor) produced killing behavior (Fig. 9; ref. 37).

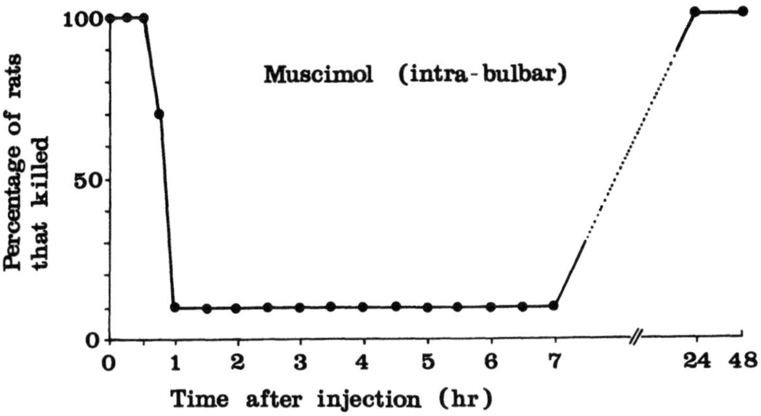

Fig. 7. Change in muricidal behavior of a group of ten killer rats after intra-bulbar injection of 100 μg of muscimol-HBr.

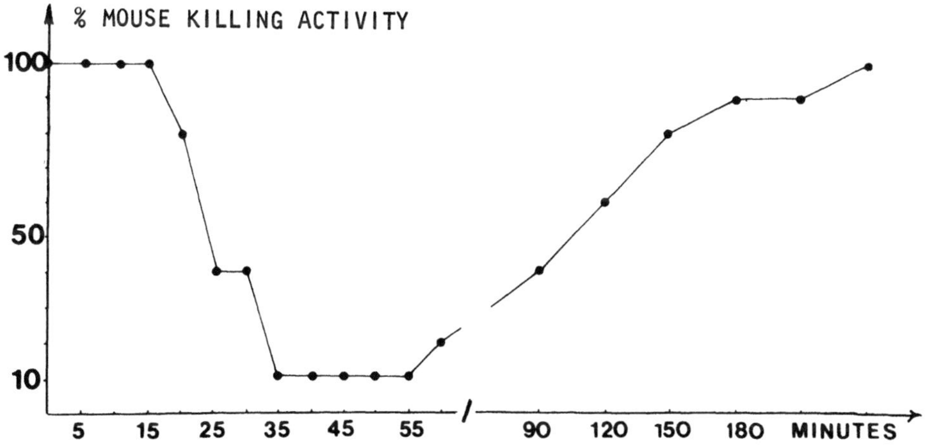

Fig. 8. Effect of L-2,4-diaminobutyric acid (184 μg; intra-bulbar injection) on the muricidal activity of a group of 10 rats.

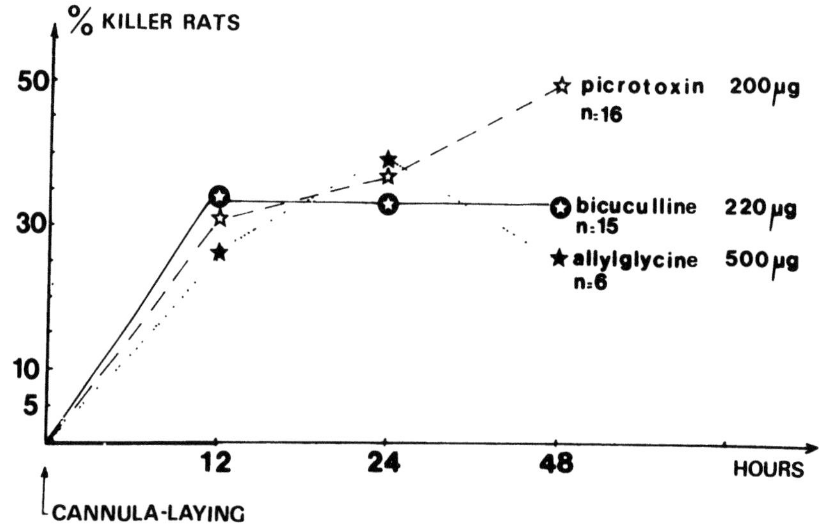

Fig. 9. Effects of intra-bulbar administration of the solid forms of allylglycine (a GAD inhibitor), picrotoxin and bicuculline on muricidal behavior. Note that substances that interfere with the action of GABA provoke muricidal behavior.

In mice which had become aggressive after isolation, the behavioral change was found to be correlated with decreases in GABA binding to cerebral synaptosomal fractions (38) and in GAD activities of whole forebrain (39) and perhaps olfactory bulb and striatum (see also Blindermann et al., this volume). Moreover, in DBA mice the aggressive behavior was suppressed by DPA which increases brain GABA content (40). Taken together, these results have revealed that: (a) killer rats and isolated mice both possess deficient GABA-ergic inhibitory systems, and potentiation of GABA-ergic systems can reduce or abolish aggressive responses; (b) a deficiency in GABA or in the activity of the GABA-ergic system in specific areas of the CNS might be responsible for aggressive behavior.

DISCUSSION

The possibility of preventing some convulsive states and some model aggressive behaviors by potentiation of central inhibitory mechanisms was studied using some GABA-analogues which inhibit GABA-T and thus produce increases in GABA in certain brain regions. DPA was the compound most frequently used. In view of the rather high K_i for DPA action on GABA-T, it was questioned whether DPA can inhibit GABA-T *in vivo*. In this regard, we should recall that the only way in which an increase in brain GABA content could be produced by DPA is by inhibition of GABA-T, since DPA is ineffective against GAD and since inhibition of SSADH cannot lead to such an increase via SSA. We should also consider that inhibition of GABA-T in a certain brain area (or its subcellular fractions or pools) depends upon the ratio of DPA/GABA which differs among various brain areas, and the only criterion required for determining whether or not an event can occur is the observation of whether or not it exists; in this case the increase in brain GABA is critical. Moreover, a good correlation has been found between GABA-T inhibition and anticonvulsant efficiency, but not between SSADH inhibition and anticonvulsant efficiency. A good correlation also existed between the increase in brain GABA and protection against audiogenic seizures. Finally, intraperitoneal or oral administration of DPA and a derivative of nipecotic acid which crosses the blood-brain barrier produce a synergistic effect, thus confirming that DPA acts through the GABA system (unpublished results). It has also been questioned whether DPA which binds to plasma proteins can penetrate into the brain. Determinations of DPA which revealed that it was present in different amounts in different rat brain areas and that its presence reflected relative increases in GABA content permitted us to overcome this criticism (unpublished results). Furthermore, it is well known that several compounds which bind to plasma proteins also penetrate into the brain, perhaps the best example being tryptophan. It was also questioned whether the time course of the brain GABA increase always coincides with the anticonvulsant effect of DPA. In this regard, we should remember the long-lasting effects observed with several other neurotropic drugs.

As we have shown, DPA might also produce secondary alterations in 5-HT, acetylcholine or dopamine in some brain areas. However, such alterations cannot explain the anticonvulsant effect of DPA since anticonvulsant effects could not be produced by drugs that act directly on these other neurotransmitter systems. It also seems worthwhile to draw attention to the fact that some anticonvulsants (e.g., barbiturates) depress cellular respiration whereas branched-chain fatty acids do not (41). Finally, certain doses of some branched-chain fatty acids produce an enhancement of learning in various animal species (42-44). In this regard, let us remember the favorable effect of DPA on the behavior of children, which is altered markedly by barbiturates (45).

REFERENCES

(1) Stryer, L. (1975) Biochemistry, Vol. 1, W. H. Freeman and Co., San Francisco, 877 pp.
(2) Krnjević, K. (1974) Physiol. Rev. 54, 418-540.
(3) Curtis, D. R. and Johnston, G. A. R. (1974) Ergebn. Physiol. 69, 97-188.
(4) Mandel, P. and Pasantes-Morales, H. (1978) In: Reveiws of Neuroscience (S. Ehrenpreis and I. Kopin, eds.) Vol. 3, pp. 157-163. Raven Press, New York.
(5) DeFeudis, F. V. (1977) Acta Physiol. Latinoamer. 27, 131-145.
(6) Tower, D. B. (1969) In: Basic Mechanisms of the Epilepsies (H.H. Jasper, A. A. Ward and A. Pope, eds.) pp. 611-638, Churchill, London.
(7) Strasberg, P., Krnjević, K., Schwartz, S. and Elliott, K. A. C. (1967) J. Neurochem. 14, 755-760.
(8) Barbeau, A., Tsukada, Y. and Inoue, N. (1976) In: Taurine (R. Huxtable and A. Barbeau, eds.) pp. 253-266, Raven Press, New York.
(9) van Gelder, N. M. (1976) In: Taurine (R. Huxtable and A. Barbeau, eds.) pp. 293-302, Raven Press, New York.
(10) Recasens, M. and Mandel, P. (1979) In: CIBA Fdn. Symp., London (in press).
(11) Burton, B. S. (1882) Amer. Chem. J. 3, 385-395.
(12) Meunier, G., Carraz, G., Meunier, Y., Eymard, P. and Aimard, M. (1963) Thérapie, 18, 435-438.

(13) Cohen-Addad, C. and Cohen-Addad, J. P. (1977) Spectrochim. Acta, 33A, 821-832.
(14) Simler, S., Randrianarisoa, H., Lehmann, A. and Mandel, P. (1968) J. Physiol. (Paris) 60, 547.
(15) Godin, Y., Heiner, L., Mark, J. and Mandel, P. (1969) J. Neurochem. 16, 869-873.
(16) Schechter, P. J., Tranier, Y. and Grove, J. (1978) J. Neurochem. 31, 1325-1327.
(17) Mandel, P., Ciesielski, L. and Maitre, M. (1977) In: Medicinal Chemistry (J. Mathieu, ed.) Vol. 1, pp. 155-172. Elsevier, Amsterdam.
(18) Maitre, M., Ciesielski, L. and Mandel, P. (1974) Biochem. Pharmacol. 23, 2363-2368.
(19) Blindermann, J. M., Maitre, M. and Mandel, P. (this volume).
(20) Blindermann, J. M., Maitre, M., Ossola, L. and Mandel, P. (1978) Eur. J. Biochem. 86, 143-152.
(21) Cash, C., Maitre, M., Ciesielski, L. and Mandel, P. (1974) FEBS Lett. 47, 199-203.
(22) Maitre, M., Ciesielski, L., Cash, C. and Mandel, P. (1975) Eur. J. Biochem. 52, 157-169.
(23) Schousboe, A. (1973) Biochemistry, 12, 2868-2873.
(24) Cash, C., Maitre, M., Ossola, L. and Mandel, P. (1978) Biochim. Biophys. Acta, 524, 26-36.
(25) Cash, C., Maitre, M., Ciesielski, L. and Mandel, P. (1977) Biochimie, 59, 257-268.
(26) Cash, C., Maitre, M., Ciesielski, L. and Mandel, P. (this volume).
(27) Maitre, M., Ossola, L. and Mandel, P. (this volume).
(28) Cash, C., Ciesielski, L., Maitre, M. and Mandel, P. (1975) C.R. Soc. Biol. Paris, 169, 884-887.
(29) Cash, C., Maitre, M. and Mandel, P. (1979) J. Neurochem. (in press).
(30) Simler, S., Randrianarisoa, H., Lehmann, L. and Mandel, P. (1971) In: Int. Symp. Golden Sands, Varna, 7-10 October, 1969, pp. 59-70. Publ. House Bulgarian Acad. Sci., Sofia.
(31) Martin del Rio, R., Garcia-Gracia, M., Orensanz Muñoz, L. M. and DeFeudis, F. V. (1978) Gen. Pharmacol. 9, 45-47.
(32) Simler, S., Gensburger, C., Ciesielski, L. and Mandel, P. (1978) Commun. Psychopharmacol. 2, 123-130.
(33) Ciesielski, L. et al. (this volume).
(34) Baxter, C. F. and Roberts, E. (1961) J. Biol. Chem. 236, 3287-3294.
(35) Maitre, M., Ossola, L. and Mandel, P. (1976) FEBS Lett. 72, 53-57.
(36) Karli, P. (1956) Behaviour, 10, 81-103.
(37) Mandel, P., Mack, G., Kempf, E., Ebel, A. and Simler, S. (1978) In: Interactions Between Putative Neurotransmitters in the Brain (S. Garattini, J. F. Pujol and R. Samanin, eds.) Vol. 1, pp. 285-304, Raven Press, New York.

(38) DeFeudis, F. V., Madtes, P., Ojeda, A. and DeFeudis, P. A. (1976) Exp. Neurol. 52, 285-294.
(39) Blindermann, J. P., DeFeudis, F. V., Maitre, M., Misslin, R., Wolff, P. and Mandel, P. (1979) J. Neurochem. 32, 1357-1359.
(40) Puglisi-Allegra, deS., Mack, G., Oliverio, A. and Mandel, P. (1979) C. R. Acad. Sci., Paris (in press).
(41) Ciesielski, L., Maitre, M., Cash, C. and Mandel, P. (1975) Biochem. Pharmacol. 24, 1055-1058.
(42) Will, B., Misslin, R., Ropartz, P. and Mandel, P. (1972) C. R. Acad. Sci., Paris (série D) 275, 2921-2923.
(43) Misslin, R., Ropartz, P. and Mandel, P. (1975) Psychopharmacologia, 44, 263-265.
(44) Misslin, R., Ropartz, P. and Mandel, P. (1975) C. R. Acad. Sci., Paris, 281, 1175-1178.
(45) Salzarulo, P. and Lairy, G. C. (1976) EEG Clin. Neurophysiol. 41, 329 (Abstr.).

SUBJECT INDEX

GPSR Compliance

The European Union's (EU) General Product Safety Regulation (GPSR) is a set of rules that requires consumer products to be safe and our obligations to ensure this.

If you have any concerns about our products, you can contact us on ProductSafety@springernature.com

In case Publisher is established outside the EU, the EU authorized representative is:

Springer Nature Customer Service Center GmbH
Europaplatz 3
69115 Heidelberg, Germany

Batch number: 10838062

Printed by Printforce, the Netherlands